Australian Signpost MATHS

5

Alan McSeveny
Alan Parker
Diane McSeveny-Foster
Rachel McSeveny
Bob Collard

Short Contents

Pearson Australia
(a division of Pearson Australia Group Pty Ltd)
459–471 Church Street, Level 1, Building B, Richmond, Victoria 3121
PO Box 23360, Melbourne, Victoria 8012
www.pearson.com.au

First published 2018 by Pearson Australia
2025 2024 2023 2022
10 9 8 7 6 5 4 3 2 1

Content and Learning Specialist: Julian Lumb
Project Manager: Shelly Wang
Production Manager: Elizabeth Gosman & Aptara
Editor: Aptara
Designer: Anne Donald & Jennifer Johnston
Cover Designer: Jennifer Johnston
Desktop operator: Aptara
Rights & Permissions Editor: Samantha Russell-Tulip
Senior Publishing Services Analyst: Rob Curulli
Cover art: Michael Barter
Illustrators: Michael Barter, Wendy Gorton, Nives Porcellato, Andrew Craig
Printed in Malaysia (CTP-VVP)

ISBN 978 1 4886 2190 1
Pearson Australia Group Pty Ltd ABN 40 004 245 943

Acknowledgements
All material identified by ACARA is material subject to copyright under the *Copyright Act 1968* and is owned by the Australian Curriculum, Assessment and Reporting Authority 2018. ACARA neither endorses nor verifies the accuracy of the information provided and accepts no responsibility for incomplete or inaccurate information. In particular, ACARA does not endorse or verify that:
- The content descriptions are solely for a particular year and subject;
- All the content descriptions for that year and subject have been used; and
- The author's material aligns with the Australian Curriculum content descriptions for the relevant year and subject.

You can find the unaltered and most up to date version of this material at http://www.australiancurriculum.edu.au/ This material is reproduced with the permission of ACARA.

Every effort has been made to trace and acknowledge copyright. However, if any infringement has occurred, the publishers tender their apologies and invite the copyright holders to contact them.

Some of the images used in *Australian Signpost Maths 5* might have associations with deceased Indigenous Australians. Please be aware that these images might cause sadness or distress in Aboriginal or Torres Strait Islander communities.

What is Australian Signpost Maths?

Australian Signpost Maths is a mathematics activity book series for students from Foundation to Year 6. The series has been written to meet the requirements of the Australian Curriculum.

The components of the series include Student Books, Teacher's Books, Mentals Books and an interactive Website. Teachers can select an appropriate program for every student from the rich and varied material provided.

The content has been carefully sequenced within each year level and across the series to take into account students' likely mathematical development.

Student Books   Teacher's Books   Mentals Books   Website

Structure of Australian Signpost Maths

Australian Signpost Maths emphasises the curriculum's syllabus content as well as problem-solving strategies, language development and the use of technology.

The syllabus is organised into three content strands and four proficiency strands:

Content Strands

- Number and Algebra
- Measurement and Geometry
- Statistics and Probability

Proficiency Strands (see page iv)

- Understanding
- Fluency
- Problem Solving
- Reasoning

The curriculum's **general capabilities** are developed throughout the Australian Signpost Maths program. These are:

- literacy
- numeracy
- information and communication technologies (ICT)
- critical and creative thinking.

Australian Signpost Maths also provides opportunities to develop other general capabilities, such as personal and social competence and intercultural understanding.

The cross-curriculum dimensions of the syllabus – 'Aboriginal and Torres Strait Islander histories and cultures', 'Asia and Australia's engagement with Asia' and 'Sustainability' – are embedded in the program.

To maximise the benefits of the program, the Student Book, Teacher's Book, Mentals Book and Website should be used together.

The structure of the **Student Book** allows teachers to determine both the order and the extent of content covered. Strands are organised separately so that the teacher, not the Student Book, decides the content of the next lesson. However, a suggested term program (see page xii of this book) and a detailed program (see the Teacher's Book and Website) are also provided.

The **Teacher's Book** also provides lesson plans for each page of the Student Book and blackline masters to assist teachers in implementing the program.

The **Mentals Book** mixes examples from all strands. It revises the content covered in the Student Book. Each content strand is thoroughly covered, with the proficiency strands incorporated within each section. A special feature woven throughout the Mentals Book is the tables program in the four operations.

The innovative **Website** help teachers to bring mathematics alive with technology. The website provides interactive maths tools, games and practice opportunities as well as relevant resource masters and worksheets for all year levels. These can be used for whole-class, small-group and individual learning. The website also includes **Concept Check-In** a new diagnostic screener.

Student Book pages are colour-coded by section.

Number and Algebra A | Measurement and Geometry A | Statistics and Probability
Number and Algebra B | Measurement and Geometry B | Answers

Australian Curriculum Proficiency Strands

The proficiency strands of the Australian Curriculum describe how content is explored or developed – that is, the 'thinking and doing' of mathematics.

Understanding

Learning the concepts

*Students build a robust knowledge of adaptable and transferable mathematical concepts. They make connections between related concepts and progressively apply the familiar to develop new ideas. They develop an understanding of the relationship between the 'why' and the 'how' of mathematics.**

Conceptual understanding of maths ideas includes the explanation of a concept using text and diagrams. This occurs throughout Australian Signpost Maths at the top of many pages and is indicated by the Concepts icon.

Fluency

Using the concepts

*Students develop skills in choosing appropriate procedures, carrying out procedures flexibly, accurately, efficiently and appropriately, and recalling factual knowledge and concepts readily.**

The practice of maths skills to build fluency occurs on every page of Australian Signpost Maths.

Problem Solving

Applying concepts and strategies to develop solutions to problems

*Students develop the ability to make choices, interpret, formulate, model and investigate problem situations, and communicate solutions effectively.**

Problem solving provides opportunities for students to use strategies and skills such as investigating and questioning, to collaborate with others and to communicate their findings to different audiences. Such activities are often indicated throughout Australian Signpost Maths by the Activity and Investigation icons.

Reasoning

Coherent and logical thought

*Students develop an increasingly sophisticated capacity for logical thought and actions, such as analysing, proving, evaluating, explaining, inferring, justifying and generalising.**

Students require opportunities to explain their mathematical thinking and can do so through both diagrams and written explanations. Reasoning questions are located throughout Australian Signpost Maths.

* The Australian Curriculum: Mathematics, v8.3 – Content structure

Special Features of Australian Signpost Maths

- **Traffic Light** system allows students to reflect on their work and highlight any units that they are having trouble understanding. They tick the red for units they feel they still don't understand, and green for those they feel they understand fully.

- Exercises are **well graded**. New work is reinforced in the Mentals Book.
- **Answers** are supplied in the back of this book as well as in the Teacher's Book.
- **Concept Check-In** diagnostic screener (on the Website) provides a snapshot of the class' conceptual understandings to aid in classroom management. It also allows teachers to measure progress over time.
- The eight **Diagnostic Tests** (now also in the back of this book) allow the teacher to discover each student's strengths and weaknesses, and the cross-references direct students to the pages where that work is introduced. Answers are supplied in the Teacher's Book.
- The **Dictionary** at the beginning of this Student Book will help students to learn the language of mathematics.
- **ID Cards** (in the Mentals Book, Teacher's Book and Website) review the language of mathematics by asking students to identify common terms, shapes and symbols.
- Important **rules and concepts** are clearly highlighted.
- **Worked examples** and explanations are given throughout the Student Book where new ideas are introduced.
- The use of **colour** makes emphasis clear and is highly motivating.
- **Cartoons** give instruction and friendly advice.
- **Interactive activities** are provided on the website for whole-class, small-group and individual learning.

Australian Signpost Icons

Signpost icons are used throughout the book as cues to the essential nature of exercises and activities, and as a guide to ways of engaging with them. These icons often indicate alternative or more concrete approaches to dealing with concepts.

This icon highlights **important rules and concepts** occurring throughout the book. It often appears with worked examples.

Activities provide **applications and enrichment**. These activities usually involve the use of concrete materials and partner or group work.

These enjoyable activities are used to **motivate and involve** students in mathematical pursuits. They include games and puzzles.

Investigations allow students to explore and discover maths concepts.

This icon indicates the use of computers, calculators or other **information and communications technology**.

5 Contents and Syllabus Overview

KEY

■ (blue)	Number and Algebra
■ (green)	Measurement and Geometry
■ (orange)	Statistics and Probability

Number and Algebra A

Page	Unit	Title	Sub-strand	Number and place value	Fractions and decimals	Patterns and algebra	Content	Counting and numeration	Place value	Fractions	Decimals	Number patterns	Suggested progress
1	1:01	Numbers to One Million		■				●	●				Term 1
2	1:02	Numbers Above One Million		■				●	●				
3	1:03	Powers of Ten		■				●	●				
4	1:04	Hundredths			■					●	●		
5	1:05	Fractions			■					●			
6	1:06	Unit Fractions			■					●			T1, T2*
7	1:07	Tenths			■					●	●		Term 2
8	1:08	Decimals			■					●	●		
9	1:09	Place Value in Decimals		■	■				●	●	●		
10	1:10	Place Value to Thousandths		■	■				●	●	●		
11	1:11	Reading and Writing Decimals		■	■				●		●		T3, T4*
12	1:12	Comparing Decimals		■	■				●		●		
13	1:13	Place Value to Thousandths		■	■				●	●	●		
14	1:14	Addition of Fractions			■					●			Term 3
15	1:15	Subtraction of Fractions			■					●			
16	1:16	Addition and Subtraction of Fractions			■					●			
17	1:17	Addition and Subtraction of Fractions			■					●			
18	1:18	Addition and Subtraction of Fractions			■					●			
19	1:19	Comparing Decimal Measurements			■						●		T5, T6*
20	1:20	Using Decimals			■	■					●	●	
21	1:21	Patterns with Fractions and Decimals			■	■				●	●	●	
22	1:22	Equivalent Fractions			■					●			
23	1:23	Equivalent Fractions			■					●			
24	1:24	Equivalent Fractions			■					●			Term 4
25	1:25	Percentages			■					●			T7, T8*
26	1:26	Using Percentages			■					●			

* Suggested progress for Diagnostic Tests 1 to 8 is found in the Teacher's Book.
The first of each pair of tests covers the first half of the period.
It is assumed that there are 10 weeks in each term.

Number and Algebra B

Page	Unit	Title	Sub-strand	Number and place value	Money and financial mathematics	Patterns and algebra	Content	Addition	Subtraction	Multiplication	Division	Place value	Number patterns	Suggested progress
27	2:01	Number Facts, ×2, ×3, ×4, ×5, ×10		■						●				Term 1
28	2:02	Number Facts, ×6, ×7, ×8, ×9		■						●				
29	2:03	Learning your Multiplication Tables		■						●				
30	2:04	Division Facts		■		■					●			
31	2:05	Rounding		■								●		
32	2:06	Addition to 999		■				●						
33	2:07	Addition to 999		■				●						
34	2:08	Subtraction Without Trading to 999		■					●					
35	2:09	Writing the Addition Algorithm		■				●						
36	2:10	Subtraction with Trading to 999		■					●					
37	2:11	Subtraction with Trading to 999		■					●					
38	2:12	Multiples		■						●				
39	2:13	Factors		■						●	●			T1, T2*
40	2:14	Addition of Money		■	■			●						
41	2:15	Subtraction of Money		■	■				●					
42	2:16	Shopping		■	■			●	●					Term 2
43	2:17	Using Strategies to Solve Problems		■		■				●	●		●	
44	2:18	Division with Remainders		■							●			
45	2:19	Division of 2-Digit Numbers		■							●			
46	2:20	Using Division Facts		■							●			
47	2:21	Number Patterns				■		●	●				●	
48	2:22	Subtraction with Trading to 999		■	■				●					
49	2:23	Subtraction from Hundreds		■	■				●					
50	2:24	Addition to 9999		■	■			●						
51	2:25	Addition to 9999		■	■			●						
52	2:26	Subtraction with Trading to 9999		■					●					
53	2:27	Four-Digit Subtraction from Thousands		■					●					
54	2:28	Subtraction from Thousands Strategy		■	■				●					
55	2:29	Mental Strategies		■				●	●					T3, T4*
56	2:30	Factors and Multiples		■						●	●			
57	2:31	Factors and Multiples		■						●	●			
58	2:32	Dividing 2-Digit Numbers		■							●			
59	2:33	Dividing 2-Digit Numbers		■							●			
60	2:34	Dividing 2-Digit Numbers		■							●			
61	2:35	Dividing 3-Digit Numbers		■							●			

Number and Algebra B

Page	Unit	Title	Sub-strand: Number and place value	Sub-strand: Money and financial mathematics	Sub-strand: Patterns and algebra	Content: Addition	Content: Subtraction	Content: Multiplication	Content: Division	Content: Place value	Content: Number patterns	Suggested progress
62	2:36	Multiplying Tens	■	■				●				Term 3
63	2:37	Multiplying Tens or Hundreds	■					●				
64	2:38	Dividing 3-Digit Numbers by 10	■						●			
65	2:39	Division Involving Zeros in Answers	■						●			
66	2:40	Divisibility	■		■				●			
67	2:41	Factors and Multiples	■					●	●			
68	2:42	Averages	■			●			●			
69	2:43	Averages	■			●			●			
70	2:44	Using Factors in Multiplication	■					●				
71	2:45	Mental Strategies for Multiplication	■					●				
72	2:46	Number Patterns			■						●	
73	2:47	Number Patterns			■						●	
74	2:48	Multiplying 2-Digit Numbers	■					●				
75	2:49	Introducing Extended Multiplication	■					●				T5, T6*
76	2:50	The Extended Form of Multiplication	■					●				
77	2:51	The Extended Form of Multiplication	■					●				
78	2:52	Estimating by Rounding	■			●	●			●		
79	2:53	Estimating Products	■					●				
80	2:54	The Contracted Form of Multiplication	■	■				●				
81	2:55	The Contracted Form of Multiplication	■	■				●				
82	2:56	Using Algorithms to Solve Problems	■			●	●					Term 4
83	2:57	Problems Involving Change of Units	■			●	●	●	●			
84	2:58	Estimation by Rounding	■			●	●	●	●			
85	2:59	Estimating Products	■					●				
86	2:60	Making a Budget	■	■		●		●				
87	2:61	Shopping	■	■		●	●					
88	2:62	Using Operations to Solve Problems	■	■		●	●	●	●			T7, T8*
89	2:63	Strategies for Multiplication	■					●				
90	2:64	Finding Missing Numbers	■		■	●	●	●	●		●	
91	2:65	Finding Missing Numbers	■		■	●	●	●	●		●	
92	2:66	Using Strategies to Solve Problems	■								●	
93	2:67	Problem Solving	■	■		●					●	
94	2:68	Number Machines	■		■	●	●	●			●	

Measurement and Geometry A

Page	Unit	Title	Sub-strand	Using units of measurement	Content	Length	Area	Volume	Capacity	Mass	Time	Temperature	Suggested progress
95	3:01	Time Units									●		Term 1
96	3:02	Kilometres				●							
97	3:03	Kilometres and Metres				●							
98	3:04	Perimeter				●							
99	3:05	Perimeter				●							T1, T2*
100	3:06	Calculating Area					●						Term 2
101	3:07	Square Metres					●						
102	3:08	Hectares					●						
103	3:09	Area					●						
104	3:10	Cubic Centimetres						●					T3, T4*
105	3:11	Cubic Centimetres						●					
106	3:12	24-Hour Time									●		Term 3
107	3:13	Using 12- and 24-Hour Time									●		
108	3:14	Cubic Metres						●					
109	3:15	Volume of Rectangular Prisms						●					
110	3:16	Millimetres				●							
111	3:17	Perimeter				●							T5, T6*
112	3:18	24-Hour Time									●		
113	3:19	Problems Involving Time									●		
114	3:20	Tonnes								●			Term 4
115	3:21	Grams and Kilograms								●			
116	3:22	Using Measurement Scales				●			●	●		●	
117	3:23	Converting Measurements				●							T7, T8*
118	3:24	Stopwatches									●		
119	3:25	Exploring Perimeter, Area and Volume				●	●	●					

* Suggested progress for Diagnostic Tests 1 to 8 is found in the Teacher's Book.
The first of each pair of tests covers the first half of the period.

Measurement and Geometry B			Sub-strand	Shape	Location and transformation	Geometric reasoning	Content	2D Space	3D Space	Location	Transformations	Angles	Suggested progress
Page	**Unit**	**Title**											
120	4:01	Review of 3D Space		■					●				Term 1
121	4:02	Prisms and Pyramids		■					●				
122	4:03	Translations, Reflections and Rotations			■						●		
123	4:04	Translations, Reflections and Rotations			■			●			●		
124	4:05	Nets		■					●				T1, T2*
125	4:06	Describing Position			■			●		●			
126	4:07	Measuring Angles Using a Protractor				■						●	
127	4:08	Angle Types in Degrees				■						●	Term 2
128	4:09	Using a Protractor				■						●	
129	4:10	Classifying Angles				■						●	T3, T4*
130	4:11	Compass Directions			■					●			
131	4:12	Reading a Street Directory			■					●			
132	4:13	Rotational Symmetry			■			●			●		
133	4:14	Measuring Angles of Rotation			■	■					●		
134	4:15	Rotational Symmetry			■			●			●		Term 3
135	4:16	Views and Nets of 3D Objects		■					●				
136	4:17	Reading a Street Directory			■					●			T5, T6*
137	4:18	Using Coordinates			■					●			
138	4:19	Drawing Angles				■						●	Term 4
139	4:20	Measuring Angles Greater Than 180°				■						●	
140	4:21	Enlargements and Reductions			■						●		
141	4:22	Enlargements and Reductions			■						●		T7, T8*
142	4:23	Enlargements and Reductions			■						●		

* Suggested progress for Diagnostic Tests 1 to 8 is found in the Teacher's Book.
The first of each pair of tests covers the first half of the period.

Statistics and Probability			Sub-strand: Chance	Sub-strand: Data representation and interpretation	Content: Chance	Content: Investigation	Content: Data representation	Suggested progress
Page	**Unit**	**Title**						
143	5:01	Reading Graphs		■			●	Term 1
144	5:02	Drawing Graphs		■			●	
145	5:03	Drawing Picture Graphs		■			●	
146	5:04	Reading Line Graphs		■			●	T1, T2*
147	5:05	Drawing Line Graphs		■			●	
148	5:06	Dot Plots		■			●	Term 2
149	5:07	Drawing Dot Plots		■			●	
150	5:08	The Likelihood of an Event	■		●			
151	5:09	Chance	■		●			
152	5:10	Divided Bar Graphs		■			●	T3, T4*
153	5:11	Sector (or Pie) Graphs		■			●	
154	5:12	Questionnaires/Surveys		■		●	●	
155	5:13	Data Investigation		■		●		Term 3
156	5:14	More Line Graphs		■			●	
157	5:15	Information Collected Over Time		■			●	T5, T6*
158	5:16	Chance	■		●			
159	5:17	Chance Events	■		●			
160	5:18	Comparing Types of Graphs		■			●	Term 4
161	5:19	Selecting the Best Graph to Present Data		■			●	
162	5:20	Collecting Data from Experiments	■			●		
163	5:21	Collecting Data	■			●		
164	5:22	Reasoning with Graphs		■		●	●	T7, T8*
165	5:23	Comparing Mobile Phone Plans		■		●	●	

Suggested Program

	Weeks 1–10	Weeks 11–20	Weeks 21–30	Weeks 31–end
Number and Algebra A	1:01–1:06	1:07–1:13	1:14–1:23	1:24–1:26
Number and Algebra B	2:01–2:15	2:16–2:35	2:36–2:55	2:56–2:68
Measurement and Geometry A	3:01–3:05	3:06–3:11	3:12–3:19	3:20–3:25
Measurement and Geometry B	4:01–4:07	4:08–4:14	4:15–4:18	4:19–4:23
Statistics and Probability	5:01–5:05	5:06–5:12	5:13–5:17	5:18–5:23

The eight Diagnostic Tests are found in the Teacher's Book.
See the Contents and Syllabus Overview on pages vi–xi for suggested placement of each test.
It is assumed that there are 10 weeks in each term.

Contents Cross-reference

Numbers and Algebra

Multiplication of large numbers by 1-digit numbers	62, 63, 70, 71, 74, 75, 76, 77, 79, 80, 81	Identify and describe factors and multiples of whole numbers and use them to solve problems (ACMNA098); Use estimation and rounding to check the reasonableness of answers to calculations (ACMNA099); Solve problems involving multiplication of large numbers by one- or two-digit numbers using efficient mental, written strategies and appropriate digital technologies (ACMNA100)
Multiplication of large numbers by 2-digit numbers	70, 71, 85, 89	Identify and describe factors and multiples of whole numbers and use them to solve problems (ACMNA098); Use estimation and rounding to check the reasonableness of answers to calculations (ACMNA099); Solve problems involving multiplication of large numbers by one- or two-digit numbers using efficient mental, written strategies and appropriate digital technologies (ACMNA100)
Mental strategies	62, 63, 70, 71, 74, 75, 89	Identify and describe factors and multiples of whole numbers and use them to solve problems (ACMNA098); Solve problems involving multiplication of large numbers by one- or two-digit numbers using efficient mental, written strategies and appropriate digital technologies (ACMNA100)
Multiples, factors and square numbers	27, 28, 29, 38, 39, 43, 56, 57, 66, 67, 70, 88, 91	Recall multiplication facts up to 10 × 10 and related division facts (ACMNA075); Identify and describe factors and multiples of whole numbers and use them to solve problems (ACMNA098); Solve problems involving multiplication of large numbers by one- or two-digit numbers using efficient mental, written strategies and appropriate digital technologies (ACMNA100); Use efficient mental and written strategies and apply appropriate digital technologies to solve problems (ACMNA291)
Problem solving	39, 75, 77, 83, 84, 86, 88, 90, 91, 156, 165	Identify and describe factors and multiples of whole numbers and use them to solve problems (ACMNA098); Use estimation and rounding to check the reasonableness of answers to calculations (ACMNA099); Solve problems involving multiplication of large numbers by one- or two-digit numbers using efficient mental, written strategies and appropriate digital technologies (ACMNA100); Use efficient mental and written strategies and apply appropriate digital technologies to solve problems (ACMNA291); Create simple financial plans (ACMNA106); Use equivalent number sentences involving multiplication and division to find unknown quantities (ACMNA121)
5 Division		
Division facts	30, 46, 66	Solve problems involving division by a one digit number, including those that result in a remainder (ACMNA101)
Linking multiplication and division	30, 66	Use equivalent number sentences involving multiplication and division to find unknown quantities (ACMNA121); Recall multiplication facts up to 10 × 10 and related division facts (ACMNA075)
Dividing by 1-digit numbers	44, 45, 46, 58, 59, 60, 61, 64, 65	Solve problems involving division by a one digit number, including those that result in a remainder (ACMNA101); Use efficient mental and written strategies and apply appropriate digital technologies to solve problems (ACMNA291)
Problem solving	44, 45, 46, 59, 60, 61, 64, 65, 68, 69, 83, 84, 88, 90, 91	Identify and describe factors and multiples of whole numbers and use them to solve problems (ACMNA098); Use estimation and rounding to check the reasonableness of answers to calculations (ACMNA099); Solve problems involving division by a one digit number, including those that result in a remainder (ACMNA101); Use efficient mental and written strategies and apply appropriate digital technologies to solve problems (ACMNA291); Use equivalent number sentences involving multiplication and division to find unknown quantities (ACMNA121)

6 Fraction and decimals	Pages	Australian Curriculum Reference
Common fractions	4, 5, 6, 7, 8, 9, 25, 26	Model and represent unit fractions including $\frac{1}{2}, \frac{1}{4}, \frac{1}{3}, \frac{1}{5}$ and their multiples to a complete whole (ACMNA058); Recognise that the place value system can be extended to tenths and hundredths, and make connections between fractions and decimal notation (ACMNA079); Compare and order common unit fractions and locate and represent them on a number line (ACMNA102)
Mixed numbers and improper fractions	9, 17, 18, 60	Count by quarters, halves and thirds, including with mixed numerals. Locate and represent these fractions on a number line (ACMNA078)
Count using fractions	7, 17, 21, 22	Compare and order common unit fractions and locate and represent them on a number line (ACMNA102); Describe, continue and create patterns with fractions, decimals and whole numbers resulting from addition and subtraction (ACMNA107)
Compare and order fractions	6, 7, 14, 15, 16	Recognise that the place value system can be extended to tenths and hundredths, and make connections between fractions and decimal notation (ACMNA079); Compare and order common unit fractions and locate and represent them on a number line (ACMNA102)
Equivalent fractions	22, 23, 24	Investigate equivalent fractions used in contexts (ACMNA077); Compare and order common unit fractions and locate and represent them on a number line (ACMNA102)
Addition and subtraction of fractions	7, 14, 15, 16, 17, 18	Investigate strategies to solve problems involving addition and subtraction of fractions with the same denominator (ACMNA103)
Decimal notation and place value	4, 7, 8, 9, 10, 11, 12, 13, 19, 25, 26	Recognise that the place value system can be extended to tenths and hundredths, and make connections between fractions and decimal notation (ACMNA079); Recognise that the place value system can be extended beyond hundredths (ACMNA104); Compare, order and represent decimals (ACMNA105)
Compare and order decimals	12, 13, 19	Recognise that the place value system can be extended beyond hundredths (ACMNA104); Compare, order and represent decimals (ACMNA105)
Addition and subtraction of decimals	20, 21	Describe, continue and create patterns with fractions, decimals and whole numbers resulting from addition and subtraction (ACMNA107)
Percentages	25, 26	Make connections between equivalent fractions, decimals and percentages (ACMNA131)
7 Money		
Operations with money	40, 41, 42, 48, 49, 50, 51, 62, 75, 80, 81, 88, 93	Use estimation and rounding to check the reasonableness of answers to calculations (ACMNA099); Solve problems involving multiplication of large numbers by one- or two-digit numbers using efficient mental, written strategies and appropriate digital technologies (ACMNA100); Use efficient mental and written strategies and apply appropriate digital technologies to solve problems (ACMNA291); Create simple financial plans (ACMNA106)
Simple financial plans	86, 87	Use efficient mental and written strategies and apply appropriate digital technologies to solve problems (ACMNA291); Create simple financial plans (ACMNA106)

8 Number patterns	Pages	Australian Curriculum Reference
Patterns and pattern rules	20, 21, 22, 47, 72, 73, 92, 93, 94	Describe, continue and create patterns with fractions, decimals and whole numbers resulting from addition and subtraction (ACMNA107)
Number sentences with missing elements	30, 66, 90, 91	Use efficient mental and written strategies and apply appropriate digital technologies to solve problems (ACMNA291); Use equivalent number sentences involving multiplication and division to find unknown quantities (ACMNA121)
Solving unusual problems	43, 92, 93, 143, 144	Identify and describe factors and multiples of whole numbers and use them to solve problems (ACMNA098); Use efficient mental and written strategies and apply appropriate digital technologies to solve problems (ACMNA291)
9 Technology		
Using a calculator	26, 56, 58, 67, 73, 84, 85, 86, 87	Use efficient mental and written strategies and apply appropriate digital technologies to solve problems (ACMNA291)
Using computers	134, 155	Use efficient mental and written strategies and apply appropriate digital technologies to solve problems (ACMNA291)

Measurement and Geometry

1 Using units of measurement	Pages	Australian Curriculum Reference
Reading scales	116, 117, 118	Use scaled instruments to measure and compare lengths, masses, capacities and temperatures (ACMMG084); Choose appropriate units of measurement for length, area, volume, capacity and mass (ACMMG108)
Length	83, 88, 96, 97, 98, 99, 110, 111, 116, 117	Choose appropriate units of measurement for length, area, volume, capacity and mass (ACMMG108); Calculate the perimeter and area of rectangles using familiar metric units (ACMMG109)
Perimeter	98, 99, 110, 111, 119	Choose appropriate units of measurement for length, area, volume, capacity and mass (ACMMG108); Calculate the perimeter and area of rectangles using familiar metric units (ACMMG109)
Area	100, 101, 102, 103, 119	Choose appropriate units of measurement for length, area, volume, capacity and mass (ACMMG108); Calculate the perimeter and area of rectangles using familiar metric units (ACMMG109)
Volume	104, 105, 108, 109, 119	Choose appropriate units of measurement for length, area, volume, capacity and mass (ACMMG108);
Capacity	83, 88, 116	Choose appropriate units of measurement for length, area, volume, capacity and mass (ACMMG108);
Mass	83, 88, 114, 115, 116	Choose appropriate units of measurement for length, area, volume, capacity and mass (ACMMG108);
Temperature	116, 155	Choose appropriate units of measurement for length, area, volume, capacity and mass (ACMMG108);
Time	83, 88, 95, 106, 107, 112, 113, 118	Choose appropriate units of measurement for length, area, volume, capacity and mass (ACMMG108); Compare 12- and 24- hour time systems and convert between them (ACMMG110)
12- and 24-hour time	95, 106, 107, 112, 113	Compare 12- and 24- hour time systems and convert between them (ACMMG110)
Time problems	83, 88, 95, 107, 112, 113, 118	Compare 12- and 24- hour time systems and convert between them (ACMMG110)
2 Shape		
2D shapes	123, 125, 132, 134	Describe translations, reflections and rotations of two-dimensional shapes. Identify line and rotational symmetries (ACMMG114)
3D shapes	120, 121, 124, 135	Connect three-dimensional objects with their nets and other two-dimensional representations (ACMMG111)
Cones, cylinders and spheres	120, 124, 135	Connect three-dimensional objects with their nets and other two-dimensional representations (ACMMG111)
Prisms and pyramids	120, 121, 124, 135	Connect three-dimensional objects with their nets and other two-dimensional representations (ACMMG111)
Angles	125, 126, 127, 128, 129, 133, 138, 139	Estimate, measure and compare angles using degrees. Construct angles using a protractor (ACMMG112)

3 Location and transformation	Pages	Australian Curriculum Reference
Maps, scales and legends	130, 131, 136, 137, 142	Use a grid reference system to describe locations and describe routes using landmarks and directional language (ACMMG113)
Using compass directions	130, 137	Use simple scales, legends and directions to interpret information contained in basic maps (ACMMG090); Use a grid reference system to describe locations and describe routes using landmarks and directional language (ACMMG113)
Using coordinates and grid references	125, 131, 136, 137	Use a grid reference system to describe locations and describe routes using landmarks and directional language (ACMMG113)
Describe routes	131, 136	Use a grid reference system to describe locations and describe routes using landmarks and directional language (ACMMG113)
Translations, reflections and rotations	122, 123, 132, 133, 134	Describe translations, reflections and rotations of two-dimensional shapes. Identify line and rotational symmetries (ACMMG114)
The enlargement transformation	140, 141, 142	Apply the enlargement transformation to familiar two-dimensional shapes and explore the properties of the resulting image compared with the original (ACMMG115)

Statistics and Probability

1 Chance	Pages	Australian Curriculum Reference
Chance experiments	149, 162, 163	List outcomes of chance experiments involving equally likely outcomes and represent probabilities of those outcomes using fractions (ACMSP116)
Describing likelihood	150, 151, 158, 159, 162, 163	List outcomes of chance experiments involving equally likely outcomes and represent probabilities of those outcomes using fractions (ACMSP116)
2 Data representation and interpretation		
Data collection	68, 69, 149, 154, 155, 157, 162, 163	List outcomes of chance experiments involving equally likely outcomes and represent probabilities of those outcomes using fractions (ACMSP116); Pose questions and collect categorical or numerical data by observation or survey (ACMSP118); Describe and interpret different data sets in context (ACMSP120)
Surveys	154	Pose questions and collect categorical or numerical data by observation or survey (ACMSP118)
Tables	145, 147, 148, 155, 156, 157, 160, 161, 165	Construct displays, including column graphs, dot plots and tables, appropriate for data type, with and without the use of digital technologies (ACMSP119); Describe and interpret different data sets in context (ACMSP120)
Drawing graphs	144, 145, 147, 152, 157, 161	Construct displays, including column graphs, dot plots and tables, appropriate for data type, with and without the use of digital technologies (ACMSP119)
Interpreting graphs	143, 146, 147, 148, 149, 152, 153, 155, 156, 157, 160, 164	Describe and interpret different data sets in context (ACMSP120)
Dot plots	148, 149	Construct displays, including column graphs, dot plots and tables, appropriate for data type, with and without the use of digital technologies (ACMSP119)

abacus
An instrument used for counting and calculating.

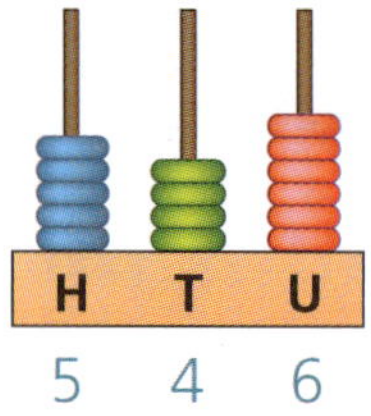

am (ante meridiem)
Any time between midnight and midday.

- The time is 25 past 7 in the morning. It is 7:25 am.

analogue time
The time shown on a clock face.

- 13 minutes to 6 is the time on this analogue clock.

angle
The amount of turning between two arms about a common point.

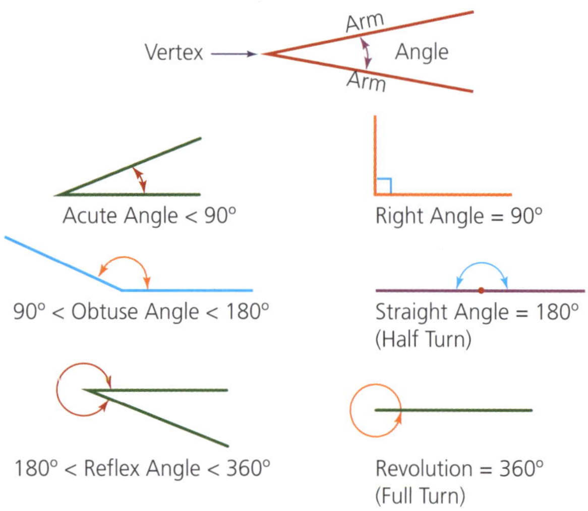

- Acute Angle
  An angle less than 90 degrees.
- Obtuse Angle
  An angle greater than 90 degrees and less than 180 degrees.
- Right Angle
  An angle that equals 90 degrees.

area
The size of a surface.

Area is measured in square units.

- **square centimetres: cm^2**
- **square metres: m^2**
- 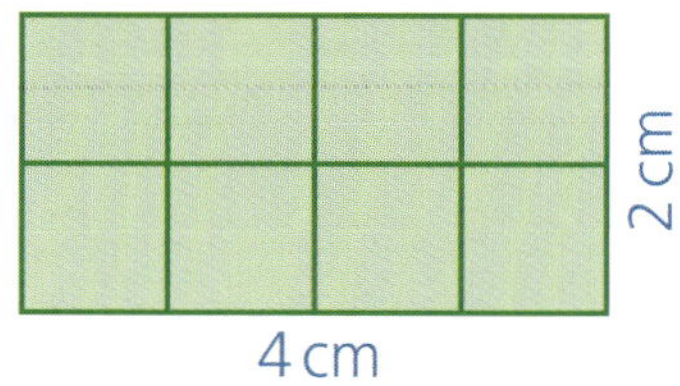

  Area = 8 cm^2

ascending order
Arranged in order from least value to greatest value.

- \$1.65 (least) \$2.30 \$4.75 \$5.96 (most)

average
A fair share.

Average = (sum of scores) ÷ (number of scores)

axis of symmetry
See *line of symmetry*.

A line that divides a picture in half so that each half is the mirror image of the other part.

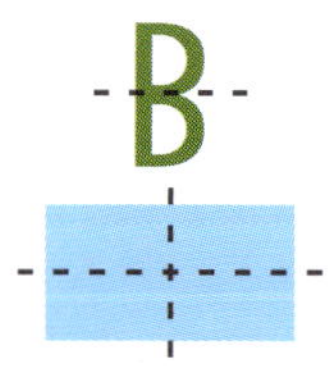

- One Axis of Symmetry
- Two Axes of Symmetry

The plural of *axis* is *axes*.

capacity
The amount that a container can hold.

- The capacity of this juice bottle is 250 mL.

centimetre (cm)

A unit of length equal to one hundredth of a metre.

- 100 cm = 1 m

chance

The chance (or probability) of something happening is its likelihood of happening.

- If you toss a coin, there is an even chance of tossing a head.

See *probability*.

compass directions

The needle of a compass points north (N).

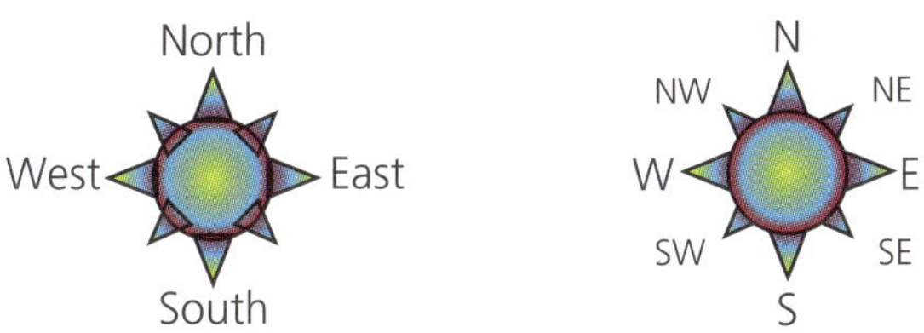

composite number

A number that has more than two factors.

- 9 is composite because it has three factors: 1, 3 and 9.

cone

A three-dimensional object with a circular base that tapers to a point.

coordinates

Pairs of letters or numbers used to show position on a grid.

- This position is D3 or (D, 3).

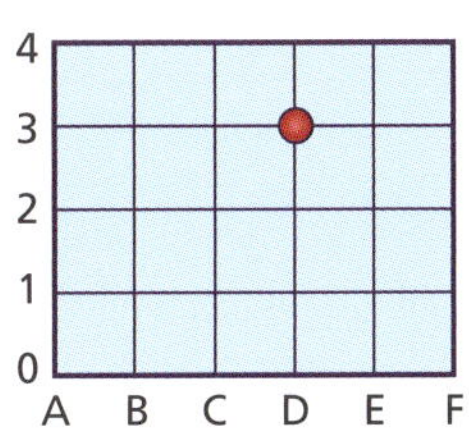

cross-section

A face that is exposed when a 3D object is cut through.

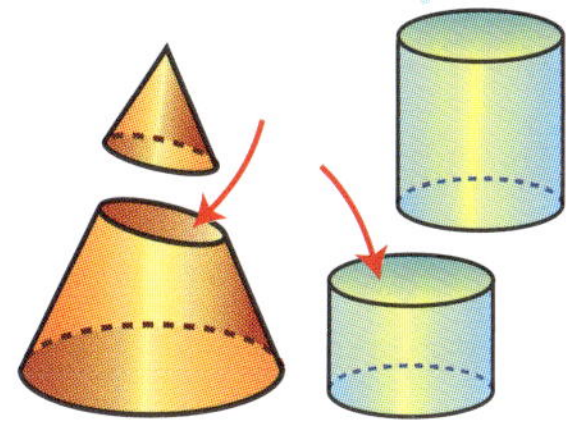

cube

A three-dimensional object that has six equal square faces, eight vertices and twelve equal edges.

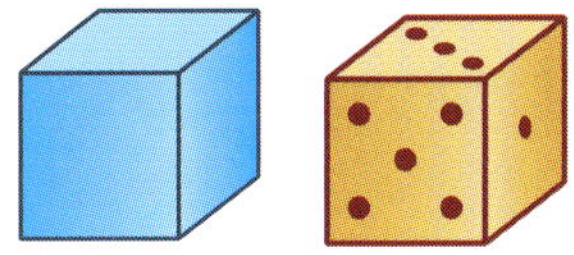

cube number

- 2 cubed = 2^3 ← Index
  = 2 × 2 × 2
  = 4 × 2
  = 8

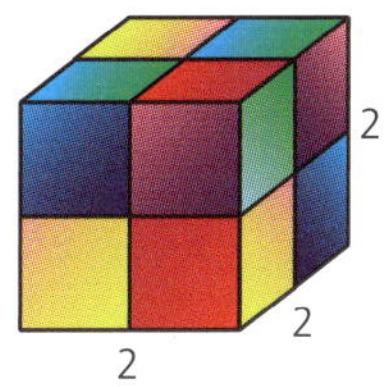

cubic centimetre (cm^3)

A unit of volume equal to the volume of a cube of side length 1 cm.

cubic metre (m^3)

A unit of volume equal to the volume of a cube of side length 1 m.

cylinder

A three-dimensional object with two equal circular faces and one curved surface.

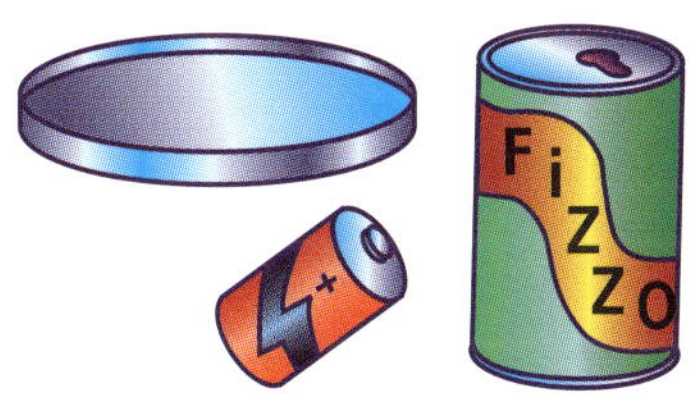

decimal

A point or dot used to separate a decimal fraction from a whole number.

- 36·9
  ↑
  Decimal Point

decrease

To make smaller or less.

denominator

The bottom number of a fraction.
It tells the number of equal parts there are in the whole.

Numerator → 7
Denominator → 8

descending order

Arranged in order from greatest value to least value.

- $5.96 (most) $4.75 $2.30 $1.65 (least)

diagonal

A line that joins any two non-adjacent corners of a polygon.

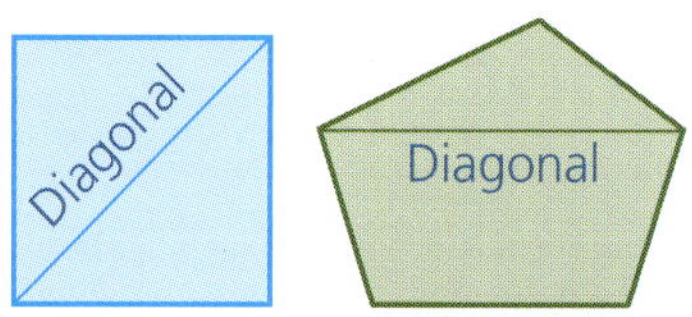

digital time

Time expressed using digits.

- This digital clock shows 24 minutes past 10.

divisible

To have no remainder when divided.

- 30 is divisible by 3.

division (÷)

Breaking up groups into equal parts.

- 6 ÷ 3
  - **a** Sharing 2 each

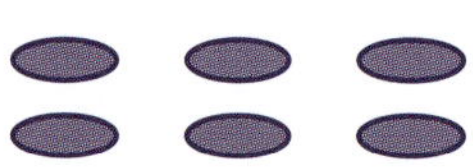

  - **b** Grouping 2 groups

enlargement

To make bigger or smaller (reduction) using a scale factor.

- A scale factor of 2 doubles lengths.

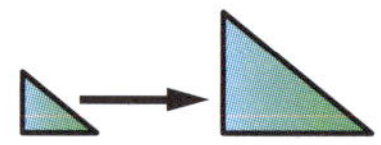

- A scale factor of $\frac{1}{2}$ halves lengths.

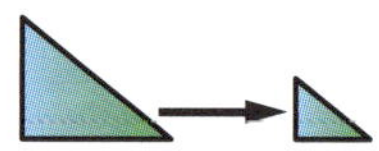

estimate (estimation)

A good guess.

expanded notation

A way of writing numerals to show the place value of each digit.

- 137 = (1 × 100) + (3 × 10) + 7

face

A flat surface of a three-dimensional object.

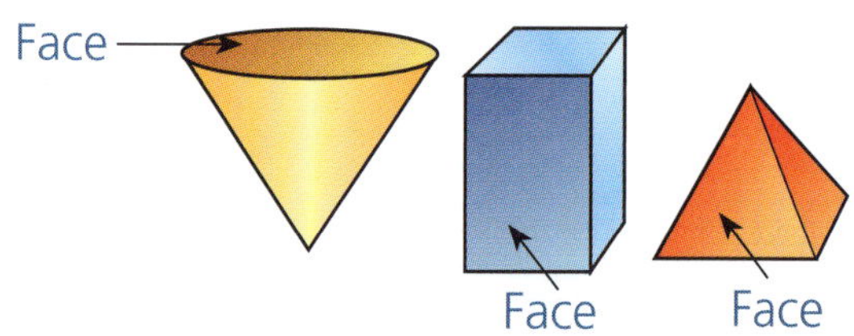

factor

A factor of a number divides the number exactly, leaving no remainder.

- The factors of 12 are 1, 12, 2, 6, 3 and 4.

flip (reflection)

To turn over.

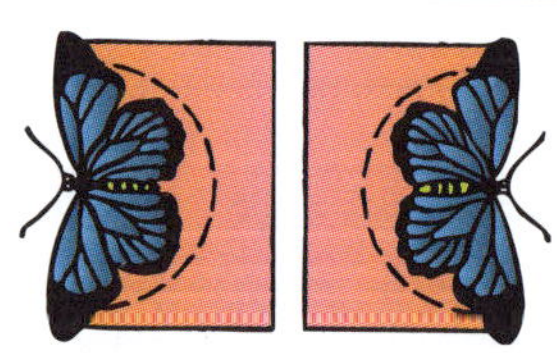

fraction

A part of a whole or group.

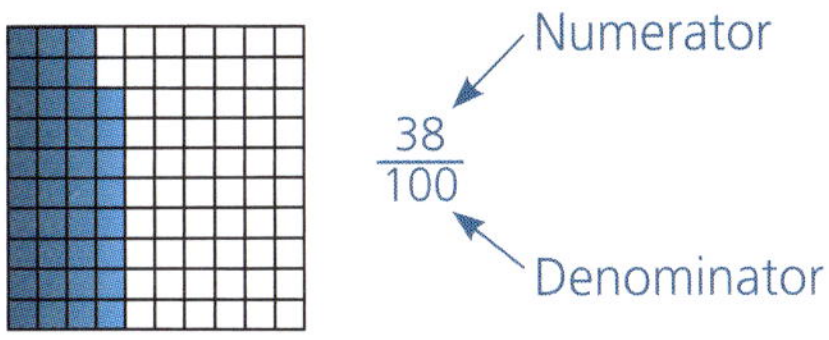

Equivalent fraction

Fractions of equal size.

- $\frac{1}{2} = \frac{5}{10} = \frac{7}{14} = \dots$

Improper fraction

A fraction which has a numerator that is bigger than the denominator.

- $\frac{9}{8}$

Mixed numeral

A numeral that has a whole number part and a fraction part.

- $1\frac{2}{3}$

gram (g)

A unit of mass.

- 1 kilogram = 1000 grams

graphs

- Bar Graph

  A graph which uses horizontal bars to compare the size of groups.

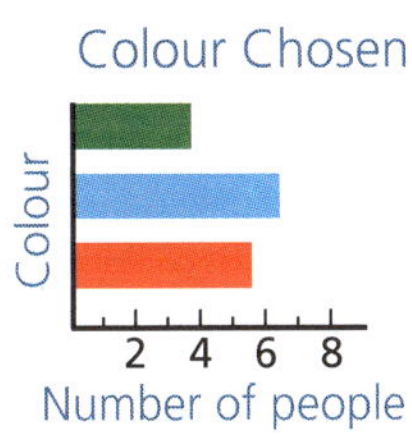

- Column Graph

  Groups are compared using the heights of columns.

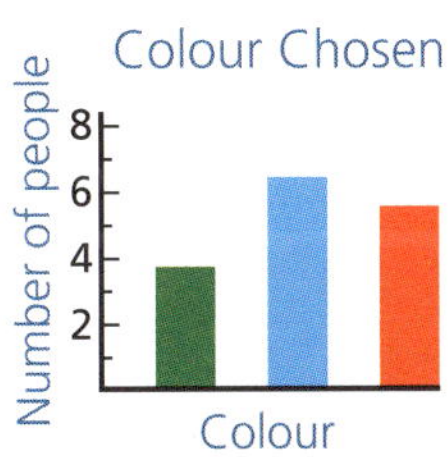

- Divided Bar Graph

  A bar is divided to show the make-up of the data.

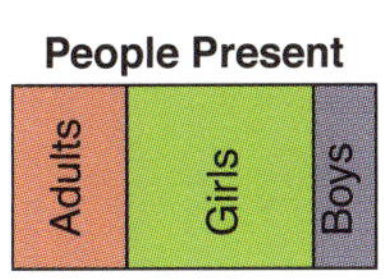

- Dot Plot

  A graph which uses dots to compare the size of groups.

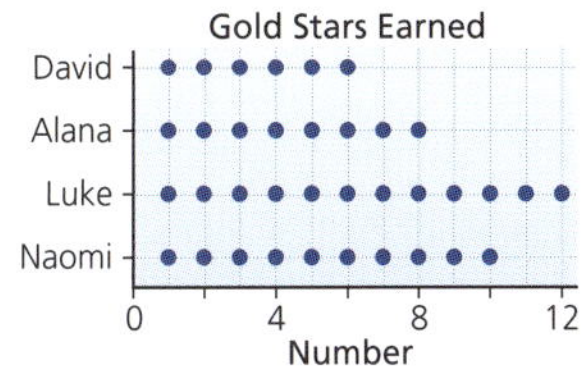

- Line Graph

  A continuous line shows the connection between variables.

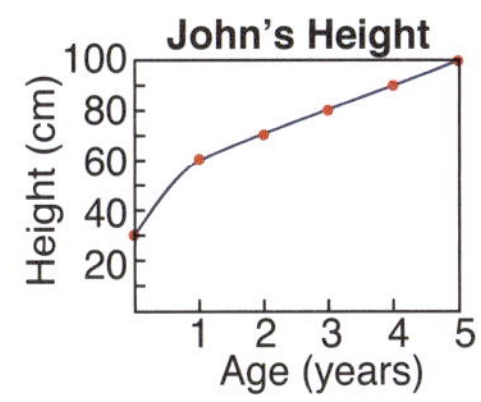

- Picture Graph

  A picture is used as a unit to show how many.

- Sector Graph

  A circle is cut into sectors to show the parts of a whole.

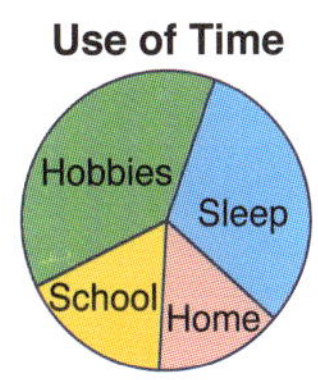

greater than (>)

A way of showing that a number is larger than another number.

- $7 > 3$

See also *less than* (<).

hectare (ha)

A unit of area equal to a square with sides of 100 m.

- 1 ha = 10 000 m^2

horizontal

- Parallel to the horizon.
- Level or flat.
- Any direction at right angles to the vertical.

increase

To make larger or more.

inverse operations

Adding 8 is the opposite (the inverse) of subtracting 8.

- 100 + 8 – 8 = 100

Multiplying by 2 is the opposite (the inverse) of dividing by 2.

- 4 × 2 ÷ 2 = 4

jump strategy

Adding or subtracting numbers, jumping by tens and ones.

- 52 – 14 = 38

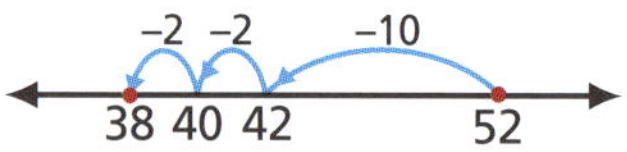

kilo (k)

Kilo means 1000.

kilogram (kg)

The basic unit of mass, equal to 1 000 grams.

- 1 kg = 1 000 g

kilometre (km)

A unit of length equal to one thousand metres.

- 1 km = 1 000 m

less than (<)

A way of showing that a number is smaller than another number.

- 3 < 7

See also *greater than (>)*.

line of symmetry

A line that divides something in half so that each half is a mirror image of the other part.

Line of Symmetry

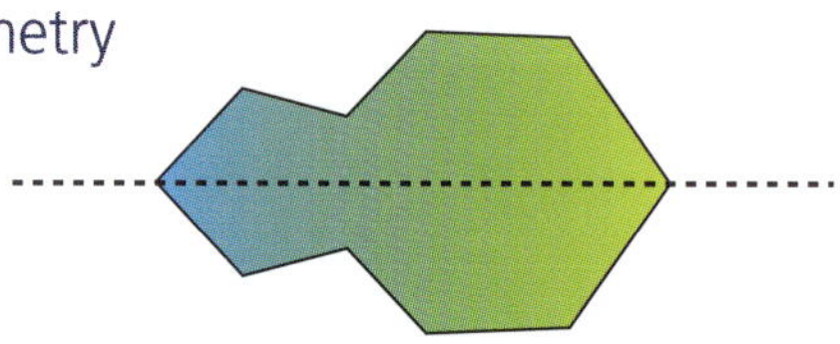

litre (L)

A unit of capacity (or volume) used for the measurement of liquids.

- 1 L = 1 000 mL

map or plan

A picture of an area viewed from above.

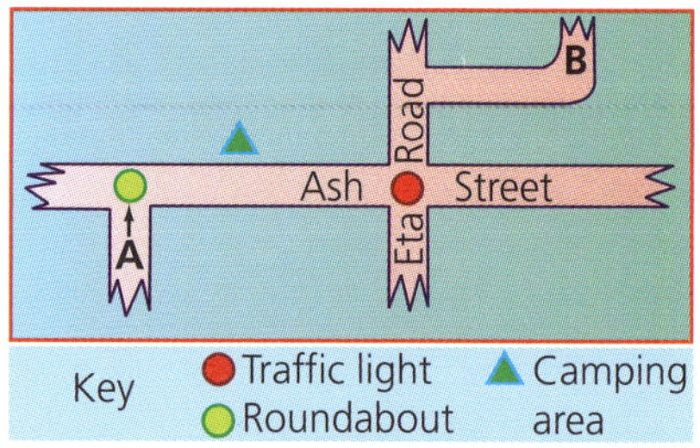

mass

The amount of matter in an object, a measure of how heavy something is.

mean

The arithmetic average.

- $\text{mean} = \dfrac{\text{sum of scores}}{\text{number of scores}}$

See also *average*.

median

The middle number of a set of numbers.

metre (m)

The basic unit of length, equal to 100 centimetres.

- 1 m = 100 cm

millilitre (mL)

A unit of capacity (or volume) equal to one thousandth of a litre.

- 1 000 mL = 1 L

millimetre (mm)

A unit of length equal to one tenth of a centimetre, or one thousandth of a metre.

- 10 mm = 1 cm
- 1 000 mm = 1 m

mixed numeral

A numeral that has a whole number part and a fraction part.

- $4\frac{1}{8}$

mode

The number that occurs the most often in a set of numbers.

- 2, 3, 3, 3, 4, 4, 5, 7
  The mode is 3.

multiple

The result of multiplying a counting number by another counting number.

- The multiples of 5 are 5, 10, 15, 20, …

net

A flat shape that can be folded to make a three-dimensional object.

Net of a Cube

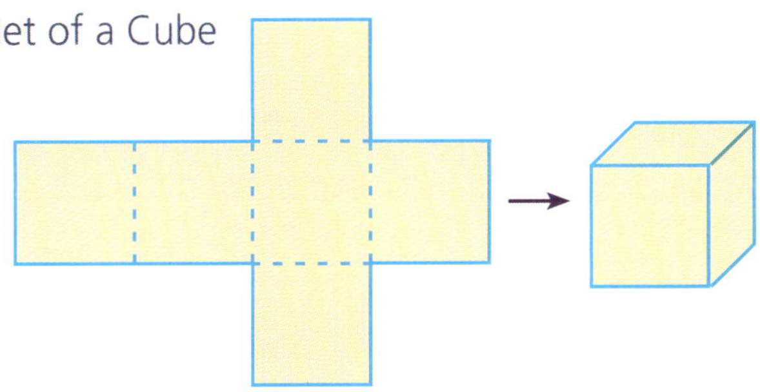

numeral expander

Breaks up a number into hundreds, tens and ones.

- 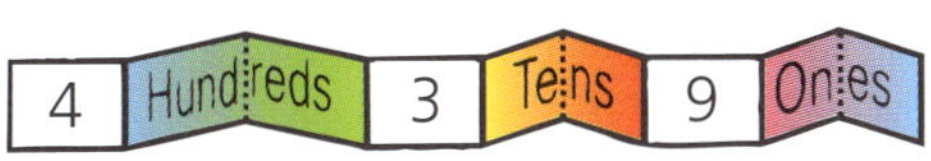

  This numeral expander shows 439.

object

The term used to describe a three-dimensional shape.

Hexagonal Prism   Cone

octagon

A polygon with eight sides.

Regular Octagon   Irregular Octagon

See also *polygon*.

parallel lines

Straight lines on the same flat surface that do not meet.

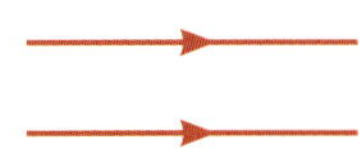

parallelogram

A shape with 4 sides such that the pairs of opposite sides are parallel and equal.

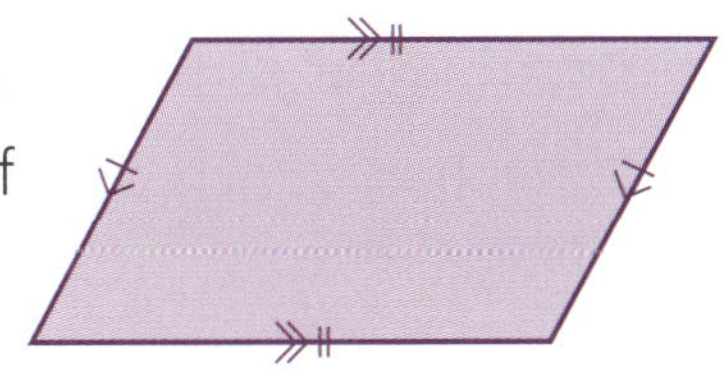

pattern (number)

Set of numbers in order, made using a rule.

- The rule for 5, 9, 13, 17, … is **Add 4**.

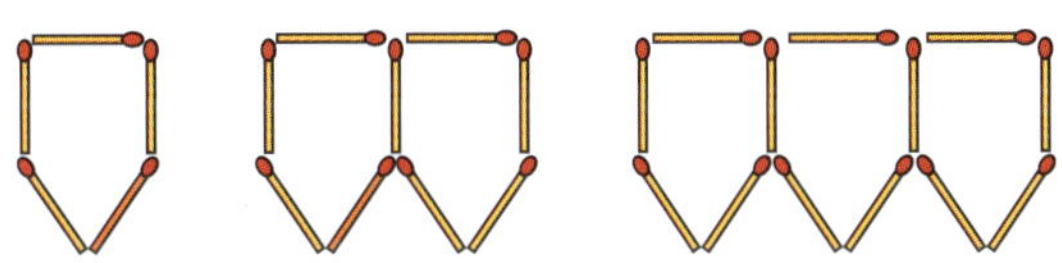

pentagon

A polygon with five sides.

Regular Pentagon Irregular Pentagon

See also *polygon*.

per cent (%)

Out of one hundred.

- $\frac{37}{100} = 0{\cdot}37 = 37\%$

perimeter

The distance around the outside of a shape; the boundary.

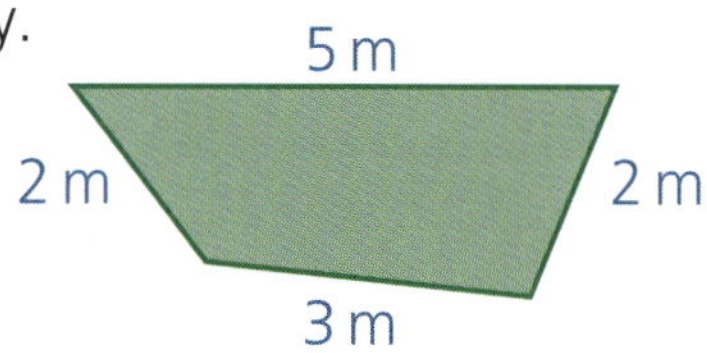

- Perimeter = 2 m + 3 m + 2 m + 5 m
  = 12 m

perpendicular lines

Lines that meet at right angles.

place value

The column value of a digit.

- 396 =

Hundreds	Tens	Ones
3	9	6

pm (post meridiem)

Any time between midday and midnight.

- The time is 20 past 1 in the afternoon. It is 1:20 pm.

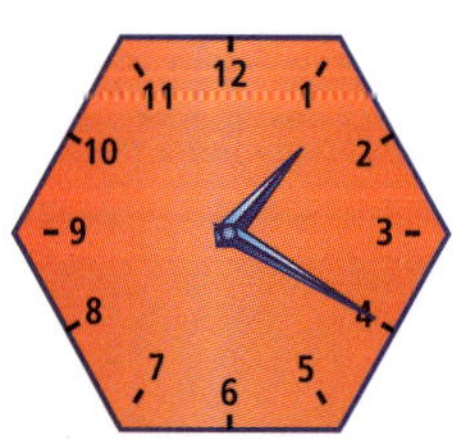

Afternoon

polygon

A two-dimensional shape with three or more straight sides, such as a triangle, quadrilateral, pentagon etc.

power

A power (or index) is a way of showing that a number has been multiplied by itself a number of times.

- $3^4 = 3 \times 3 \times 3 \times 3$
- $10^3 = 10 \times 10 \times 10$

prism

A three-dimensional object with a uniform cross-section. The ends are identical shapes and all other faces are rectangles. Prisms are named by the shape of their ends.

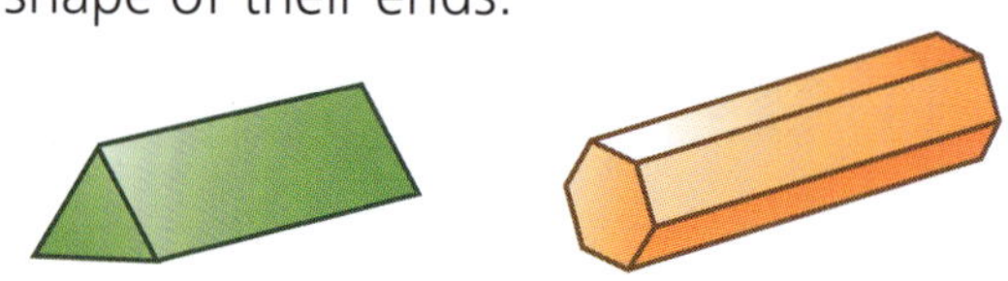

Triangular Prism Hexagonal Prism

probability

The probability (or chance) of something happening is its likelihood of happening.

- The probability of rolling an even number on a dice is 50%.

product

The answer to a multiplication question.

- The product of 8 and 9 is 72.

protractor

An instrument used for measuring and drawing angles.

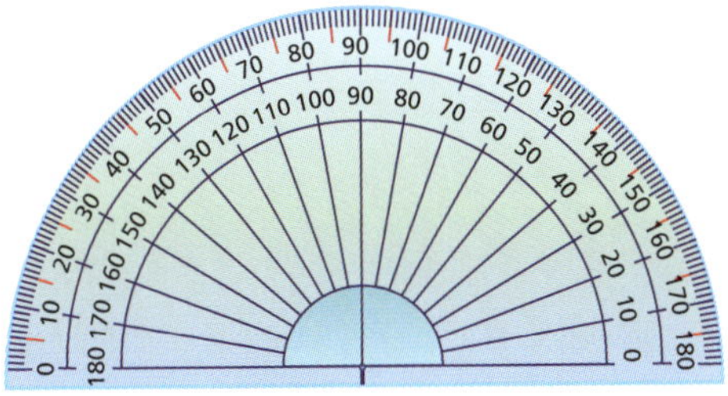

pyramid

A three-dimensional object that has a polygon for a base and triangles for all other faces. Pyramids are named by the shape of their base.

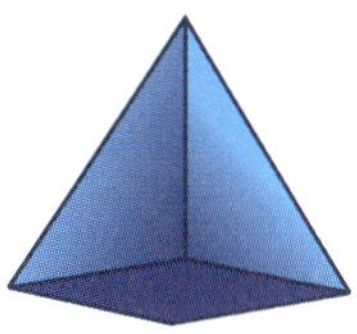

Square Pyramid

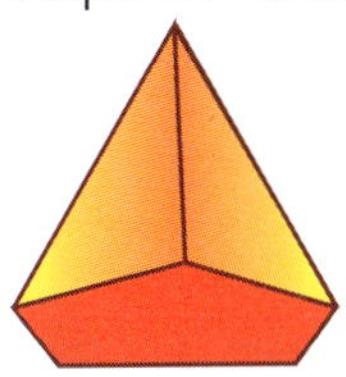

Pentagonal Pyramid

quadrilateral

A two-dimensional shape with four straight sides.

questionnaire or survey

A list of questions used to obtain statistical information or opinions from people.

quotient

The answer when one number is divided by another.

random selection

Choosing without looking.
Each item has an equal chance of being chosen.

range

The difference between the highest and lowest numbers in a set.

- 2, 3, 5, 8, 9, 10, 10
  Range = 10 – 2 = 8

reflection

See *flip*.

regular and irregular shapes

Regular shapes have all sides and all angles equal. Irregular shapes do not.

Regular Shape

Irregular Shape

remainder

The number that is left over after sharing or dividing.

- 22 cups shared among 5 people gives 4 cups each, remainder 2.

rhombus

A shape with 4 sides, opposite sides parallel, all sides equal.

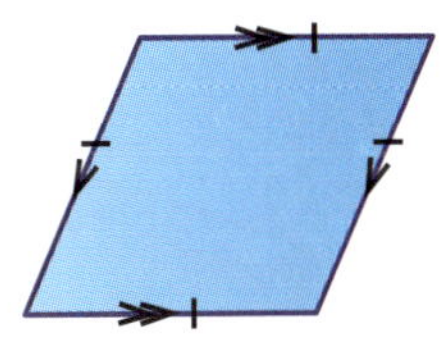

rigid shape

A model that cannot be pushed out of shape because triangles have been used in its construction.

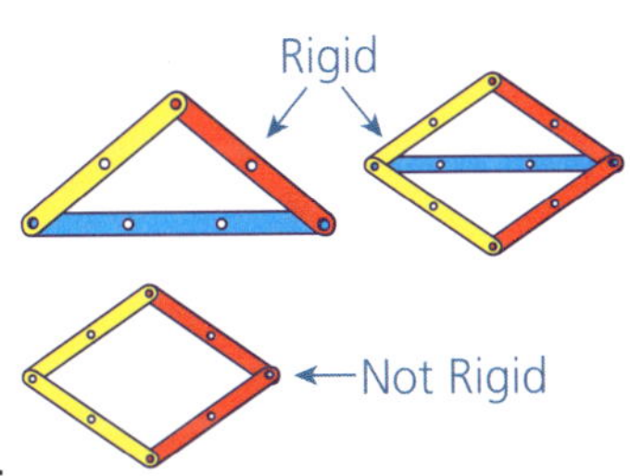

Roman numerals

A number system devised by the ancient Romans.

Roman numerals use letters for numbers:

I	V	X	L	C	D	M
1	5	10	50	100	500	1000

- XXVIII = 28

rounding

Rewriting a number to the nearest 5, 10 or 100.

skip counting

Counting on, adding the same number each time.

- 5, 10, 15, 20, 25, … is skip counting by 5.

slide (translation)

To move a shape in any direction without changing its orientation.

solid

A term used to describe a three-dimensional shape.

sphere

A three-dimensional object that is ball-shaped and round. All points on the surface of a sphere are the same distance from its centre.

split strategy

Adding numbers by splitting them into their parts.

- 36 + 52 = 30 + 6 + 50 + 2
  = (30 + 50) + (6 + 2)
  = 80 + 8
  = 88

square centimetre (cm^2)

A unit of area equal to a square with sides of 1 cm.

square kilometre (km^2)

A unit of area equal to a square with sides of 1 km.

- 1 km^2 = 1 000 000 m^2, 1 km^2 = 100 ha

square metre (m^2)

A unit of area equal to a square with sides of 1 m.

- 10 000 m^2 = 1 ha

sum

The answer when you add numbers.

surface

The outside layer of a three-dimensional object. A surface can be flat or curved.

See also *face*.

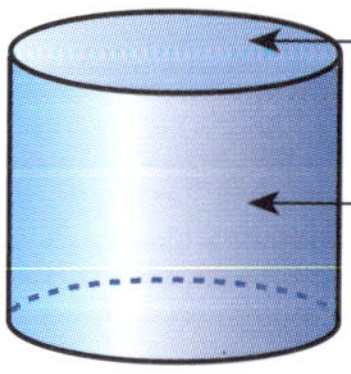

survey or questionnaire

A list of questions used to discover information.

symmetry

A balanced arrangement.

- Line Symmetry

  A property of a figure where one half is the mirror image of the other.

- Line (or Axis) of Symmetry

  A line that divides a figure into two parts that are mirror images of each other.
- Rotational Symmetry

  A property of a figure where it can be spun about a point so that it repeats its shape more than once in a rotation.

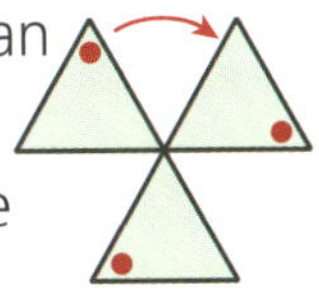

table

A simple way to display information in rows and columns.

Birds	12
Lions	2
Monkeys	6

tally

To keep count by making a mark for each item. To make counting easy, the marks are drawn in groups of five with each fifth mark crossed over the other four marks.

-  = 18

tangram

A traditional Chinese puzzle. A square is cut into seven pieces that can be rearranged to make different pictures.

- 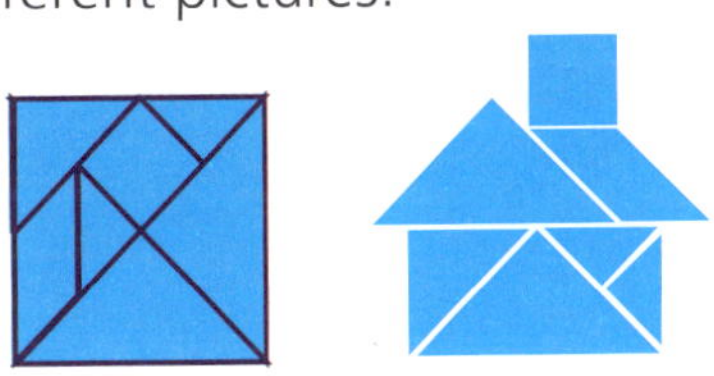

temperature

A measure of how hot or cold something is. Temperature is usually measured in degrees Celsius (°C).

- Water freezes at 0°C.
- Water boils at 100°C.

tessellation

A pattern of identical shapes that fit together without gaps or overlaps.

thermometer

An instrument used for measuring temperature.

three-dimensional (3D) object

Objects are three-dimensional. They have length, width and height.

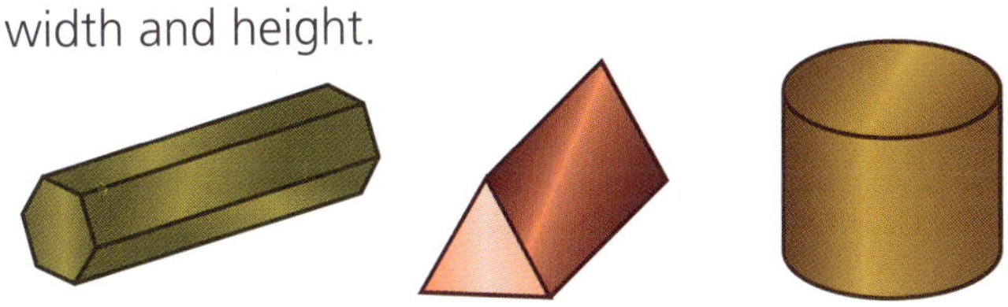

tonne (t)

A unit of mass equal to 1000 kg.

translation

See *slide*.

trapezium

A quadrilateral with one pair of parallel sides.

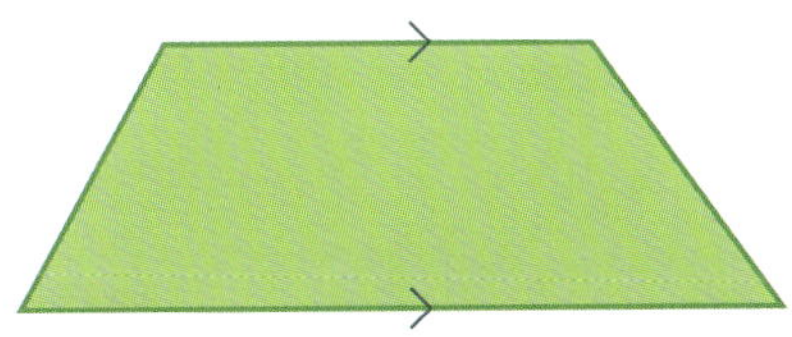

triangle

A two-dimensional shape with three straight sides and three angles.

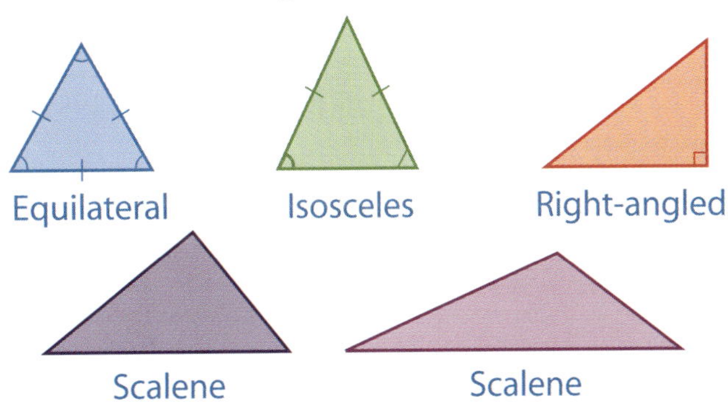

Scalene triangles have no sides equal.

See also *polygon*.

turn (rotation)

To rotate a shape about a given point.

twenty-four hour time

Time shown as a 4-digit number, the first two digits indicating the hour and the second two digits indicating minutes.

- 13:20 is 20 past 1 in the afternoon, or 1:20 pm.

two-dimensional (2D) shapes

Plane shapes are two-dimensional. They have length and width.

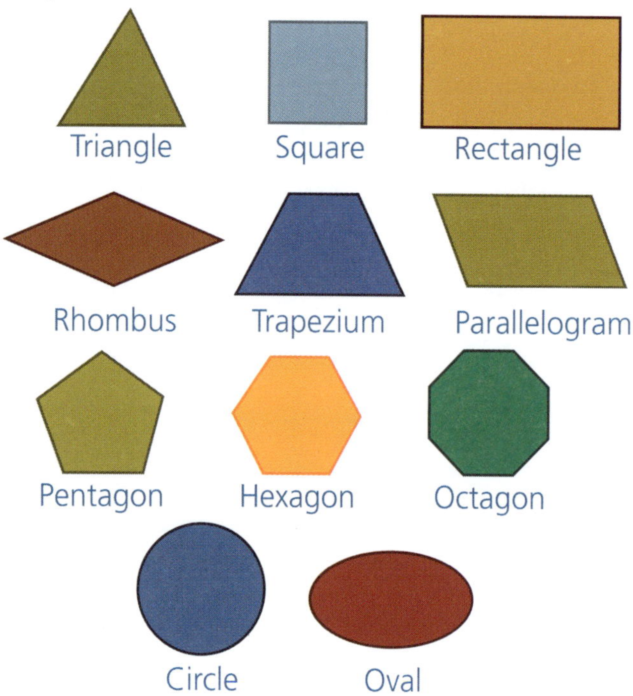

two-way tables

Tables of data that show categories down the side as well as across the top.

	Blue	Red	White
Bus	2	5	0
Car	8	10	21
Truck	1	6	2

vertex

A point at which two or more lines meet to form a corner on a 2D shape or 3D object.

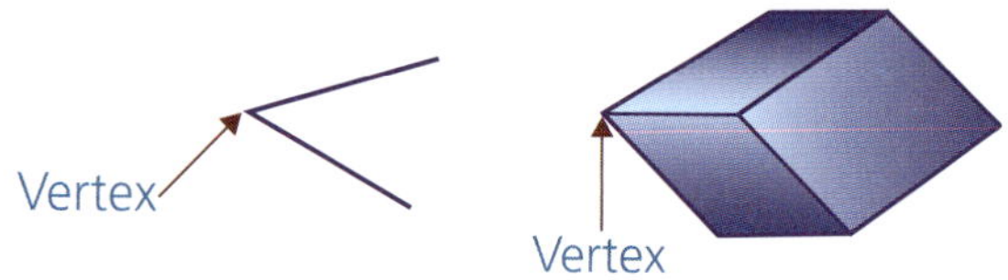

The plural of *vertex* is *vertices*.

vertical

- At right angles to the horizontal.
- Straight up and down.
- The direction in which an object falls under gravity.

volume

The amount of space occupied by a 3D object.

-  Volume = 10 cubic units

width or breadth

The distance from side to side.

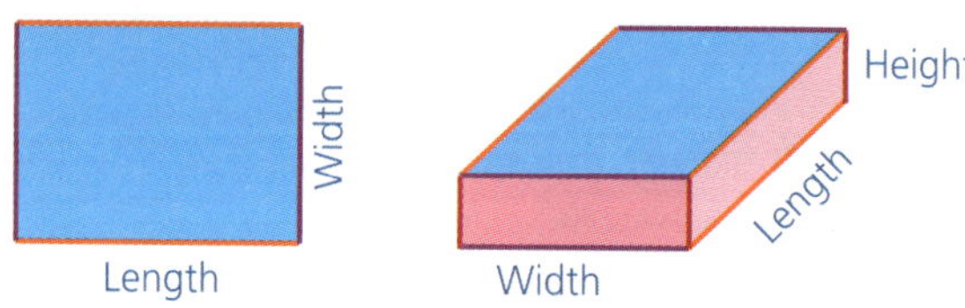

NUMBER & ALGEBRA

1:01 Numbers to One Million

Hund Thous	Ten Thous	Thous	Hund	Tens	Ones
● ● ● ●	● ● ● ● ● ● ●	● ● ●	● ● ● ● ● ●	● ● ● ● ● ● ● ●	● ●
4	7	3	6	8	2

473 682

CONCEPT

Four hundred and seventy-three thousand six hundred and eighty-two

Use counters and a place-value chart to model:

a two hundred and thirty-nine thousand four hundred and fifty-three
b six hundred and twenty-five thousand seven hundred and sixty-four
c three hundred and forty-two thousand one hundred and twenty-nine
d one hundred and fifty-eight thousand two hundred and seventy-one
e 734 396 **f** 253 817 **g** 814 536 **h** 471 239 **i** 526 949

1 Read these numbers aloud and then write them in figures on the place-value chart.

a three hundred and forty-six thousand nine hundred and seventy-five
b five hundred and twenty-three thousand four hundred and eighty-two
c seven hundred and sixteen thousand eight hundred and fifty-nine
d six hundred and seventy-eight thousand nine hundred and sixty
e two hundred and fifty-one thousand six hundred and three

	Hund Thous	Ten Thous	Thousands	Hundreds	Tens	Ones
a						
b						
c						
d						
e						

2 Write the value for each coloured digit.

a 394 128 ☐ **b** 726 403 ☐ **c** 152 776 ☐ **d** 430 275 ☐
e 863 410 ☐ **f** 507 395 ☐ **g** 483 721 ☐ **h** 273 496 ☐

3 Write the numeral for:

a 100 000 + 40 000 + 7 000 + 600 + 20 + 4 ☐
b 400 000 + 90 000 + 6 000 + 500 + 30 + 9 ☐
c 800 000 + 60 000 + 9 000 + 300 + 70 + 6 ☐
d 600 000 + 50 000 + 8 000 + 100 + 90 + 2 ☐
e 200 000 + 70 000 + 5 000 + 400 + 6 ☐

999 999 + 1
= 1 000 000
(one million)

1:02 Numbers Above One Million

CONCEPT

Millions	Hundred thousands	Ten thousands	Thousands	Hundreds	Tens	Units
2	9	5	3	6	7	4

2 953 674

Two million nine hundred and fifty-three thousand six hundred and seventy-four

1 Use numerals to write:

a two million three hundred and eighty-six thousand five hundred and thirty-one

b seven million eight hundred and forty-three thousand two hundred and sixty-six

c three million five hundred and twenty-one thousand six hundred and fifty-three

2 Write the numeral for:

a 600 000 + 50 000 + 9000 + 800 + 70 + 4

b 700 000 + 60 000 + 3000 + 700 + 50 + 9

c 3 000 000 + 800 000 + 40 000 + 8000 + 600 + 30 + 1

d 5 000 000 + 900 000 + 10 000 + 6000 + 500 + 90 + 2

3 Write the value for each coloured digit.

a 15723**9**2 **b** 46**3**1762 **c** 3**8**46724 **d** 627**5**648

e **6**334704 **f** 5435**2**46 **g** 2**1**65424 **h** 923**4**619

4 Arrange each group of numbers in ascending order.

a 3 654 761 5 814 903 4 607 519

b 7 651 411 7 323 916 7 135 976

c 4 238 175 4 962 345 4 572 391

5 Round each number to the nearest million.

a 2 469 725 **b** 6 243 915 **c** 1 385 476

d 4 517 219 **e** 5 172 403 **f** 8 319 647

Make the Number

FUN SPOT

- Use a set of playing cards marked with the digits 0 to 9. (Use four of each type.)
- One student is asked to nominate a task from the task board.
- Each player is dealt six cards.
- The players arrange their cards to best fit the task.
- The player who best meets the task receives one point.

Task Board	
Task	**Score**
biggest number	
smallest number	
closest to 1 000 000	
less than 500 000	
Total score	

 • *AUSTRALIAN SIGNPOST MATHS 5* • ISBN 978 1 4886 2190 1

1:03 Powers of Ten

CONCEPT

Thousands 1000	Hundreds 100	Tens 10	Ones 1
10^3 $10 \times 10 \times 10$	10^2 10×10	10^1 10	1 1
6	9	4	3

$6943 = (6 \times 1000) + (9 \times 100) + (4 \times 10) + 3$
$= (6 \times 10^3) + (9 \times 10^2) + (4 \times 10) + 3$

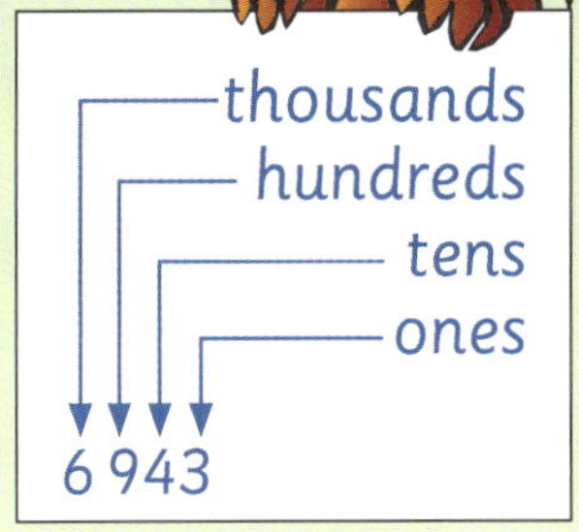

1 Write the numeral for:

a $(4 \times 10^3) + (2 \times 10^2) + (8 \times 10) + 7$ ☐

b $(8 \times 10^3) + (6 \times 10^2) + (9 \times 10) + 5$ ☐

c $(7 \times 10^3) + (4 \times 10^2) + (5 \times 10) + 4$ ☐

d $(5 \times 10^3) + (1 \times 10^2) + (3 \times 10) + 8$ ☐

e $(9 \times 10^3) + (3 \times 10^2) + (7 \times 10) + 5$ ☐

f $(2 \times 10^3) + (7 \times 10^2) + (6 \times 10) + 1$ ☐

g $(6 \times 10^3) + (5 \times 10^2) + (2 \times 10) + 9$ ☐

h $(1 \times 10^3) + (9 \times 10^2) + (1 \times 10) + 6$ ☐

2 Write the following using powers of ten.

a 3742 ☐

b 8463 ☐

c 9529 ☐

d 7385 ☐

e 6956 ☐

3 Write the value of each coloured digit using a power of ten.

a 46**3**1 ☐ b 9**5**78 ☐ c **6**752 ☐ d 69**7**3 ☐

e 3**2**76 ☐ f **4**230 ☐ g 2**8**65 ☐ h **7**624 ☐

4 Write the numeral for the number shown on each abacus.

a
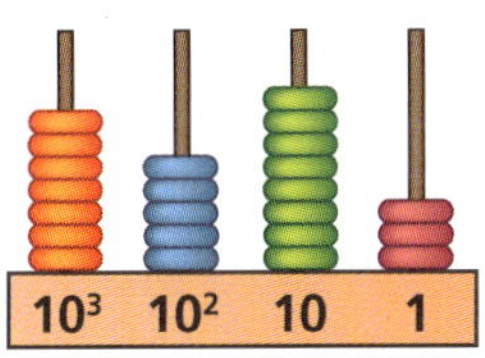

☐

b
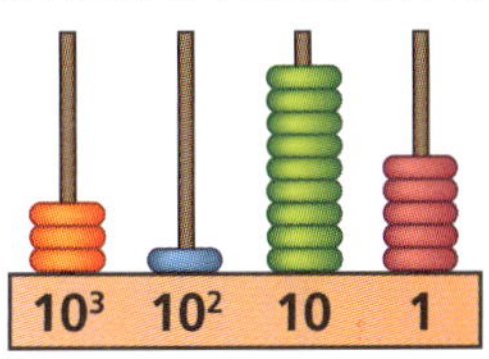

☐

c
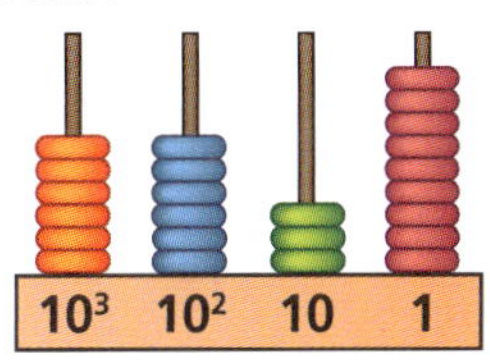

☐

d
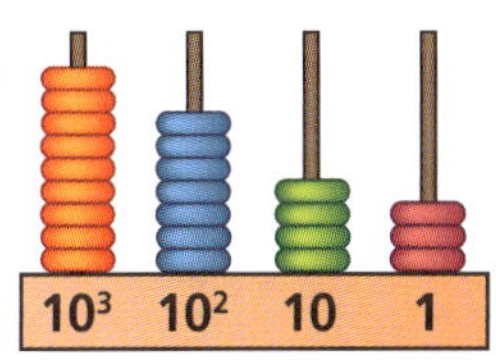

☐

e
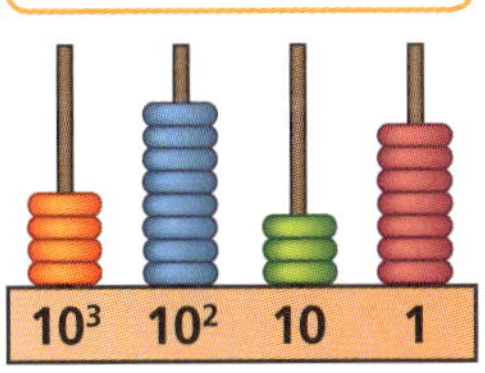

☐

f
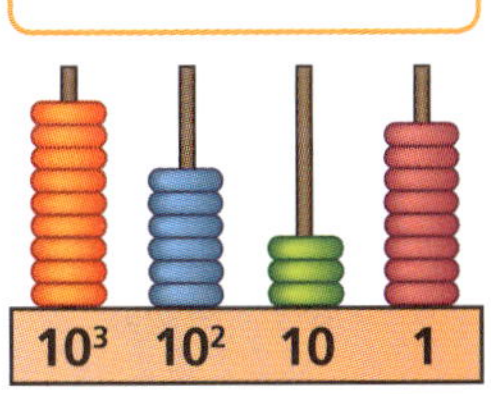

☐

g
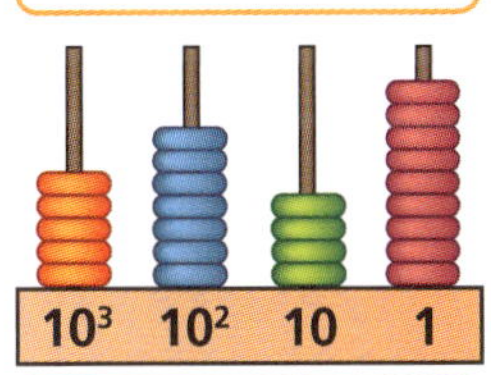

☐

h
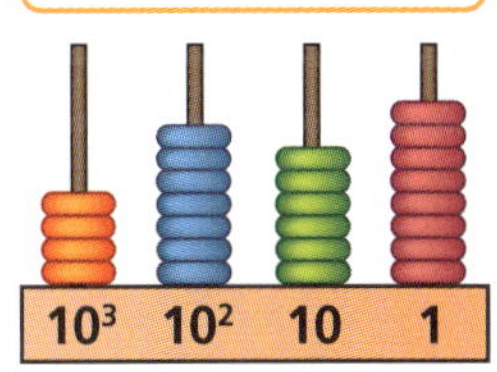

☐

1:04 Hundredths

1 What fraction of each group has been shaded?

a  ☐

b 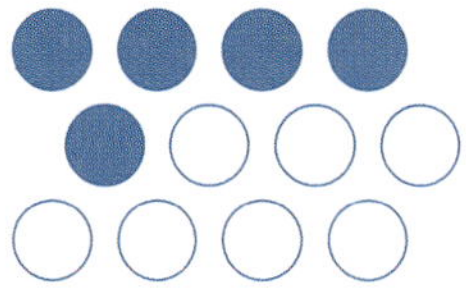 ☐

c  ☐

d 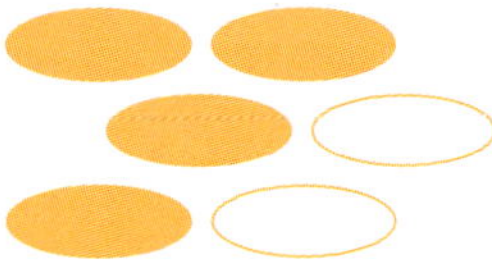 ☐

e  ☐

f  ☐

CONCEPT

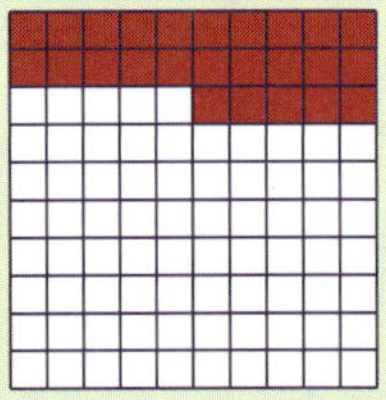

25 hundredths

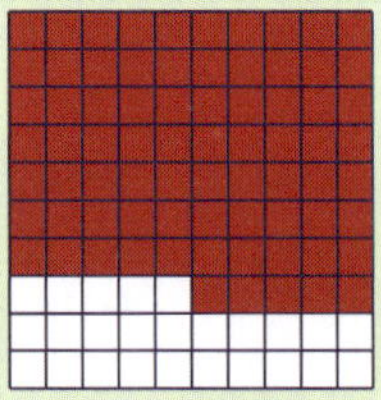

75 hundredths

75 out of 100 is $\frac{75}{100}$ or 0·75

Hundredths are also called percentages.

2 Write the fraction and the decimal shown in each hundred square.

a 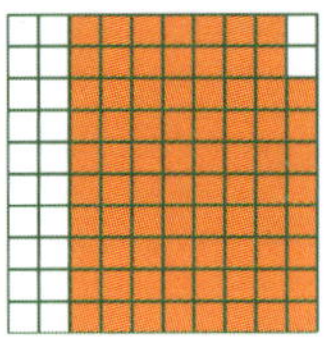 ☐ or ☐

b 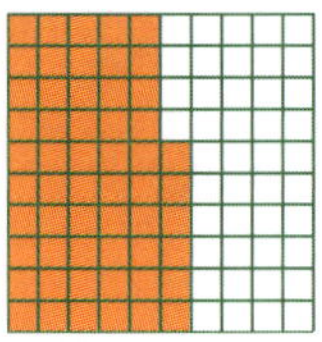 ☐ or ☐

c 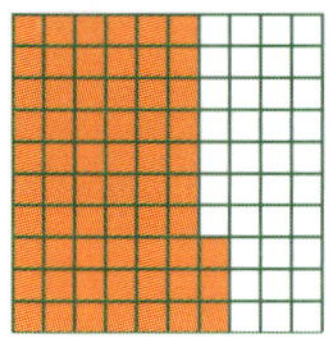 ☐ or ☐

d 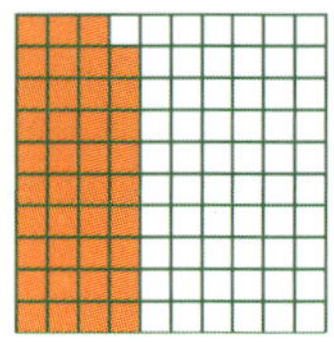 ☐ or ☐

e 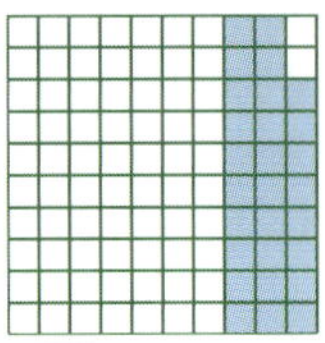 ☐ or ☐

f 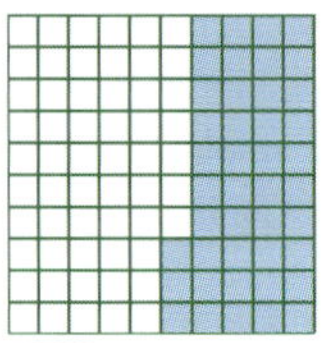 ☐ or ☐

g 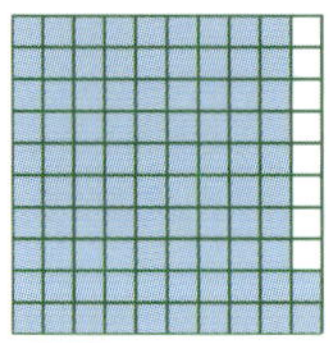 ☐ or ☐

h 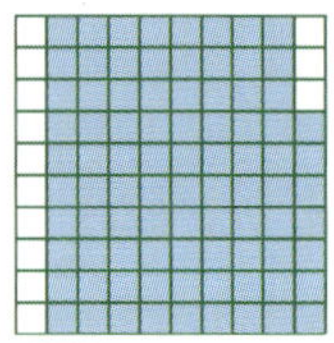 ☐ or ☐

3 For each square, colour the fraction shown.

a 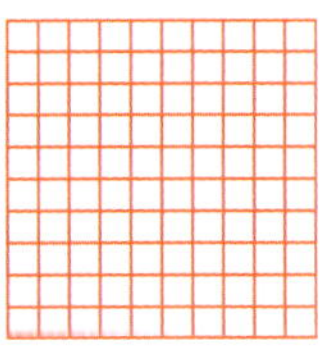 $\frac{24}{100}$

b 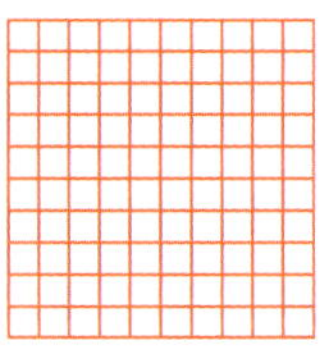 $\frac{16}{100}$

c 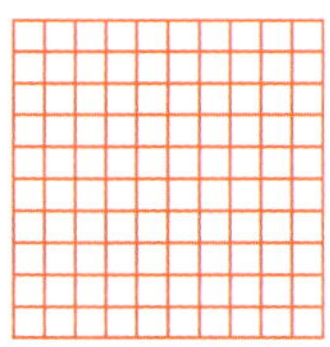 $\frac{33}{100}$

d 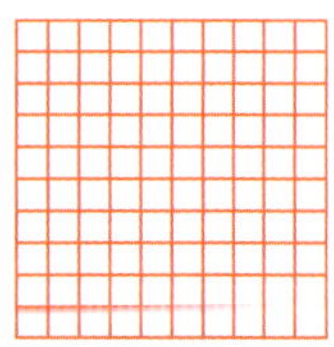 $\frac{89}{100}$

4 What fraction of each square in Questions 2**a** to 2**d** has not been coloured?

a ☐ b ☐ c ☐ d ☐

1:05 Fractions

CONCEPT

$\frac{7}{10}$ is coloured red.

$\frac{3}{10}$ is coloured blue.

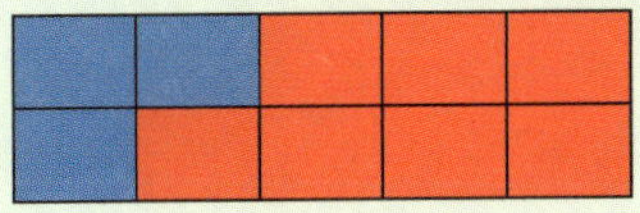

$\frac{7}{10}$ and $\frac{3}{10}$ makes 1 whole.

1 Complete this exercise.

a

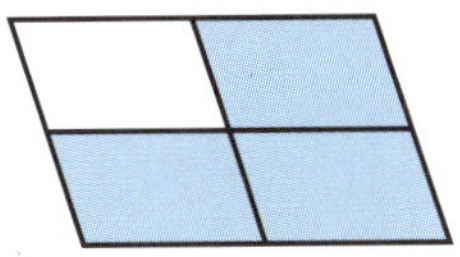

$\square$ is coloured.

$\square$ is not coloured.

$\square$ and $\square$ makes 1 whole.

b

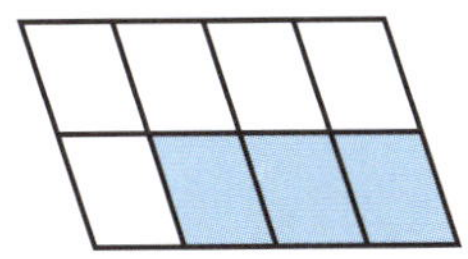

$\square$ is coloured.

$\square$ is not coloured.

$\square$ and $\square$ makes 1 whole.

2 Complete.

a $\frac{1}{2}$ and $\frac{1}{2}$ makes $\square$ whole.

b $\frac{1}{4}$ and $\square$ makes 1 whole.

c $\square$ and $\frac{2}{3}$ makes 1 whole.

d $\frac{2}{5}$ and $\frac{3}{5}$ makes $\square$ whole.

e $\frac{3}{8}$ and $\square$ makes 1 whole.

f $\square$ and $\frac{4}{10}$ makes 1 whole.

3 Complete these fraction problems.

a If $\frac{3}{4}$ of our class come to school by bus, what fraction does not come by bus? $\square$

b If a water tank is $\frac{5}{8}$ full, what fraction of the water tank is empty? $\square$

c $\frac{3}{5}$ of a pizza is left. What fraction has been eaten? $\square$

d $\frac{9}{10}$ of my pavers have arrived. What fraction still needs to arrive? $\square$

e If $\frac{3}{8}$ of a cake has been eaten, what fraction is left? $\square$

f If $\frac{7}{10}$ of the class is present, what fraction is absent? $\square$

FUN SPOT

4 What fraction of the pieces of fruit is:

watermelons?	$\frac{1}{20}$	apples?	
pears?		bananas?	
oranges?		cherries?	

1:06 Unit Fractions

1 Use dots to show the fractions on the number line.
Use < or > to complete the sentence.

a $\frac{1}{2}$ and $\frac{1}{4}$

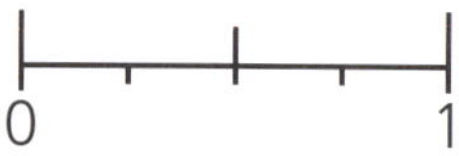

$\frac{1}{2}$ ☐ $\frac{1}{4}$

b $\frac{1}{10}$ and $\frac{1}{3}$

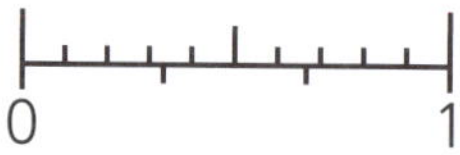

$\frac{1}{10}$ ☐ $\frac{1}{3}$

c $\frac{1}{2}$ and $\frac{1}{8}$

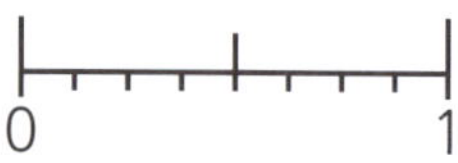

$\frac{1}{2}$ ☐ $\frac{1}{8}$

d $\frac{1}{6}$ and $\frac{1}{3}$

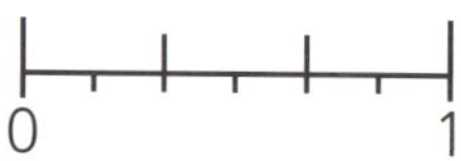

$\frac{1}{6}$ ☐ $\frac{1}{3}$

The Order of Unit Fractions

CONCEPT

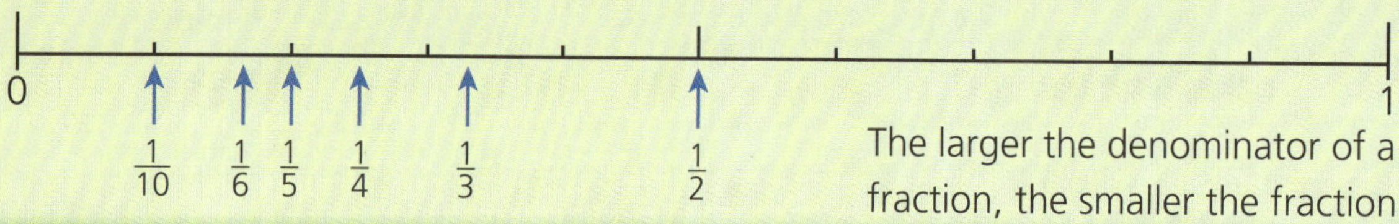

The larger the denominator of a unit fraction, the smaller the fraction is.

2 Put each group of fractions in order, from smallest to largest.

a $\frac{1}{2}$, $\frac{1}{5}$, $\frac{1}{4}$ ☐

b $\frac{1}{100}$, $\frac{1}{10}$, $\frac{1}{20}$ ☐

c $\frac{1}{3}$, $\frac{1}{8}$, $\frac{1}{2}$ ☐

d $\frac{1}{4}$, $\frac{1}{8}$, $\frac{1}{2}$ ☐

3 Match each fraction with a part of the circle.

$\frac{1}{6}$

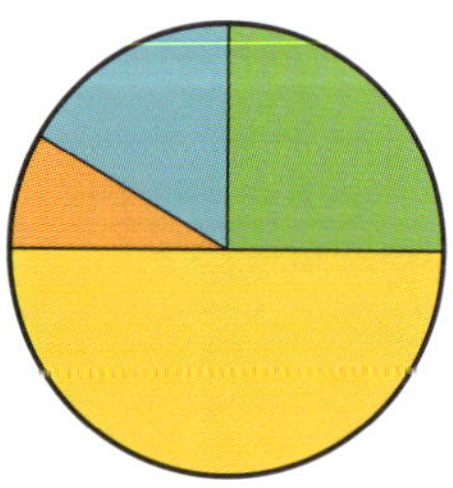

$\frac{1}{2}$

$\frac{1}{4}$

4 Match each fraction with a part of the decagon.

$\frac{1}{2}$

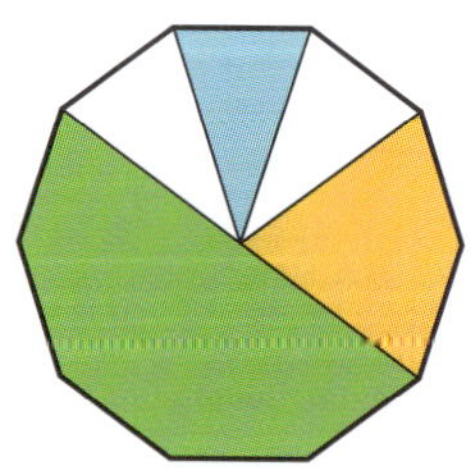

$\frac{1}{10}$

$\frac{1}{5}$

5 For unit fractions, the greater the denominator, the ☐ the fraction.

1:07 Tenths

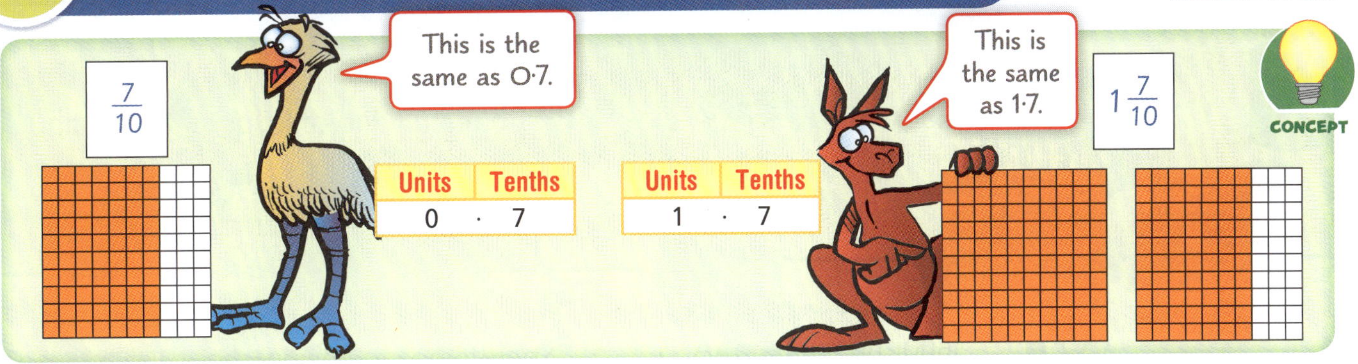

1 Write the decimal for:

a $\frac{9}{10}$ ____ b $\frac{5}{10}$ ____ c $\frac{3}{10}$ ____ d $\frac{7}{10}$ ____

e $\frac{4}{10}$ ____ f $1\frac{6}{10}$ ____ g $2\frac{8}{10}$ ____ h $1\frac{1}{10}$ ____

i $2\frac{8}{10}$ ____ j $3\frac{2}{10}$ ____ k $6\frac{9}{10}$ ____ l $2\frac{8}{10}$ ____

2 Match each fraction with the correct decimal.

a

$\frac{5}{10}$	0·5
$\frac{8}{10}$	0·2
$\frac{2}{10}$	0·8

b

$2\frac{3}{10}$	2·7
$2\frac{9}{10}$	2·9
$2\frac{7}{10}$	2·3

c

$4\frac{6}{10}$	4·5
$4\frac{1}{10}$	4·6
$4\frac{5}{10}$	4·1

3 Use decimals to write:

a 8 tenths ____ b 3 tenths ____ c 9 tenths ____ d 4 tenths ____

e 5 tenths ____ f 1 tenth ____ g 2 tenths ____ h 6 tenths ____

i zero point eight ____ j zero point three ____ k one point zero ____

l one point five ____ m four point two ____ n eight point nine ____

4 Complete the number lines.

a

0	$\frac{1}{10}$		$\frac{3}{10}$			$\frac{6}{10}$				1
0	0·1				0·5	0·6				1·0

b

3	$3\frac{1}{10}$	$3\frac{2}{10}$	$3\frac{3}{10}$								
3	3·1	3·2	3·3								

1:08 Decimals

CONCEPT

Writing fractions as decimals

0·82

Decimal point

or 82%

1. How many squares are grouped together? [ ]

2. What part of the group of squares:

   a is green? [ ] out of 100 $\frac{}{100}$ 0·

   b is white? [ ] out of 100 $\frac{}{100}$ 0·

   c is blue? [ ] out of 100 $\frac{}{100}$ 0·

3. What part of the group of squares is red? 0·

$0.17 is 17 cents or $\frac{17}{100}$ of a dollar.

4. Write the fraction and decimal that is coloured.

a

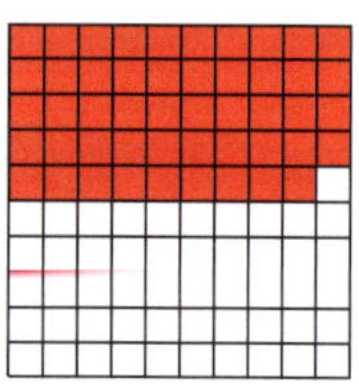

0·

b

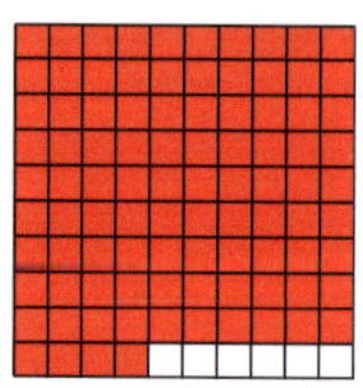

0·

c

0·

d

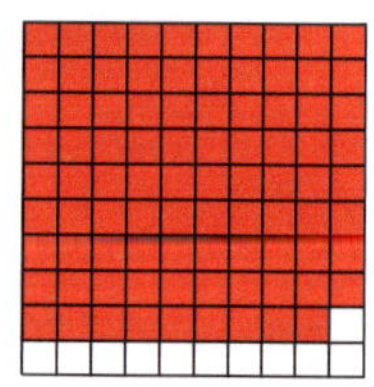

0·

e

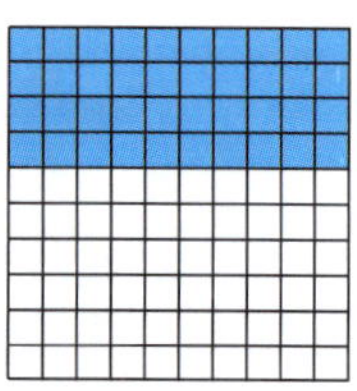

0·

f

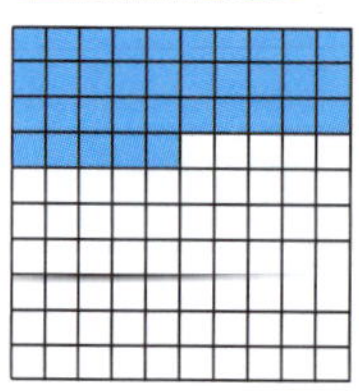

0·

g

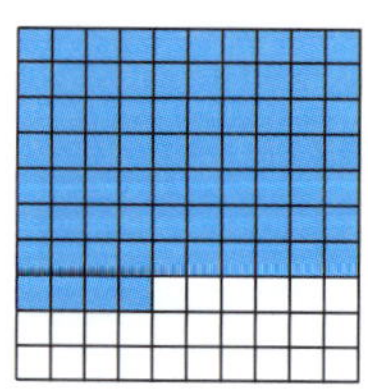

0·

h

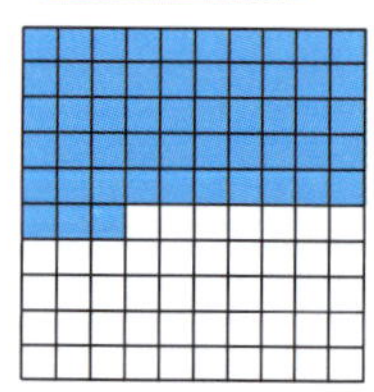

0·

1:09 Place Value in Decimals

Ones	Tenths	Hundredths
0 ·	2	7

1 Complete the label for each model.

a

Ones	Tenths	Hundredths
0 ·		

b

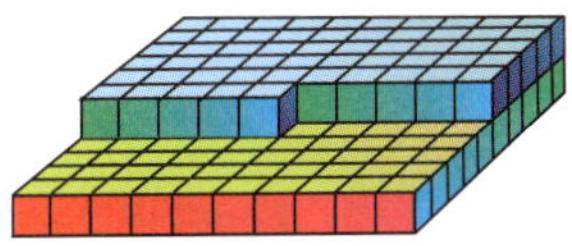

Ones	Tenths	Hundredths
0 ·		

c

Ones	Tenths	Hundredths
·		

2 Draw a line from each decimal to the correct fraction.

a

$1\frac{73}{100}$	0·17
$\frac{17}{100}$	1·73
$1\frac{53}{100}$	1·53

b

$2\frac{51}{100}$	1·35
$1\frac{35}{100}$	2·15
$2\frac{15}{100}$	2·51

c

$\frac{10}{100}$	3·49
$2\frac{49}{100}$	0·1
$3\frac{49}{100}$	2·49

- Use place-value blocks to show your answers to Question 2.

3 Complete the numeral expanders for each decimal.

a 0·64

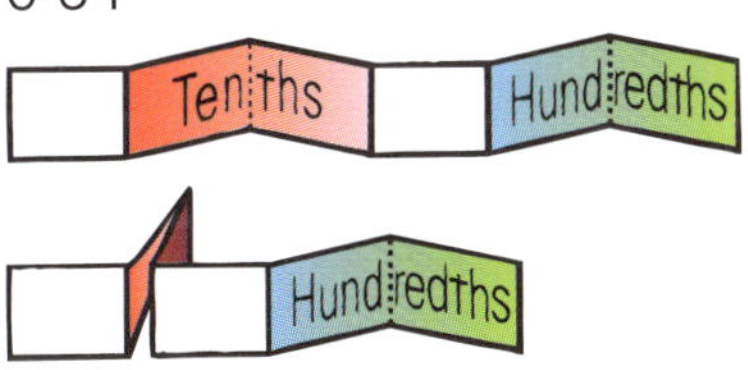

b 0·32

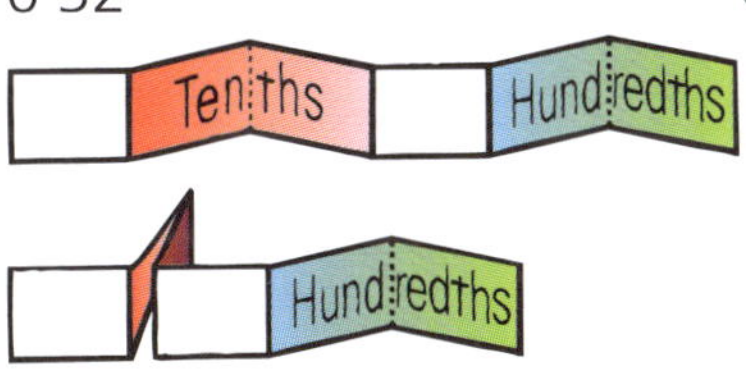

c 0·19

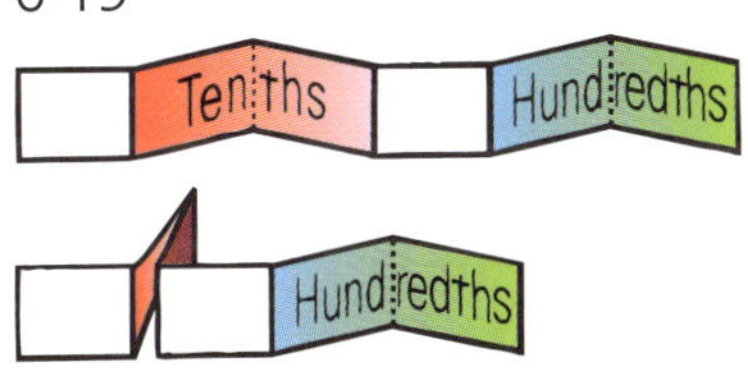

d 0·08

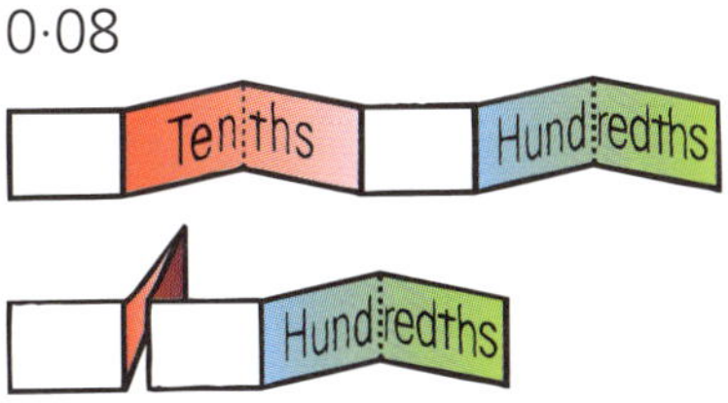

e 0·7

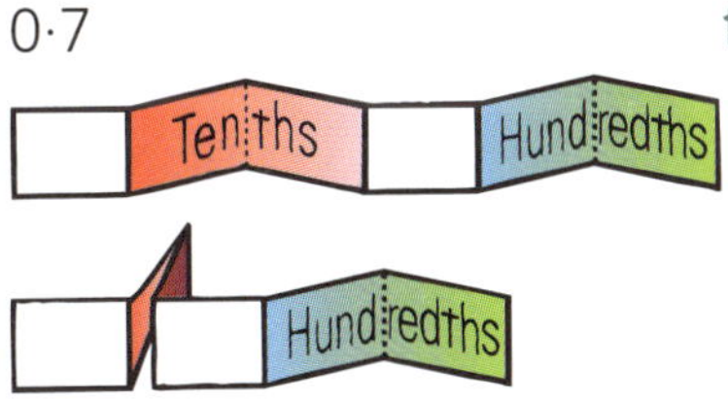

f 0·93

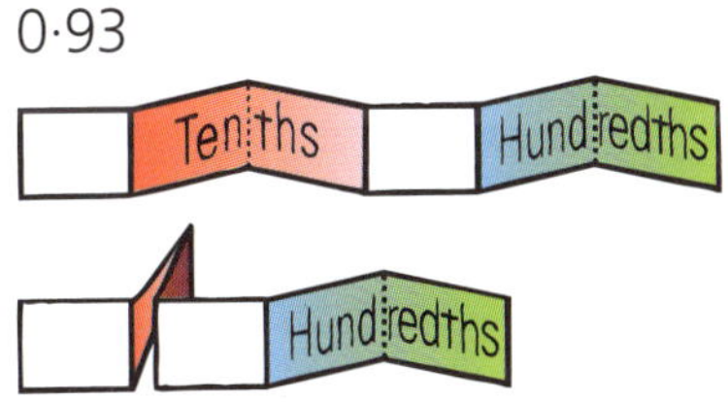

 • *AUSTRALIAN SIGNPOST MATHS 5* • ISBN 978 1 4886 2190 1

1:10 Place Value to Thousandths

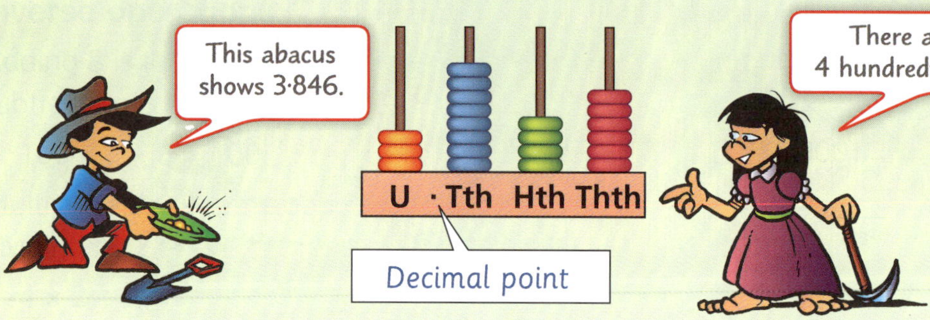

1 = 10 tenths or 100 hundredths or 1000 thousandths.

1 Write the numeral for the number shown on each abacus.

a 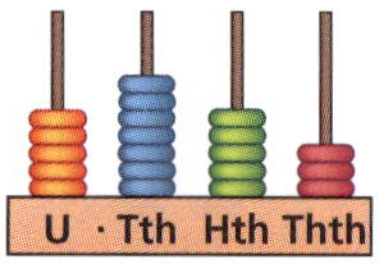

b 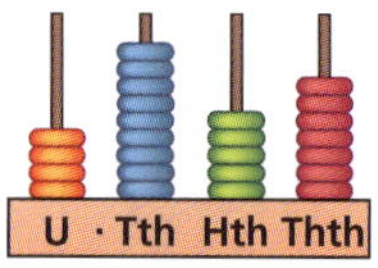

c 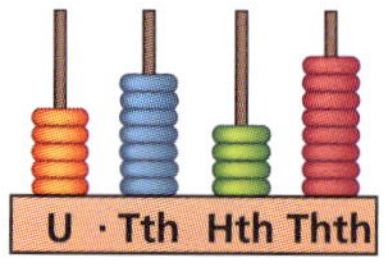

d 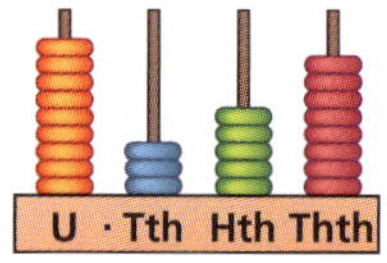

e 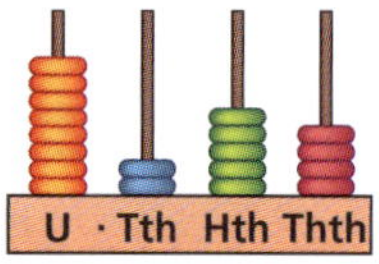

f 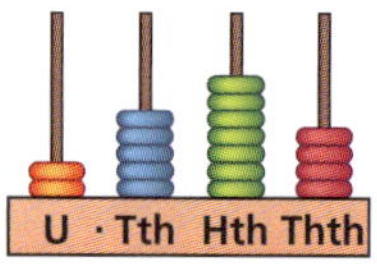

g 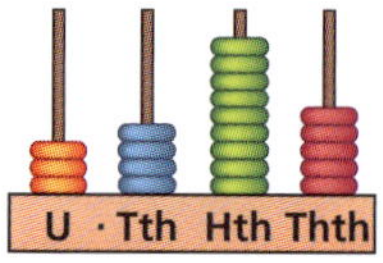

h 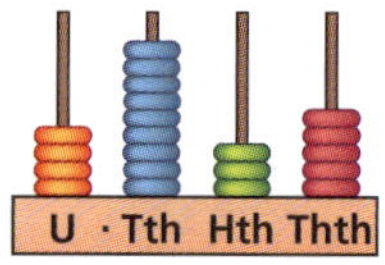

2 Write each number on the place-value chart.

a three point one nine seven
b five point six three eight
c nine point two four nine
d six point five four eight
e eight point three five two
f two point seven one nine

Units	·	10ths	100ths	1000ths
	·			
	·			
	·			
	·			
	·			
	·			

ten 1000ths = one 100th	$\frac{10}{1000} = \frac{1}{100}$
ten 100ths = one 10th	$\frac{10}{100} = \frac{1}{10}$
ten 10ths = one unit	$\frac{10}{10} = 1$

Ten of one column gives one in the column on the left.

Make the Largest Number

- Each player, in turn, rolls the dice and records the number in the column of their choice in the place-value card.
- The player rolls three more dice to fill the place-value card.
- The player with the largest 4-digit number wins the game.

Units		Tenths	Hundredths	Thousandths
6	·	4	2	1

1:11 Reading and Writing Decimals

CONCEPT

1 Write the numeral for the number shown on each abacus.

a
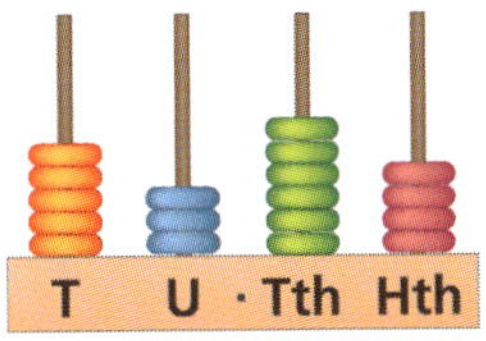

b
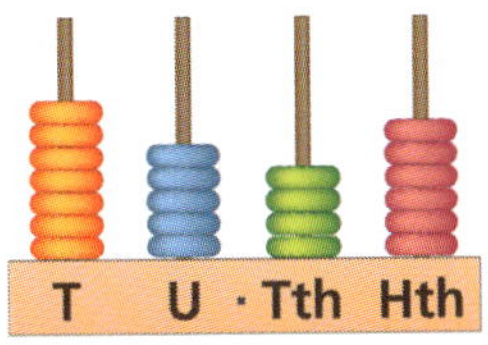

c
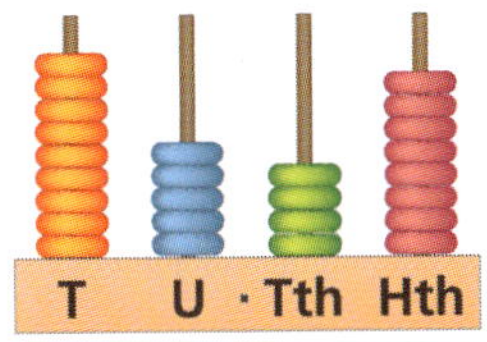

d
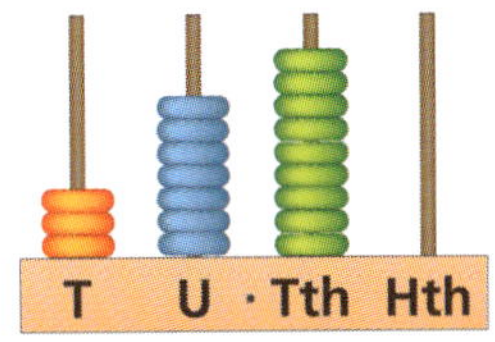

2 Use numerals to write:

a fifteen point seven five ☐
b eighty point three six ☐
c twenty-seven point nine two ☐
d ninety point zero eight ☐
e sixty-two point seven three ☐
f seventy-one point nine nine ☐

3 Complete the following.

a 0·91 = ☐ tenths and ☐ hundredth
b 0·40 = ☐ tenths and ☐ hundredths
c 0·09 = ☐ tenths and ☐ hundredths

4 Write the number represented by:

a 5 hundreds, 8 tens, 3 units, 2 tenths and 7 hundredths ☐
b 6 hundreds, 0 tens, 3 units, 0 tenths and 2 hundredths ☐
c 5 hundreds, 6 tens, 3 units and 8 tenths ☐
d 4 hundreds, 1 ten, 0 units, 3 tenths and 9 hundredths ☐
e six and nine tenths ☐
f four and two tenths ☐
g five and three tenths ☐
h nine and one tenth ☐

5 Write the value for each coloured digit.

a 273·61 ☐
b 651·93 ☐
c 911·42 ☐
d 827·04 ☐
e 129·93 ☐
f 372·14 ☐
g 740·51 ☐
h 391·72 ☐
i 692·76 ☐
j 509·73 ☐
k 233·21 ☐
l 609·49 ☐

 • *AUSTRALIAN SIGNPOST MATHS 5* • ISBN 978 1 4886 2190 1

1:12 Comparing Decimals

5·254 is bigger than 4·570.

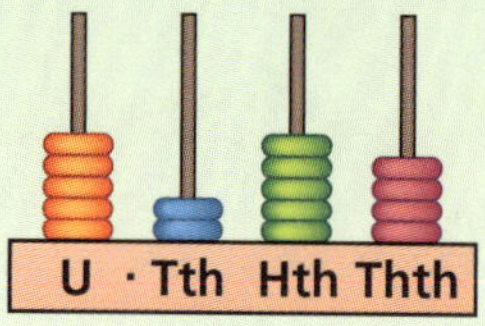

5·254

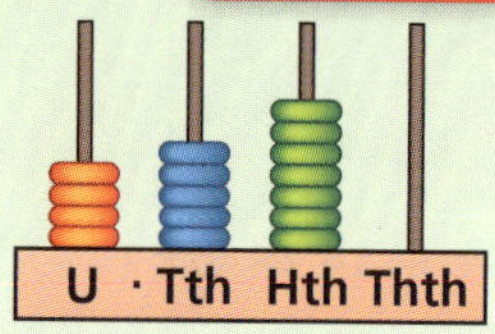

4·570 or 4·57

CONCEPT

1 Write the numeral for the number shown on each abacus.

a 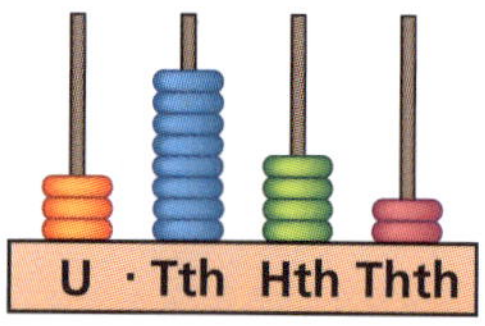

b 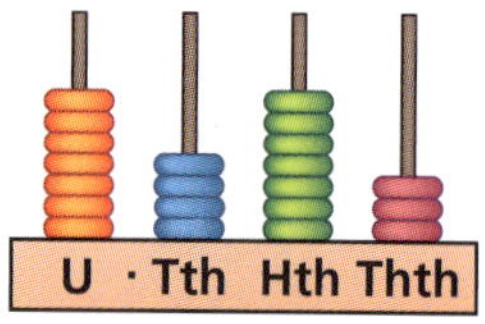

c 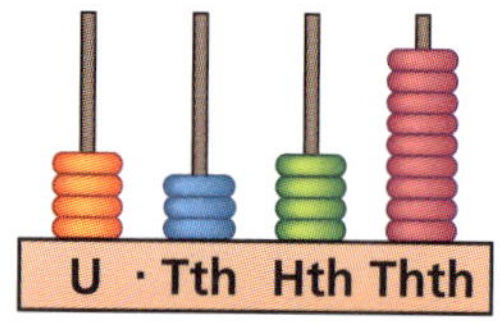

d 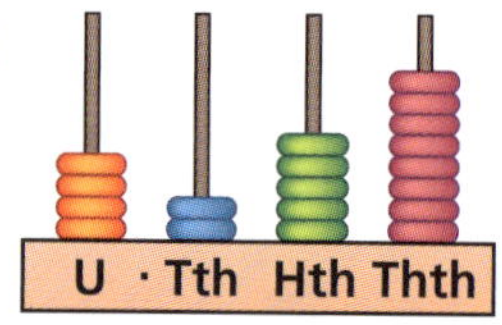

e 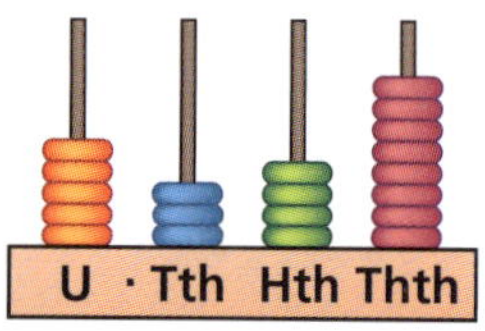

f 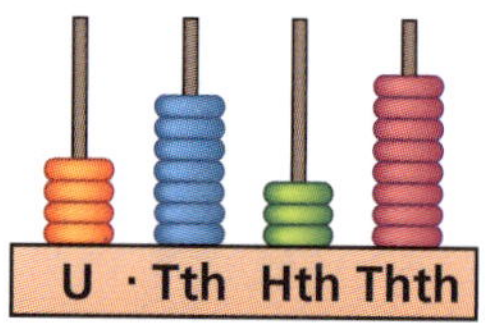

g 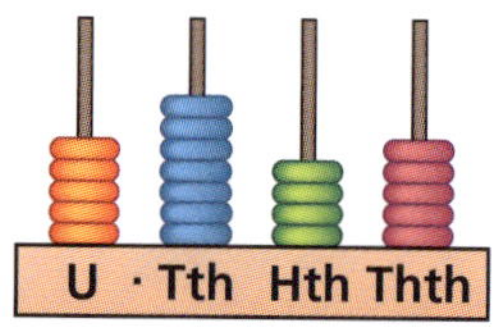

h 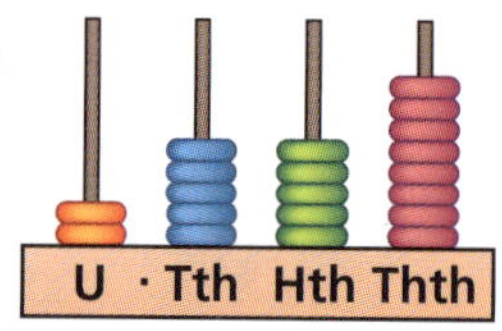

2 Use **>** or **<** to show the bigger decimal number.

a 3·619 ☐ 4·207 b 9·138 ☐ 9·56 c 4·196 ☐ 5·231 d 8·953 ☐ 7·641

e 1·403 ☐ 1·936 f 6·925 ☐ 5·863 g 5·306 ☐ 5·638 h 7·612 ☐ 7·59

3 For each statement write **true** or **false**.

a 6·735 < 6·603 ☐ b 9·345 > 9·245 ☐ c 8·238 < 8·304 ☐

4 Write the numeral for the number shown on each abacus.

a 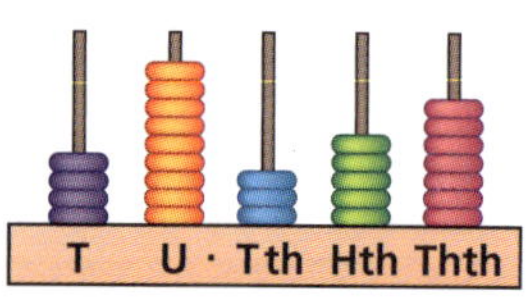

b 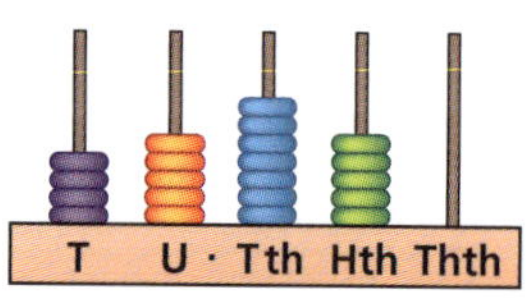

c 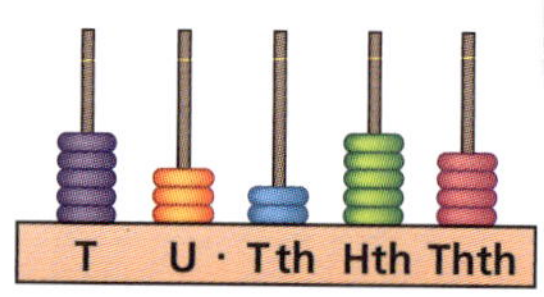

d 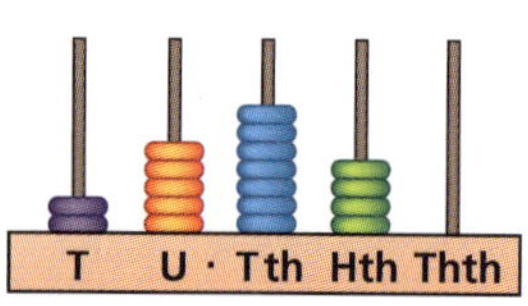

e 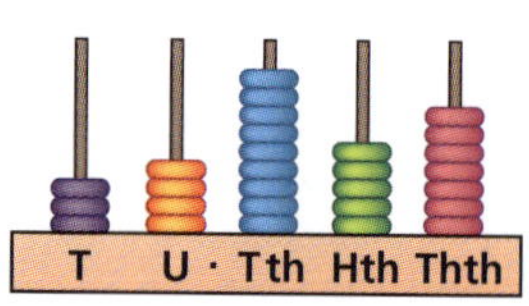

f 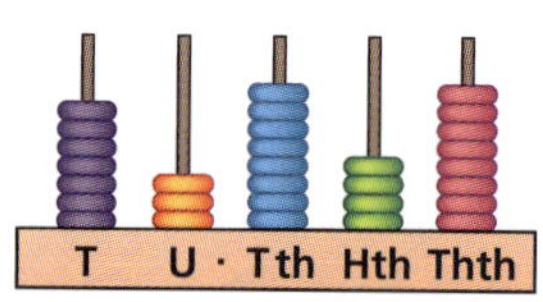

1:13 Place Value to Thousandths

CONCEPT

This abacus shows 58·254.

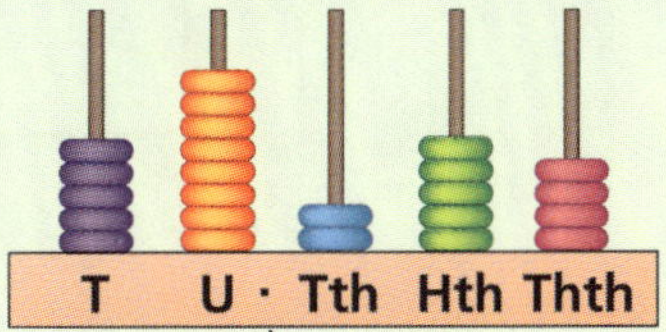

Decimal point

That's
5 tens
8 units
2 tenths
5 hundredths and
4 thousandths.

1 Write the numeral for the number shown on each abacus.

a 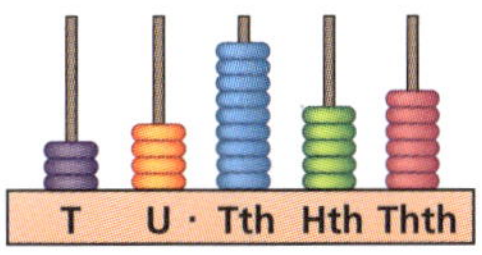

b 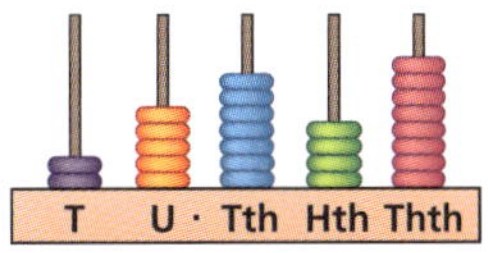

c 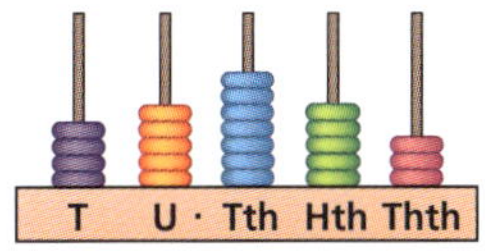

d 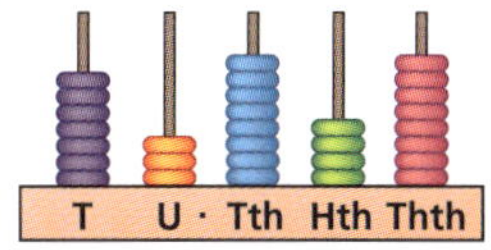

2 Write the value for each coloured digit.

a 34·**6**29 ☐　b 97·63**8** ☐　c 79·0**3**5 ☐

d 75·3**5**1 ☐　e 52·7**4**6 ☐　f 21·**9**17 ☐

g 6**4**·007 ☐　h 83·5**7**9 ☐　i 48·10**2** ☐

3 Write the decimal for each fraction.

a $\frac{47}{100}$ = ☐　b $\frac{359}{1000}$ = ☐　c $\frac{7}{10}$ = ☐

d $\frac{206}{1000}$ = ☐　e $\frac{6}{10}$ = ☐　f $\frac{8}{1000}$ = ☐

The length of a decimal does not indicate its relative size.

4 Write the fraction for each decimal.

a 0·36 = ☐　b 0·417 = ☐　c 0·5 = ☐

d 0·03 = ☐　e 0·48 = ☐　f 0·745 = ☐

5 Circle the larger number and explain why.

a 0·6 or 0·276　because $\frac{6}{10}$ is greater than $\frac{2}{10}$.

b 0·723 or 0·75　because $\frac{75}{100}$ is greater than

c 0·999 or 0·998　because

d 3·701 or 3·72　because

Compare tenths or hundredths or thousandths.

1:14 Addition of Fractions

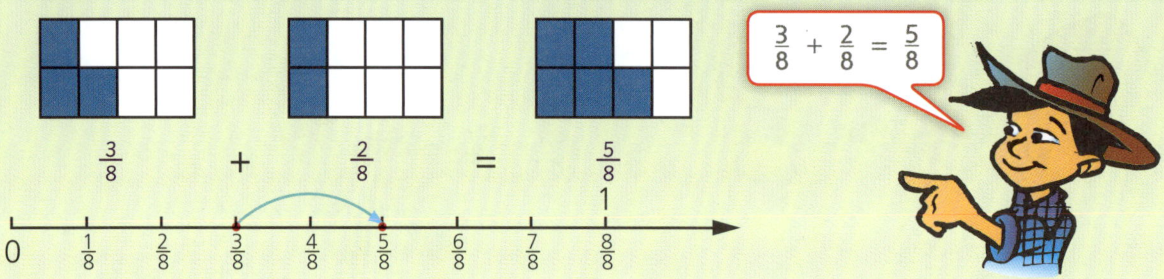

1 Add these fractions. Colour part of the last grid to match your answer.

a 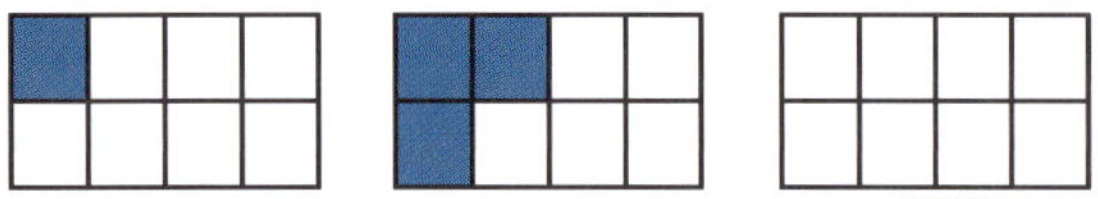

$\frac{1}{8} + \frac{3}{8} =$ ☐

b 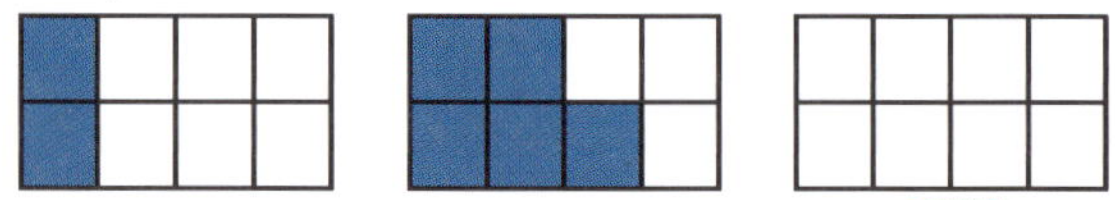

$\frac{2}{8} + \frac{5}{8} =$ ☐

c 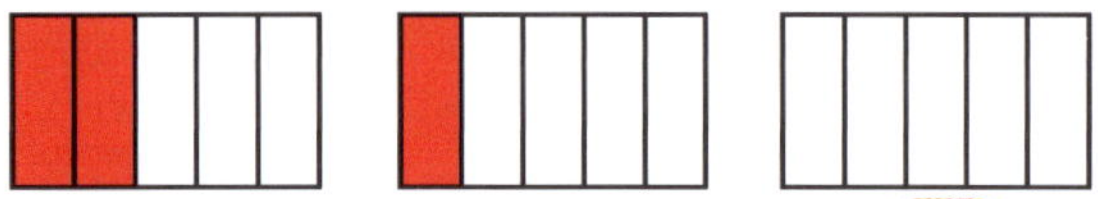

$\frac{2}{5} + \frac{1}{5} =$ ☐

d 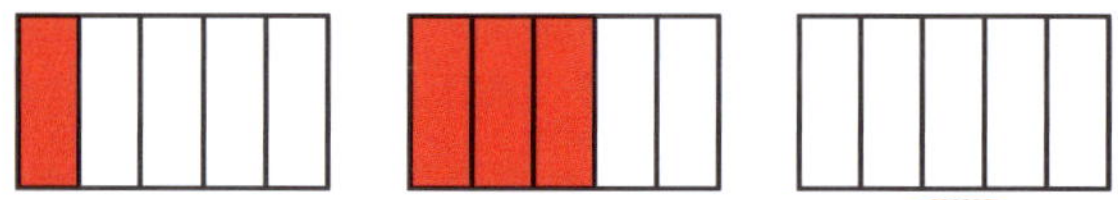

$\frac{1}{5} + \frac{3}{5} =$ ☐

e 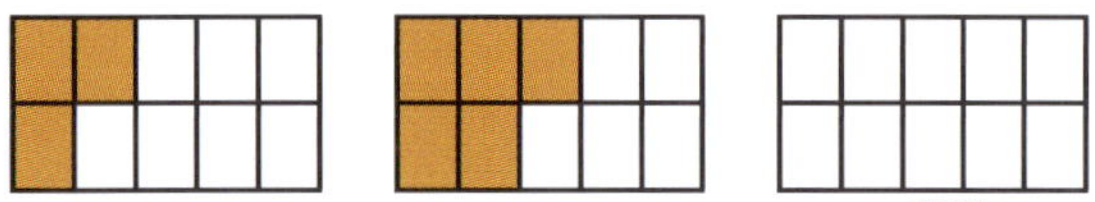

$\frac{3}{10} + \frac{5}{10} =$ ☐

f 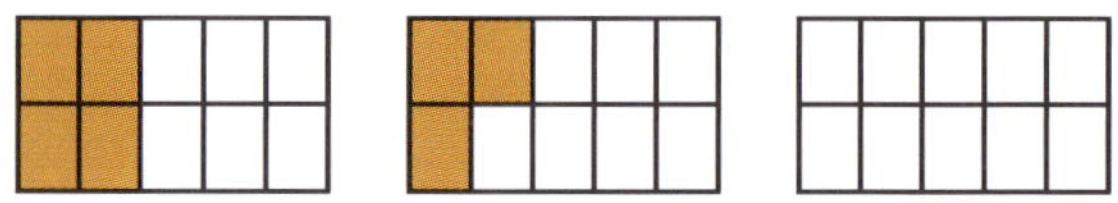

$\frac{4}{10} + \frac{3}{10} =$ ☐

2 Use the fraction card 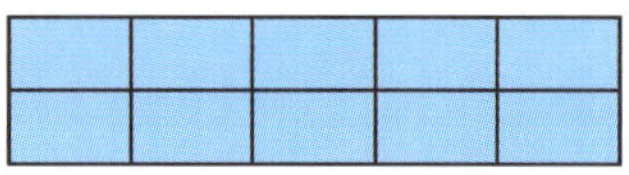 to find the answers. (Remember: $\frac{4}{10} + \frac{6}{10} = \frac{6}{10} + \frac{4}{10}$)

a $\frac{1}{10} + \frac{5}{10} =$ ☐ **b** $\frac{3}{10} + \frac{4}{10} =$ ☐ **c** $\frac{2}{10} + \frac{7}{10} =$ ☐ **d** $\frac{6}{10} + \frac{3}{10} =$ ☐

e $\frac{5}{10} + \frac{2}{10} =$ ☐ **f** $\frac{4}{10} + \frac{5}{10} =$ ☐ **g** $\frac{7}{10} + \frac{1}{10} =$ ☐ **h** $\frac{1}{10} + \frac{8}{10} =$ ☐

3 Use the fraction card 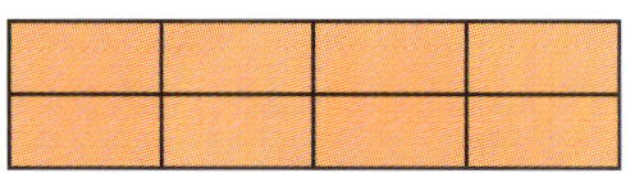 to find the answers.

a $\frac{1}{8} + \frac{6}{8} =$ ☐ **b** $\frac{2}{8} + \frac{4}{8} =$ ☐

c $\frac{3}{8} + \frac{3}{8} =$ ☐ **d** $\frac{2}{8} + \frac{3}{8} =$ ☐

It is easier to add fractions if they have the same denominator.

4 Answer **true** or **false**.

a $\frac{6}{8} > \frac{3}{8}$ ☐ **b** $\frac{7}{8} < \frac{6}{8}$ ☐ **c** $\frac{2}{8} < \frac{5}{8}$ ☐ **d** $\frac{1}{8} > \frac{7}{8}$ ☐

e $\frac{9}{10} > \frac{6}{10}$ ☐ **f** $\frac{5}{10} < \frac{9}{10}$ ☐ **g** $\frac{3}{10} > \frac{1}{10}$ ☐ **h** $\frac{8}{10} > \frac{7}{10}$ ☐

1:15 Subtraction of Fractions

CONCEPT

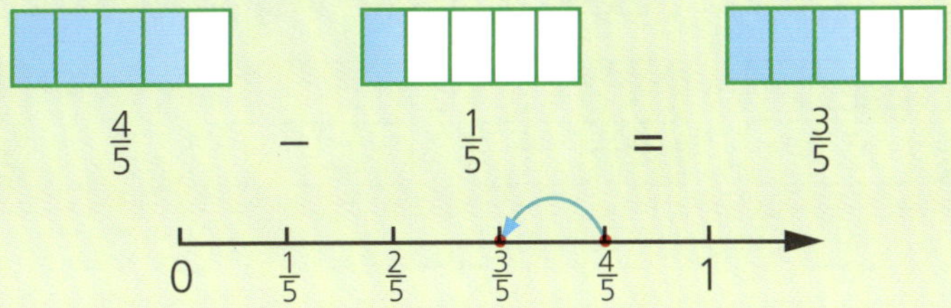

1 Subtract these fractions. Colour part of the grid to match your answer.

a 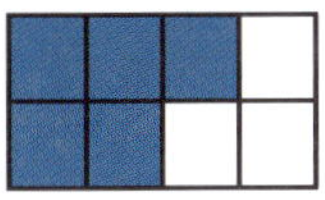 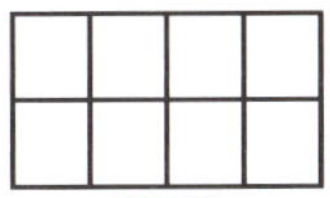

$\frac{5}{8} - \frac{1}{8} =$ ☐

b 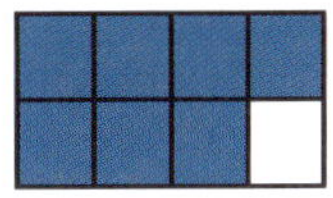 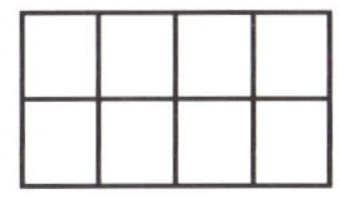

$\frac{7}{8} - \frac{2}{8} =$ ☐

c 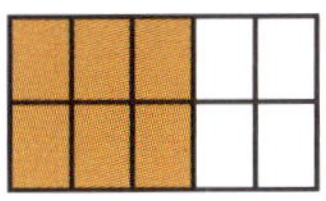 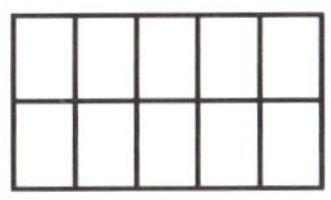

$\frac{6}{10} - \frac{3}{10} =$ ☐

d 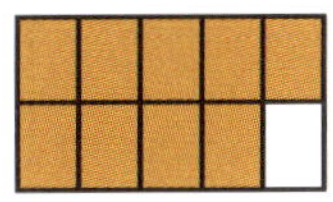 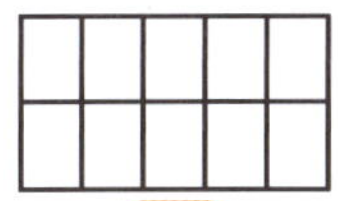

$\frac{9}{10} - \frac{4}{10} =$ ☐

e 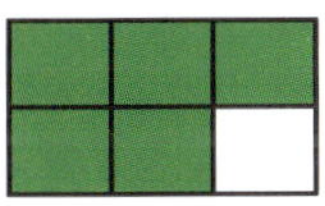 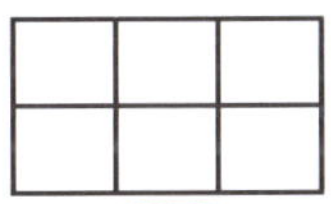

$\frac{5}{6} - \frac{2}{6} =$ ☐

f 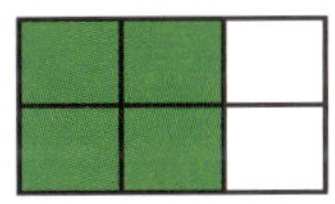 

$\frac{4}{6} - \frac{1}{6} =$ ☐

2 Use the fraction card 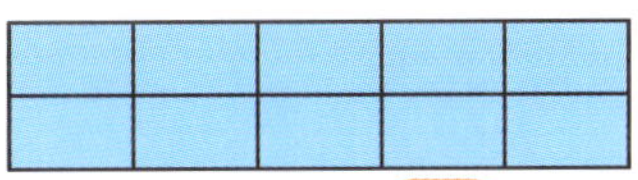 to find the answers.

a $\frac{3}{10} - \frac{1}{10} =$ ☐ **b** $\frac{5}{10} - \frac{3}{10} =$ ☐ **c** $\frac{7}{10} - \frac{2}{10} =$ ☐ **d** $\frac{9}{10} - \frac{5}{10} =$ ☐

e $\frac{8}{10} - \frac{7}{10} =$ ☐ **f** $\frac{6}{10} - \frac{1}{10} =$ ☐ **g** $\frac{4}{10} - \frac{3}{10} =$ ☐ **h** $\frac{7}{10} - \frac{4}{10} =$ ☐

3 Use the fraction card 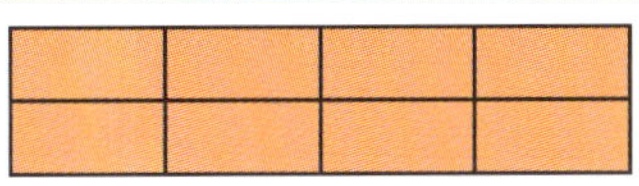 to find the answers.

a $\frac{7}{8} - \frac{3}{8} =$ ☐ **b** $\frac{5}{8} - \frac{3}{8} =$ ☐

c $\frac{6}{8} - \frac{2}{8} =$ ☐ **d** $\frac{4}{8} - \frac{1}{8} =$ ☐

It is easier to subtract fractions if they have the same denominator.

4 Answer **true** or **false**.

a $\frac{2}{8} < \frac{5}{8}$ ☐ **b** $\frac{6}{8} < \frac{5}{8}$ ☐ **c** $\frac{1}{8} < \frac{3}{8}$ ☐ **d** $\frac{7}{8} > \frac{4}{8}$ ☐

e $\frac{7}{10} > \frac{5}{10}$ ☐ **f** $\frac{9}{10} < \frac{8}{10}$ ☐ **g** $\frac{3}{10} < \frac{4}{10}$ ☐ **h** $\frac{6}{10} > \frac{3}{10}$ ☐

1:16 Addition and Subtraction of Fractions

1 Add or subtract these fractions. Colour part of the last grid to match your answer.

a
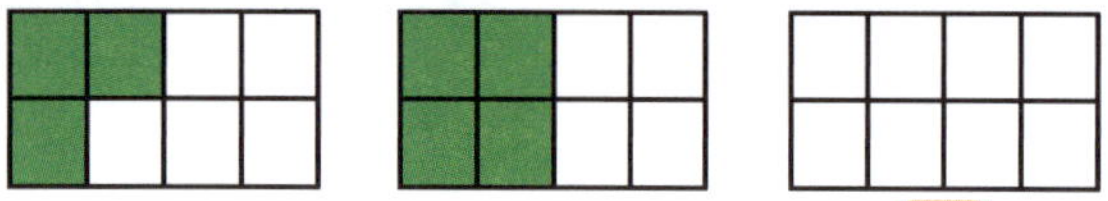
$\frac{3}{8} + \frac{4}{8} =$

b
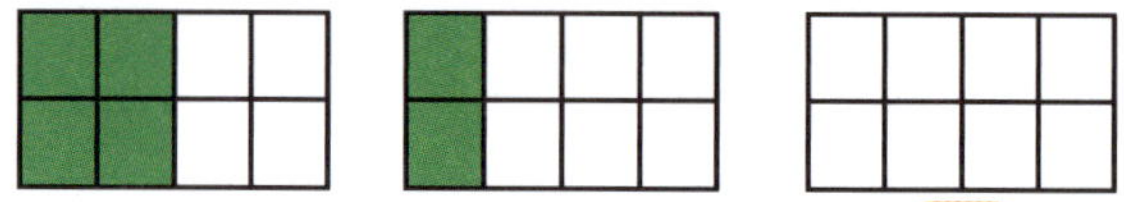
$\frac{4}{8} + \frac{2}{8} =$

c
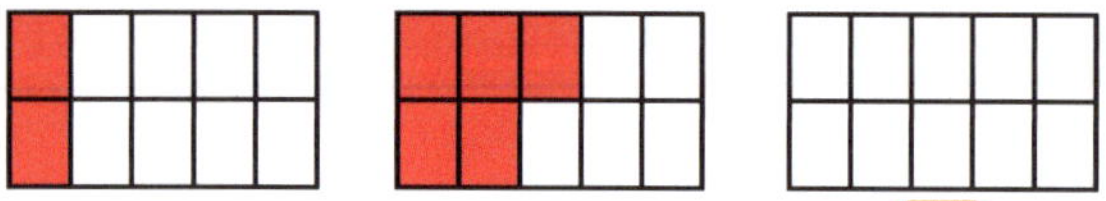
$\frac{2}{10} + \frac{5}{10} =$

d
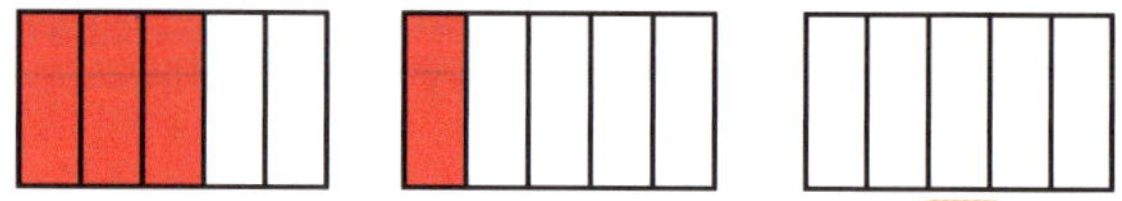
$\frac{3}{5} + \frac{1}{5} =$

e

$\frac{3}{5} - \frac{2}{5} =$

f

$\frac{5}{6} - \frac{3}{6} =$

g

$\frac{8}{10} - \frac{5}{10} =$

h

$\frac{7}{8} - \frac{5}{8} =$

2 Use the fraction card to find the answers.

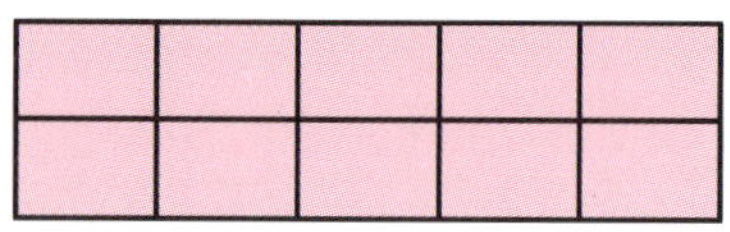

a $\frac{6}{10} - \frac{5}{10} =$ **b** $\frac{4}{10} - \frac{1}{10} =$ **c** $\frac{8}{10} - \frac{3}{10} =$ **d** $\frac{6}{10} - \frac{4}{10} =$

e $\frac{7}{10} - \frac{5}{10} =$ **f** $\frac{5}{10} - \frac{4}{10} =$ **g** $\frac{9}{10} - \frac{3}{10} =$ **h** $\frac{8}{10} - \frac{6}{10} =$

3 Use the fraction card to find the answers.

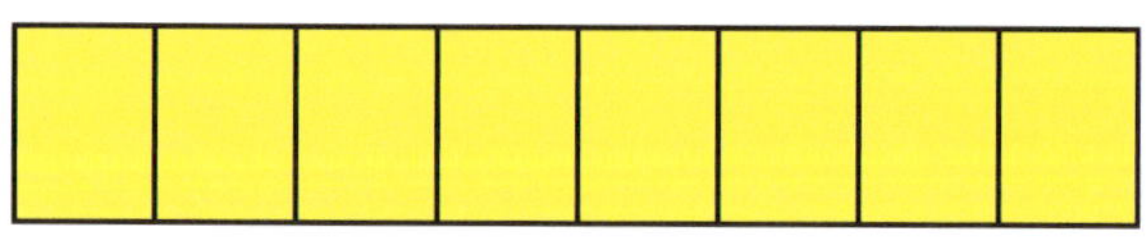

a $\frac{7}{8} - \frac{3}{8} =$ **b** $\frac{5}{8} - \frac{3}{8} =$ **c** $\frac{6}{8} - \frac{2}{8} =$ **d** $\frac{4}{8} - \frac{1}{8} =$

e $\frac{5}{8} - \frac{2}{8} =$ **f** $\frac{3}{8} - \frac{2}{8} =$ **g** $\frac{6}{8} - \frac{3}{8} =$ **h** $\frac{5}{8} - \frac{4}{8} =$

4 Use **>** or **<** to make each number sentence true.

a $\frac{3}{8}$ ☐ $\frac{1}{8}$ **b** $\frac{7}{8}$ ☐ $\frac{3}{8}$ **c** $\frac{2}{8}$ ☐ $\frac{5}{8}$ **d** $\frac{4}{8}$ ☐ $\frac{5}{8}$

e $\frac{1}{10}$ ☐ $\frac{3}{10}$ **f** $\frac{9}{10}$ ☐ $\frac{4}{10}$ **g** $\frac{7}{10}$ ☐ $\frac{2}{10}$ **h** $\frac{3}{10}$ ☐ $\frac{4}{10}$

5 Explain your answers to Question 4 to a friend.

 ISBN 978 1 4886 2190 1

1:17 Addition and Subtraction of Fractions

We can use the number line to help us work out answers.

CONCEPT

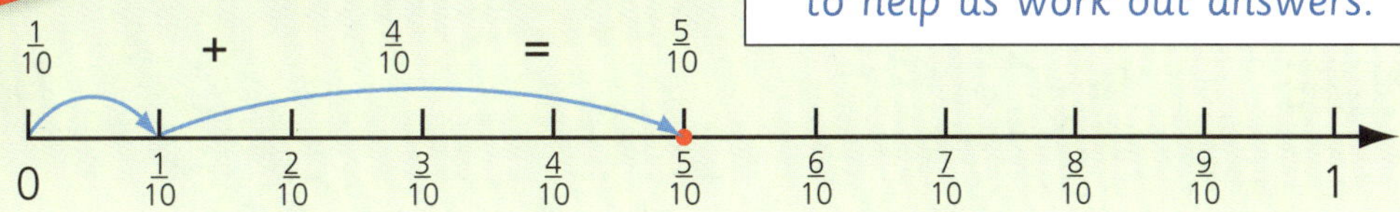

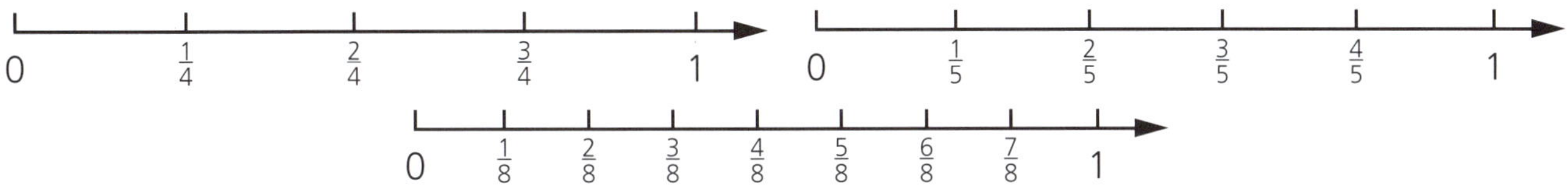

1 Add these fractions. Use the number lines above to help you.

a $\frac{1}{8} + \frac{3}{8} =$ ☐　b $\frac{1}{4} + \frac{2}{4} =$ ☐　c $\frac{2}{10} + \frac{7}{10} =$ ☐　d $\frac{1}{5} + \frac{3}{5} =$ ☐

e $\frac{2}{5} + \frac{2}{5} =$ ☐　f $\frac{5}{8} + \frac{2}{8} =$ ☐　g $\frac{1}{4} + \frac{2}{4} =$ ☐　h $\frac{3}{10} + \frac{4}{10} =$ ☐

2 Subtract these fractions. Use the number lines above to help you.

a $\frac{3}{4} - \frac{1}{4} =$ ☐　b $\frac{4}{5} - \frac{2}{5} =$ ☐　c $\frac{5}{10} - \frac{2}{10} =$ ☐　d $\frac{5}{8} - \frac{3}{8} =$ ☐

e $\frac{3}{5} - \frac{1}{5} =$ ☐　f $\frac{7}{8} - \frac{4}{8} =$ ☐　g $\frac{7}{10} - \frac{3}{10} =$ ☐　h $\frac{9}{10} - \frac{4}{10} =$ ☐

i $\frac{6}{10} - \frac{5}{10} =$ ☐　j $\frac{4}{5} - \frac{3}{5} =$ ☐　k $\frac{3}{4} - \frac{2}{4} =$ ☐　l $\frac{8}{10} - \frac{3}{10} =$ ☐

3 Add these fractions. Use the number lines to help you.

a $\frac{3}{8} + \frac{1}{8} + \frac{2}{8} =$ ☐　b $\frac{1}{5} + \frac{1}{5} + \frac{2}{5} =$ ☐　c $\frac{1}{10} + \frac{2}{10} + \frac{2}{10} =$ ☐　d $\frac{3}{10} + \frac{2}{10} + \frac{4}{10} =$ ☐

4 Write the improper fraction for the mixed numbers:

a $1\frac{2}{8} =$ ☐　b $1\frac{4}{5} =$ ☐　c $1\frac{3}{4} =$ ☐

d $3\frac{1}{2} =$ ☐　e $2\frac{3}{5} =$ ☐　f $1\frac{9}{10} =$ ☐

5 Write the mixed number for the improper fractions:

a $\frac{8}{5} =$ ☐　b $\frac{4}{3} =$ ☐

c $\frac{9}{4} =$ ☐　d $\frac{9}{8} =$ ☐

Note: $\frac{6}{4} = 1\frac{2}{4}$, $\frac{7}{4} = 1\frac{3}{4}$

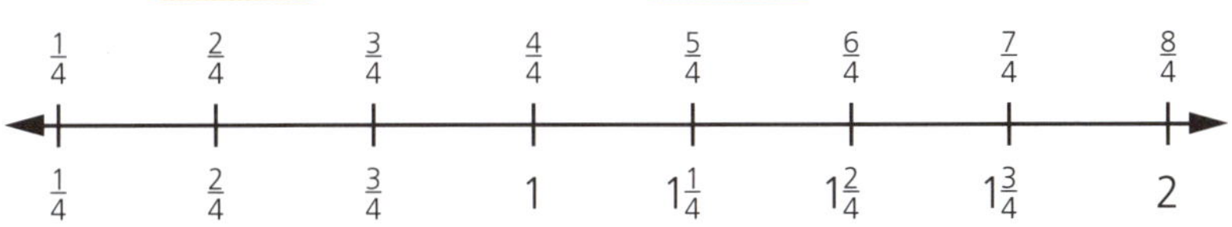

Improper fraction: $\frac{11}{8}$
Mixed number: $1\frac{3}{8}$

1:18 Addition and Subtraction of Fractions

1 Complete:

a $1 - \frac{1}{6} =$ ☐ b $1 - \frac{1}{10} =$ ☐ c $1 - \frac{1}{8} =$ ☐ d $1 - \frac{1}{12} =$ ☐

e $1 - \frac{1}{5} =$ ☐ f $1 - \frac{3}{4} =$ ☐ g $1 - \frac{7}{10} =$ ☐ h $1 - \frac{2}{5} =$ ☐

i $1 - \frac{2}{3} =$ ☐ j $1 - \frac{5}{6} =$ ☐ k $1 - \frac{3}{8} =$ ☐ l $1 - \frac{5}{12} =$ ☐

2 Complete:

a $\frac{7}{8} - \frac{3}{8} =$ ☐ b $\frac{5}{6} - \frac{2}{6} =$ ☐ c $\frac{2}{3} - \frac{1}{3} =$ ☐ d $\frac{3}{4} - \frac{1}{4} =$ ☐

e $\frac{3}{6} - \frac{1}{6} =$ ☐ f $\frac{5}{8} - \frac{2}{8} =$ ☐ g $\frac{9}{10} - \frac{3}{10} =$ ☐ h $\frac{4}{5} - \frac{2}{5} =$ ☐

i $\frac{11}{12} - \frac{6}{12} =$ ☐ j $\frac{7}{10} - \frac{4}{10} =$ ☐ k $\frac{3}{5} - \frac{1}{5} =$ ☐ l $\frac{7}{12} - \frac{3}{12} =$ ☐

3 Complete:

a $3 - \frac{1}{2} =$ ☐ b $2 - \frac{1}{3} =$ ☐ c $2 - \frac{1}{6} =$ ☐

d $3 - \frac{1}{5} =$ ☐ e $4 - \frac{1}{10} =$ ☐ f $4 - \frac{1}{12} =$ ☐

g $3 - \frac{1}{8} =$ ☐ h $4 - \frac{1}{4} =$ ☐ i $2 - \frac{3}{4} =$ ☐

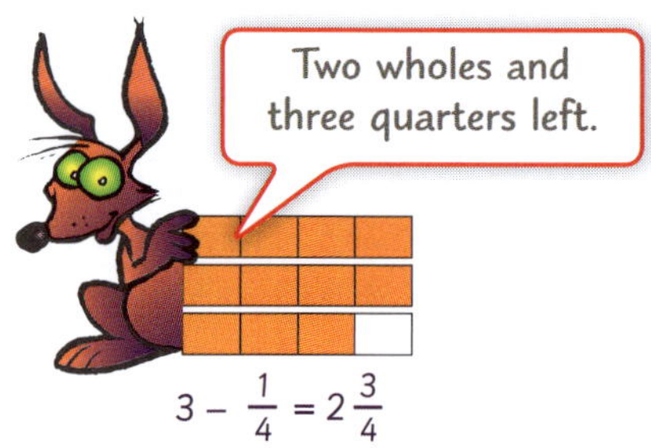

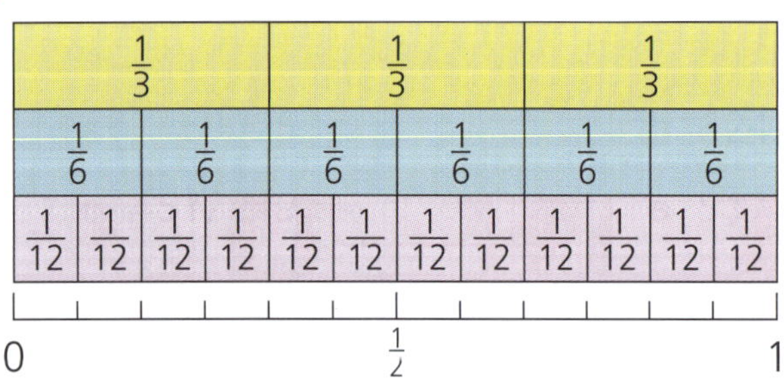

4 Use the diagram to show an equivalent fraction for:

a $\frac{2}{12} =$ ☐ b $\frac{10}{12} =$ ☐ c $\frac{4}{6} =$ ☐

d $\frac{2}{3} =$ ☐ e $\frac{8}{12} =$ ☐ f $\frac{3}{6} =$ ☐

5 Use the diagram in Question 4 to answer **true** or **false**.

a $\frac{1}{3} = \frac{4}{12}$ ☐ b $\frac{8}{12} = \frac{4}{6}$ ☐ c $\frac{2}{3} = \frac{6}{12}$ ☐ d $\frac{10}{12} = \frac{5}{6}$ ☐

6 Using the diagram, explain your answers to Questions 4 and 5 to a friend.

 ISBN 978 1 4886 2190 1

1:19 Comparing Decimal Measurements

15·45 s means …

15 and forty-five hundredths seconds.

9·31 s means …

9 and thirty-one hundredths seconds.

CONCEPT

1 Yes or No?

a The fastest time for a race is the smallest time. ______

b 31·14 s is faster than 31·65 s. ______

c 9·65 s is slower than 8·1 s. ______

2 Match these race times to the number line to show their order.

21·43 s | 21·38 s | 21·8 s | 21·65 s | 22·36 s | 22·6 s | 22·48 s

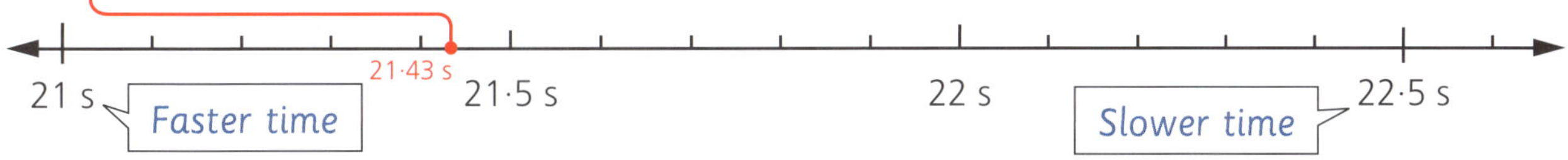

3 Circle the faster time (smallest decimal) in each pair.

a 4·75, 4·70

b 31·46, 30·40

c 39·74, 40·06

d 59·60, 59·50

e 29·56, 29·83

f 19·14, 26·99

g 36·43, 27·86

h 19·63, 19·62

A stopwatch allows accurate measurement of time intervals.

ACTIVITY

4 Examine and operate various types of stopwatches.
Discuss the following:

- Give an example of time being measured to the nearest **hundredth** of a second. ______
- What is the greatest length of time that can be measured on a stopwatch? ______
- Why does a timekeeper of a 100 m race watch for the smoke instead of listening for the sound of the gun? ______

How accurately can you measure time?
Compare and discuss your results.
With a partner, find how long it takes to hop ten metres. ______

 • *AUSTRALIAN SIGNPOST MATHS 5* • ISBN 978 1 4886 2190 1

1:20 Using Decimals

CONCEPT

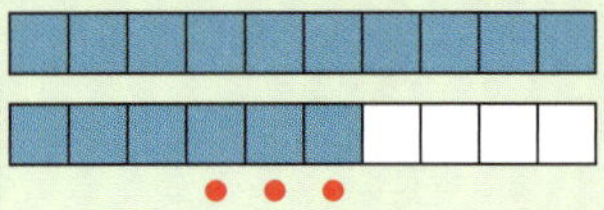

16 tenths
− 3 tenths

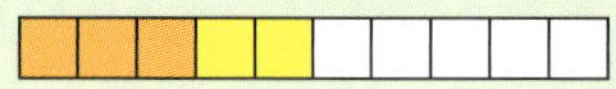

3 tenths
+ 2 tenths

Points
Under
Points

1 a 0·7 + 0·2 = ☐ b 0·8 + 0·5 = ☐ c 1·1 + 0·3 = ☐ d 1·9 + 0·1 = ☐ e 4·3 + 0·3 = ☐

f 0·4 + 0·2 = ☐ g 0·8 + 0·3 = ☐ h 1·9 + 0·4 = ☐

2 a 0·8 − 0·3 = ☐ b 1·4 − 0·6 = ☐ c 0·9 − 0·3 = ☐ d 1·8 − 0·3 = ☐ e 1·0 − 0·4 = ☐

Patterns Using Decimals

CONCEPT

1·7, 2·0, 2·3, 2·6, 2·9, …

Rule: Add 0·3.

1·5, 1·3, 1·1, 0·9, 0·7, …

Rule: Subtract 0·2.

3 Complete the patterns.

a 0·1, 0·3, 0·5, ☐, ☐
b 0·8, 1·3, 1·8, ☐, ☐
c 0·9, 0·8, 0·7, ☐, ☐
d 3·1, 2·9, 2·7, ☐, ☐
e 0·15, 0·17, 0·19, ☐, ☐
f 1·22, 1·32, 1·42, ☐, ☐
g 0·91, 0·86, 0·81, ☐, ☐
h 3·06, 3·12, 3·18, ☐, ☐

4 Use the starting number and make up your own pattern.

a 0·5, ☐, ☐, ☐
b 0·8, ☐, ☐, ☐
c 0·19, ☐, ☐, ☐
d 0·36, ☐, ☐, ☐
e 3·7, ☐, ☐, ☐
f 9·3, ☐, ☐, ☐

 • *AUSTRALIAN SIGNPOST MATHS 5* • ISBN 978 1 4886 2190 1

1:21 Patterns with Fractions and Decimals

1 Complete each pattern and write the rule.

a $\frac{7}{10}$, $\frac{9}{10}$, $\frac{11}{10}$, ___, ___, ___, ___ The rule is: ___

b 4, $3\frac{8}{10}$, $3\frac{6}{10}$, ___, ___, ___, ___ The rule is: ___

c 0·85, 0·87, 0·89, ___, ___, ___ The rule is: ___

d 1·6, 1·5, 1·4, ___, ___, ___ The rule is: ___

2 Write and continue the pattern made by the jumps on the number line. Write the rule for each.

a

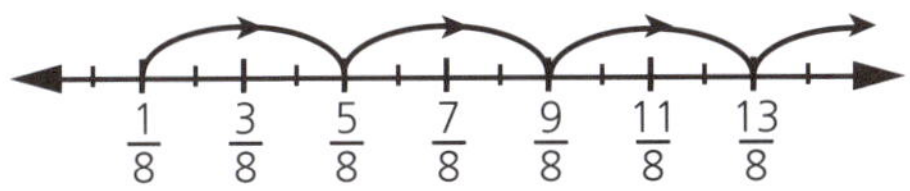

$\frac{1}{8}$, ___, ___, ___, ___, ___, ___

The rule is: ___

b

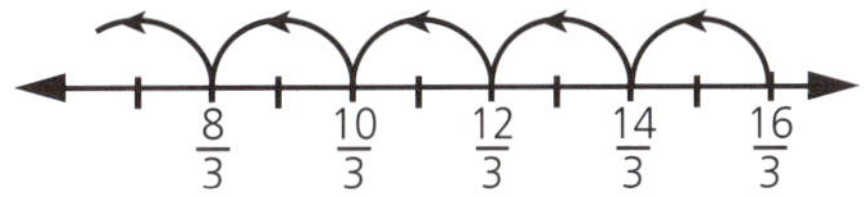

$\frac{16}{3}$, ___, ___, ___, ___, ___, ___

The rule is: ___

3 Create your own decimal number pattern using jumps on the number line. Write the rule.

a

___

___, ___, ___, ___, ___, ___, ___

The rule is: ___

4 Write and continue the pattern shown in these diagrams. Write the rule for each.

a

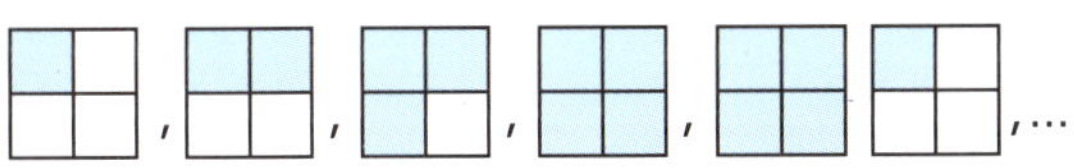

$\frac{1}{4}$, ___, ___, ___, ___, ___, ___

The rule is: ___

b

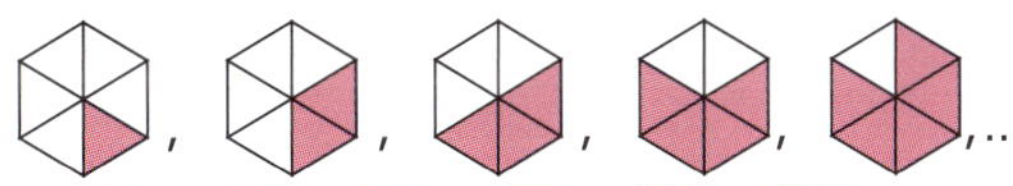

$\frac{1}{6}$, ___, ___, ___, ___, ___, ___

The rule is: ___

c

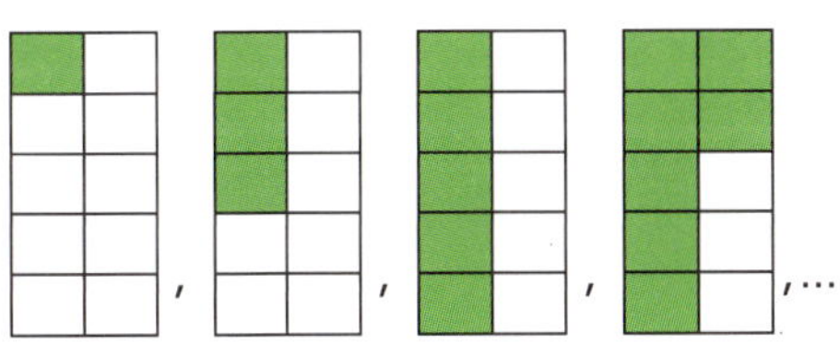

0·1, ___, ___, ___, ___, ___, ___

The rule is: ___

d

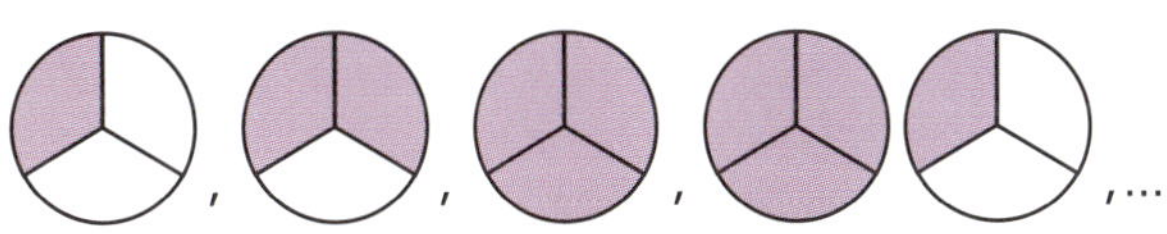

$\frac{1}{3}$, ___, ___, ___, ___, ___, ___

The rule is: ___

 • *AUSTRALIAN SIGNPOST MATHS 5* • ISBN 978 1 4886 2190 1

1:22 Equivalent Fractions

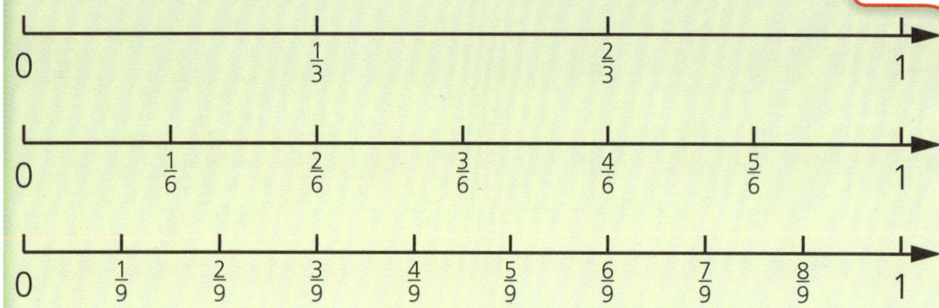

CONCEPT

$\frac{1}{3} = \frac{2}{6}$

$\frac{1}{3} = \frac{3}{9}$

1 Use the number lines to show an equivalent fraction for:

a $\frac{1}{3}$ ☐ **b** $\frac{9}{9}$ ☐ **c** $\frac{3}{9}$ ☐ **d** $\frac{2}{3}$ ☐

e $\frac{2}{6}$ ☐ **f** $\frac{4}{6}$ ☐ **g** $\frac{6}{6}$ ☐ **h** $\frac{6}{9}$ ☐

2 Use the number lines above to answer **true** or **false**.

a $\frac{1}{6} = \frac{2}{9}$ ☐ **b** $\frac{2}{3} = \frac{6}{9}$ ☐ **c** $\frac{2}{3} = \frac{4}{6}$ ☐ **d** $\frac{6}{9} = \frac{4}{6}$ ☐

e $\frac{3}{3} = \frac{9}{9}$ ☐ **f** $\frac{3}{6} = \frac{5}{9}$ ☐ **g** $\frac{1}{3} = \frac{3}{9}$ ☐ **h** $\frac{5}{6} = \frac{2}{3}$ ☐

3 Complete the number lines.

a

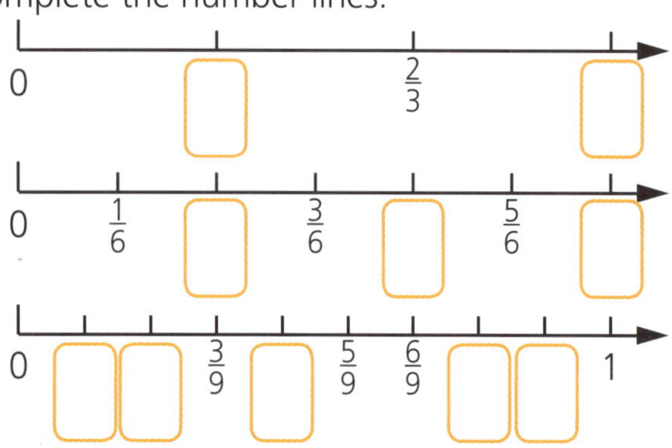

b

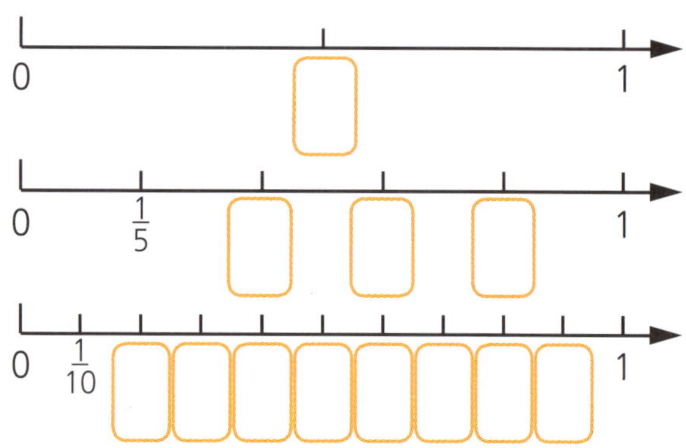

4 Use the number lines above to answer **true** or **false**.

a $\frac{1}{5} = \frac{2}{10}$ ☐ **b** $\frac{6}{9} = \frac{4}{6}$ ☐ **c** $\frac{1}{3} = \frac{1}{6}$ ☐ **d** $\frac{2}{5} = \frac{4}{10}$ ☐

e $\frac{2}{3} = \frac{4}{6}$ ☐ **f** $\frac{3}{6} = \frac{6}{9}$ ☐ **g** $\frac{4}{5} = \frac{9}{10}$ ☐ **h** $\frac{3}{5} = \frac{8}{10}$ ☐

i $\frac{1}{5} = \frac{1}{10}$ ☐ **j** $\frac{2}{3} = \frac{6}{9}$ ☐ **k** $\frac{1}{2} = \frac{5}{10}$ ☐ **l** $\frac{5}{6} = \frac{8}{9}$ ☐

5 Use the number lines above to write an equivalent fraction for:

a $\frac{1}{3}$ ☐ **b** $\frac{4}{6}$ ☐ **c** $\frac{1}{2}$ ☐ **d** $\frac{6}{9}$ ☐

e $\frac{8}{10}$ ☐ **f** $\frac{3}{9}$ ☐ **g** $\frac{2}{3}$ ☐ **h** $\frac{1}{5}$ ☐

1:23 Equivalent Fractions

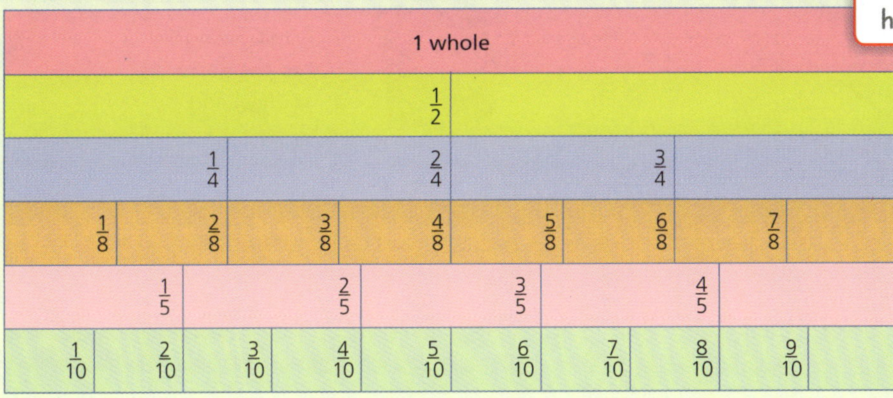

CONCEPT

We can make equivalent fractions by multiplying the numerator and denominator by the same number.

$$\frac{1}{2}\;\frac{(\times 5)}{(\times 5)} = \frac{5}{10}$$

1 Complete these to make equivalent fractions.

a $\frac{1}{4} = \frac{\square}{8}$ b $\frac{2}{5} = \frac{\square}{10}$ c $\frac{3}{4} = \frac{\square}{8}$ d $\frac{1}{5} = \frac{\square}{10}$ e $\frac{1}{2} = \frac{\square}{4}$

f $\frac{3}{5} = \frac{\square}{10}$ g $\frac{2}{4} = \frac{\square}{8}$ h $\frac{4}{5} = \frac{\square}{10}$ i $\frac{1}{2} = \frac{\square}{8}$ j $1 = \frac{\square}{10}$

2 Complete these to make equivalent fractions.

a $\frac{1}{4}\;\frac{(\times 2)}{(\times 2)} = \frac{\square}{\square}$ b $\frac{1}{2}\;\frac{(\times 4)}{(\times 4)} = \frac{\square}{\square}$ c $\frac{1}{5}\;\frac{(\times 2)}{(\times 2)} = \frac{\square}{\square}$ d $\frac{3}{5}\;\frac{(\times 2)}{(\times 2)} = \frac{\square}{\square}$

e $\frac{3}{4}\;\frac{(\times 2)}{(\times 2)} = \frac{\square}{\square}$ f $\frac{4}{5}\;\frac{(\times 2)}{(\times 2)} = \frac{\square}{\square}$ g $\frac{1}{2}\;\frac{(\times 2)}{(\times 2)} = \frac{\square}{\square}$ h $\frac{2}{5}\;\frac{(\times 2)}{(\times 2)} = \frac{\square}{\square}$

i $\frac{1}{3}\;\frac{(\times 2)}{(\times 2)} = \frac{\square}{\square}$ j $\frac{1}{6}\;\frac{(\times 2)}{(\times 2)} = \frac{\square}{\square}$ k $\frac{2}{3}\;\frac{(\times 4)}{(\times 4)} = \frac{\square}{\square}$ l $\frac{4}{6}\;\frac{(\times 2)}{(\times 2)} = \frac{\square}{\square}$

3 Complete these to make equivalent fractions.

a $\frac{1}{3}\;\frac{(\times\;\;)}{(\times\;\;)} = \frac{4}{\square}$ b $\frac{3}{6}\;\frac{(\times\;\;)}{(\times\;\;)} = \frac{6}{\square}$ c $\frac{2}{3}\;\frac{(\times\;\;)}{(\times\;\;)} = \frac{8}{\square}$

d $\frac{2}{6}\;\frac{(\times\;\;)}{(\times\;\;)} = \frac{4}{\square}$ e $\frac{4}{10}\;\frac{(\times\;\;)}{(\times\;\;)} = \frac{8}{\square}$ f $\frac{3}{5}\;\frac{(\times\;\;)}{(\times\;\;)} = \frac{9}{\square}$

g $\frac{3}{5}\;\frac{(\times\;\;)}{(\times\;\;)} = \frac{6}{\square}$ h $\frac{1}{4}\;\frac{(\times\;\;)}{(\times\;\;)} = \frac{5}{\square}$ i $\frac{2}{7}\;\frac{(\times\;\;)}{(\times\;\;)} = \frac{6}{\square}$

1:24 Equivalent Fractions

CONCEPT

Multiply the numerator and denominator by the same number.

$$\frac{1}{2}\ \frac{(\times 5)}{(\times 5)} = \frac{5}{10}$$

Multiplying by $\frac{5}{5}$ is the same as multiplying by 1.

1 Write an equivalent fraction for each of these.

a $\frac{1}{3}\ \frac{(\times 2)}{(\times 2)} = \frac{\square}{\square}$ **b** $\frac{1}{4}\ \frac{(\times 3)}{(\times 3)} = \frac{\square}{\square}$ **c** $\frac{1}{2}\ \frac{(\times 5)}{(\times 5)} = \frac{\square}{\square}$ **d** $\frac{1}{5}\ \frac{(\times 3)}{(\times 3)} = \frac{\square}{\square}$

e $\frac{2}{5}\ \frac{(\times 3)}{(\times 3)} = \frac{\square}{\square}$ **f** $\frac{2}{3}\ \frac{(\times 4)}{(\times 4)} = \frac{\square}{\square}$ **g** $\frac{3}{4}\ \frac{(\times 2)}{(\times 2)} = \frac{\square}{\square}$ **h** $\frac{3}{5}\ \frac{(\times 3)}{(\times 3)} = \frac{\square}{\square}$

2 Complete these.

a $\frac{1}{2}\ \frac{(\times 4)}{(\times 4)} = \frac{\square}{\square}$ **b** $\frac{1}{5}\ \frac{(\times 2)}{(\times 2)} = \frac{\square}{\square}$ **c** $\frac{1}{3}\ \frac{(\times 3)}{(\times 3)} = \frac{\square}{\square}$ **d** $\frac{1}{6}\ \frac{(\times 2)}{(\times 2)} = \frac{\square}{\square}$

e $\frac{2}{3}\ \frac{(\times 2)}{(\times 2)} = \frac{\square}{\square}$ **f** $\frac{1}{2}\ \frac{(\times 3)}{(\times 3)} = \frac{\square}{\square}$ **g** $\frac{1}{5}\ \frac{(\times 4)}{(\times 4)} = \frac{\square}{\square}$ **h** $\frac{1}{3}\ \frac{(\times 4)}{(\times 4)} = \frac{\square}{\square}$

3 Multiply both the numerator and the denominator by 2.

a $\frac{1}{4} = \frac{\square}{\square}$ **b** $\frac{1}{6} = \frac{\square}{\square}$ **c** $\frac{1}{2} = \frac{\square}{\square}$ **d** $\frac{1}{5} = \frac{\square}{\square}$ **e** $\frac{1}{3} = \frac{\square}{\square}$

4 Multiply both the numerator and the denominator by 3.

a $\frac{1}{6} = \frac{\square}{\square}$ **b** $\frac{2}{3} = \frac{\square}{\square}$ **c** $\frac{1}{3} = \frac{\square}{\square}$ **d** $\frac{3}{4} = \frac{\square}{\square}$ **e** $\frac{4}{5} = \frac{\square}{\square}$

5 Multiply both the numerator and the denominator by 4.

a $\frac{1}{4} = \frac{\square}{\square}$ **b** $\frac{1}{5} = \frac{\square}{\square}$ **c** $\frac{1}{3} = \frac{\square}{\square}$ **d** $\frac{3}{4} = \frac{\square}{\square}$ **e** $\frac{2}{3} = \frac{\square}{\square}$

6 What number has been used to multiply the numerator and denominator in each pair of equivalent fractions below?

a $\frac{1}{2} = \frac{4}{8}$ $\square$ **b** $\frac{1}{4} = \frac{3}{12}$ $\square$ **c** $\frac{1}{3} = \frac{3}{9}$ $\square$ **d** $\frac{1}{6} = \frac{2}{12}$ $\square$ **c** $\frac{2}{3} = \frac{4}{6}$ $\square$

f $\frac{3}{4} = \frac{6}{8}$ $\square$ **g** $\frac{1}{5} = \frac{2}{10}$ $\square$ **h** $\frac{3}{5} = \frac{6}{10}$ $\square$ **i** $\frac{1}{2} = \frac{5}{10}$ $\square$ **j** $\frac{1}{3} = \frac{4}{12}$ $\square$

7 Complete these equivalent fractions.

a $\frac{1}{2} = \frac{\square}{4} = \frac{\square}{6} = \frac{\square}{8} = \frac{\square}{10} = \frac{\square}{12}$

b $\frac{1}{3} = \frac{\square}{6} = \frac{\square}{9} = \frac{\square}{12} = \frac{\square}{15} = \frac{\square}{18}$

 ISBN 978 1 4886 2190 1

1:25 Percentages

CONCEPT

50 out of 100
50 hundredths
50%

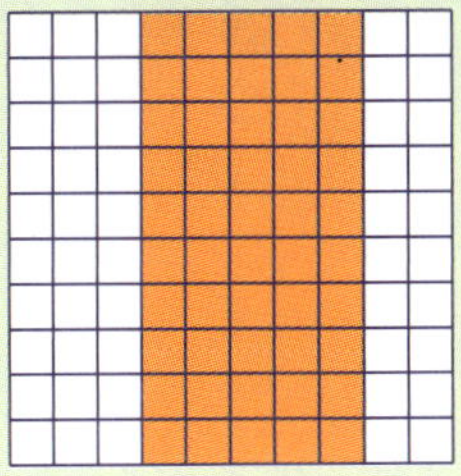

0.50
$\frac{50}{100}$

1 What percentage of each square is coloured?

a 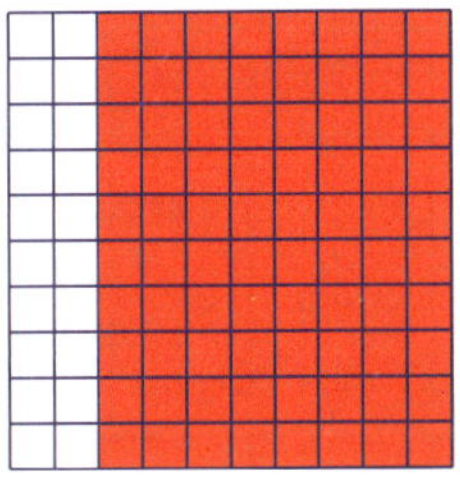

b 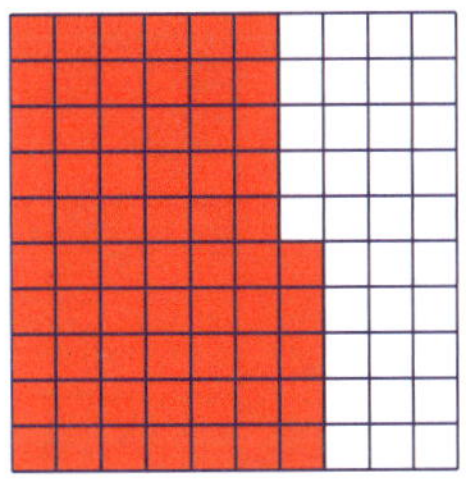

c 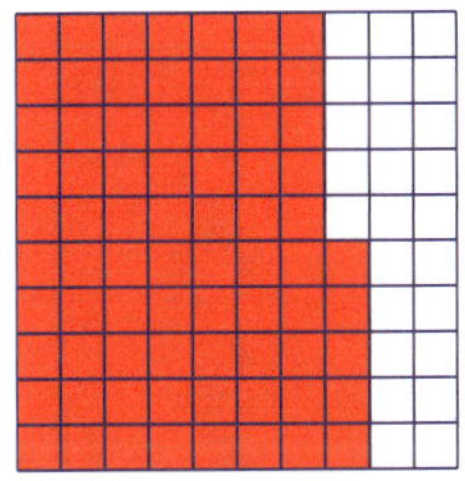

d 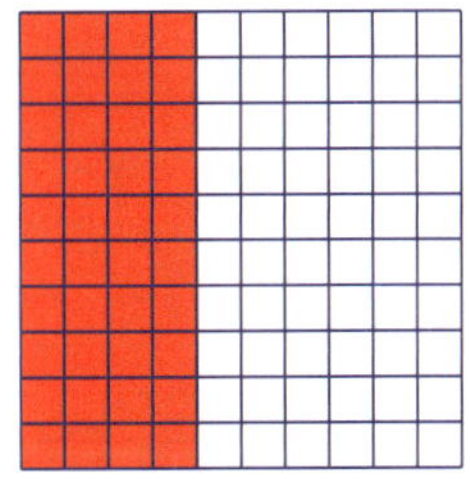

e 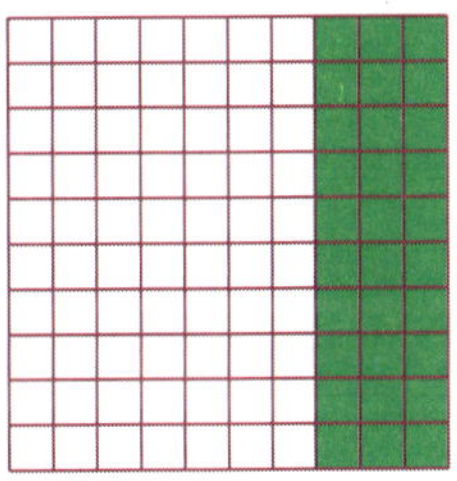

f 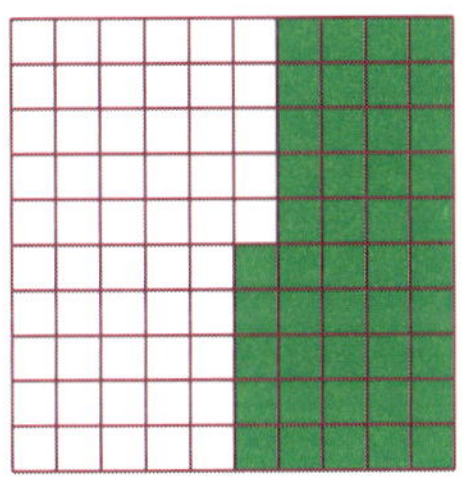

g 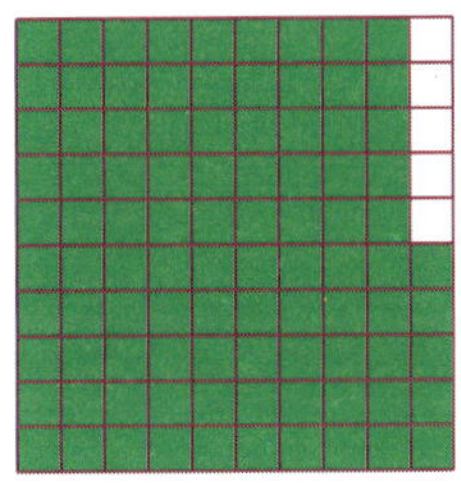

h 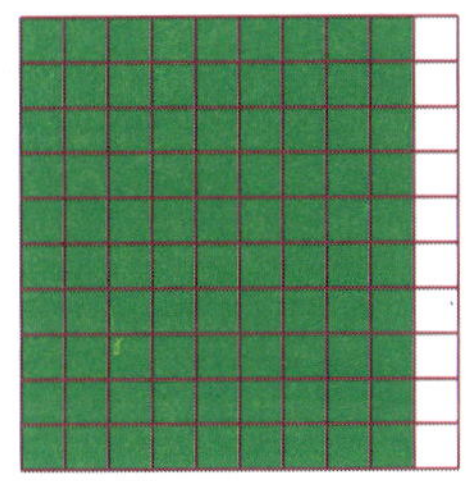

2 What percentage of each square in Question 1 is not coloured?

a ____ b ____ c ____ d ____ e ____ f ____ g ____ h ____

3 Complete these equivalents:

a	0·25	$\frac{\quad}{100}$	%
b	0·65		
c	0·45		
d	0·80		
e	0·50		
f	0·20		
g	0·60		
h	0·35		
i	0·75		
j	0·55		
k	0·95		
l	0·40		

Percentages in the Environment

ACTIVITY

- Collect examples of percentages from newspapers and food packages.
- Discuss different ways in which percentages are used.

1:26 Using Percentages

1 For each square, colour the percentage shown.

a 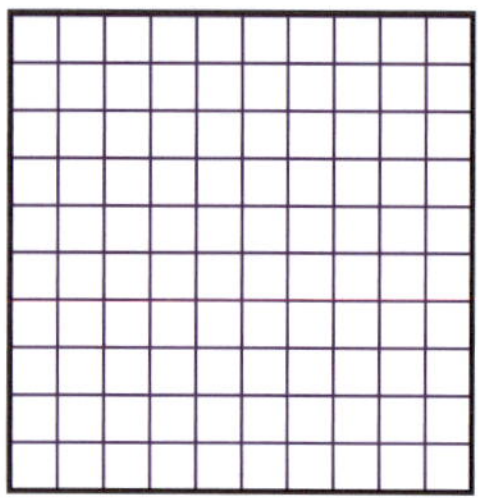 10%

b 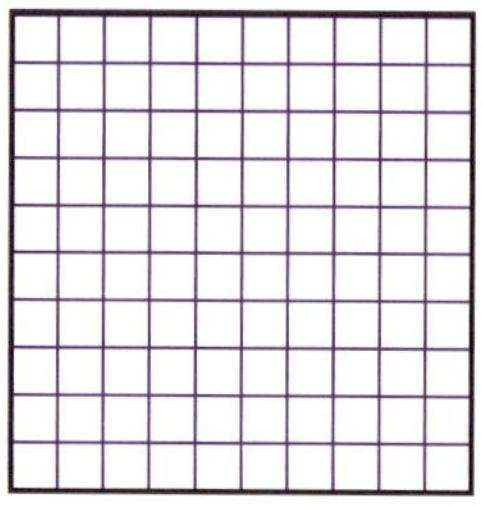 25%

c 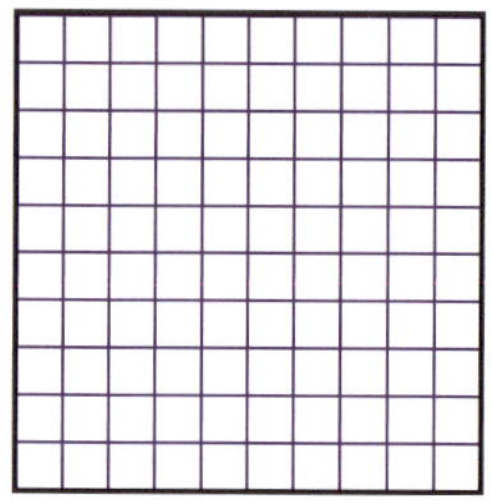 20%

d 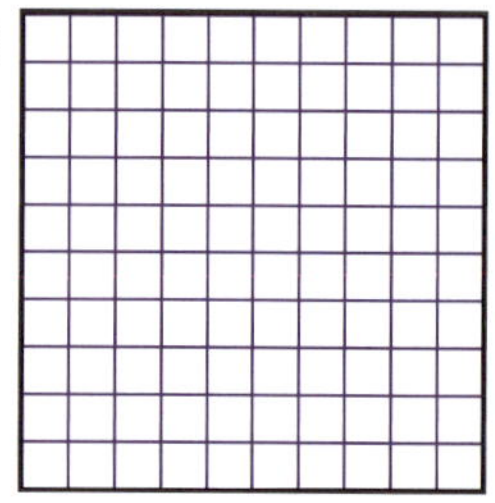 75%

e 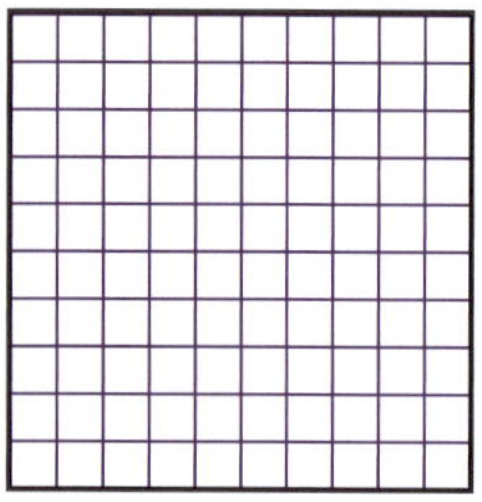 50%

f 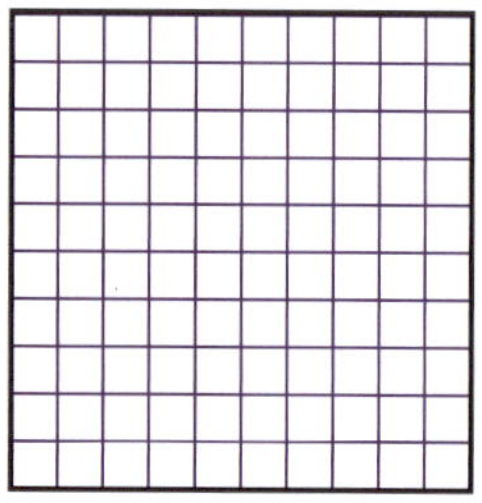 15%

g 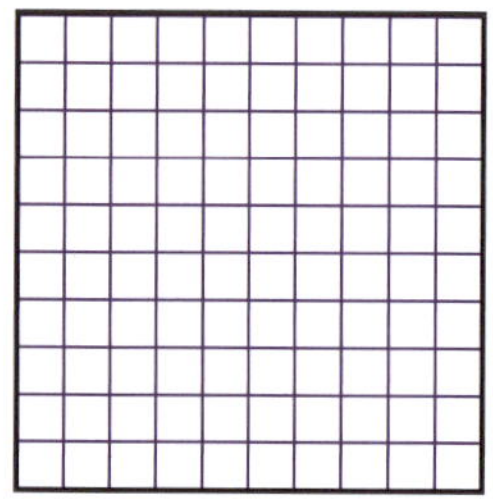 60%

h 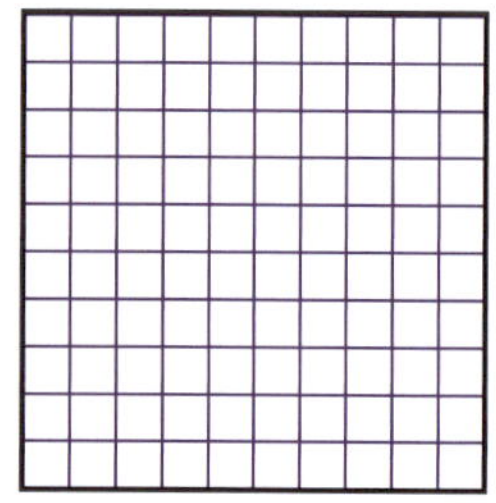 90%

2 What percentage of each square in Question 1 should be not coloured?

a ☐ b ☐ c ☐ d ☐ e ☐ f ☐ g ☐ h ☐

3 Complete the following.

	Fraction	Decimal	Percentage
a	$\frac{15}{100}$	0·	%
b	$\frac{75}{100}$	0·	%
c	$\frac{25}{100}$	0·	%
d	$\frac{45}{100}$	0·	%
e	$\frac{95}{100}$	0·	%
f	$\frac{85}{100}$	0·	%
g	$\frac{35}{100}$	0·	%
h	$\frac{10}{100}$	0·	%
i	$\frac{30}{100}$	0·	%
j	$\frac{20}{100}$	0·	%
k	$\frac{60}{100}$	0·	%
l	$\frac{90}{100}$	0·	%

Converting Fractions to Decimals

ICT

4 Use a calculator to divide the denominator into the numerator.

e.g. $\frac{35}{100}$ ← numerator, ← denominator is [3] [5] [÷] [1] [0] [0] [=] 0·35

Now calculate:

a	$\frac{65}{100}$ ☐	b	$\frac{15}{100}$ ☐	c	$\frac{95}{100}$ ☐	d	$\frac{45}{100}$ ☐	e	$\frac{75}{100}$ ☐
f	$\frac{25}{100}$ ☐	g	$\frac{5}{100}$ ☐	h	$\frac{60}{100}$ ☐	i	$\frac{80}{100}$ ☐	j	$\frac{40}{100}$ ☐
k	$\frac{10}{100}$ ☐	l	$\frac{37}{100}$ ☐	m	$\frac{91}{100}$ ☐	n	$\frac{20}{100}$ ☐	o	$\frac{100}{100}$ ☐

 ISBN 978 1 4886 2190 1

2:01 Number Facts, ×2, ×3, ×4, ×5, ×10

×	2	3	4	5
1	2	3	4	5
2	4	6	8	10
3	6	9	12	15
4	8	12	16	20
5	10	15	20	25
6	12	18	24	30
7	14	21	28	35
8	16	24	32	40
9	18	27	36	45
10	20	30	40	50

CONCEPT

The circle shows 6 × 4.

The product of 6 and 4 is 24.

$6 \times 4 = 4 \times 6$

1 Try to do these without using the table.

a $9 \times 3 =$ ___ b $8 \times 5 =$ ___ c $8 \times 4 =$ ___

d $5 \times 4 =$ ___ e $7 \times 5 =$ ___ f $8 \times 3 =$ ___

g $7 \times 2 =$ ___ h $7 \times 4 =$ ___ i $3 \times 5 =$ ___

j $6 \times 5 =$ ___ k $6 \times 3 =$ ___ l $6 \times 4 =$ ___

2 Find the product of:

a 5 and 5 ___ b 9 and 4 ___

c 3 and 3 ___ d 1 and 1 ___

e 2 and 5 ___ f 9 and 5 ___

g 2, 5 and 4 ___ h 3, 2 and 4 ___

The answer to a multiplication question is called the **product**.

3 a

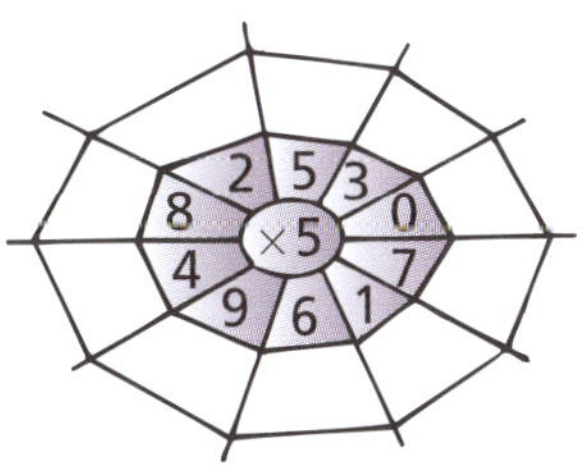

b

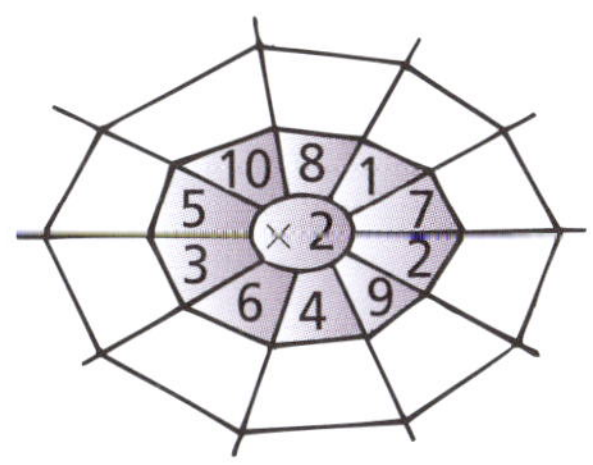

c

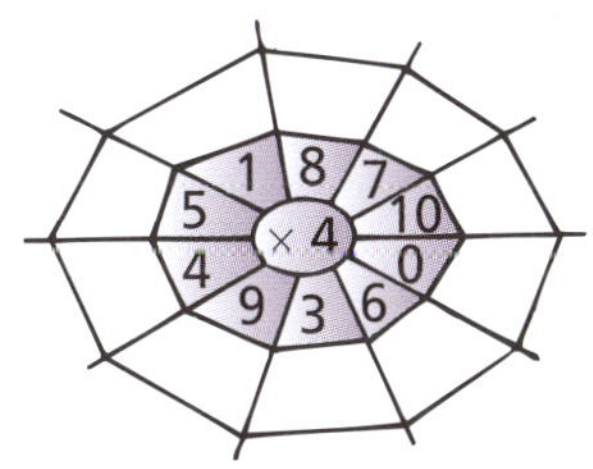

d

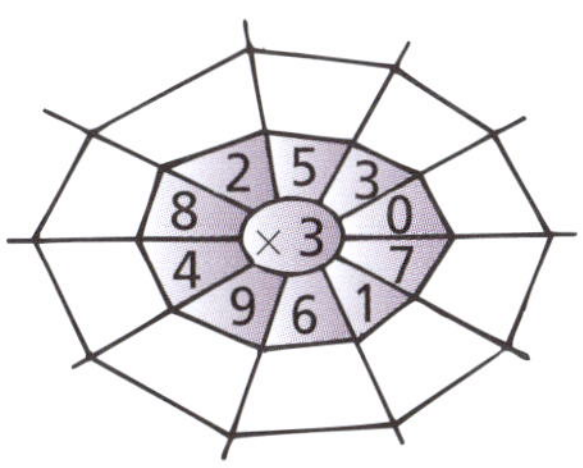

e

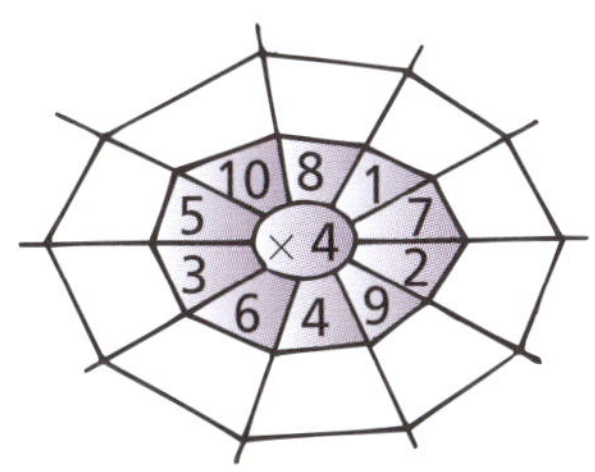

f

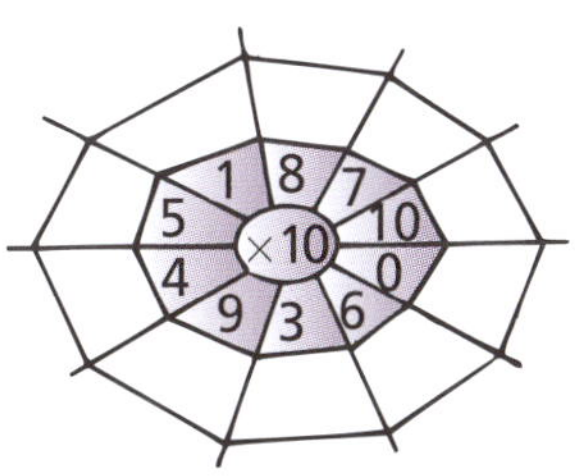

4 Write the first ten multiples of:

a 3	3									
b 4	4									
c 5	5									

 • *AUSTRALIAN SIGNPOST MATHS 5* • ISBN 978 1 4886 2190 1

2:02 Number Facts, ×6, ×7, ×8, ×9

×	6	7	8	9
1	6	7	8	9
2	12	14	16	18
3	18	21	24	27
4	24	28	32	36
5	30	35	40	45
6	36	42	48	54
7	42	49	56	63
8	48	56	64	72
9	54	63	72	81
10	60	70	80	90

CONCEPT

The circle shows 8 × 9.

The product of 8 and 9 is 72.

9×8 = 8×9

The answer to a multiplication question is called the **product**.

1 Try to do these without using the table.

a 5 × 6 = ____ **b** 3 × 8 = ____ **c** 3 × 7 = ____

d 4 × 9 = ____ **e** 5 × 8 = ____ **f** 7 × 9 = ____

g 6 × 6 = ____ **h** 7 × 7 = ____ **i** 9 × 6 = ____

j 8 × 9 = ____ **k** 6 × 8 = ____ **l** 8 × 6 = ____

m 7 × 6 = ____ **n** 8 × 7 = ____

o 9 × 8 = ____ **p** 9 × 9 = ____

q 4 × 6 = ____ **r** 8 × 8 = ____

s The product of 2, 3 and 8 = ____.

2 **a**

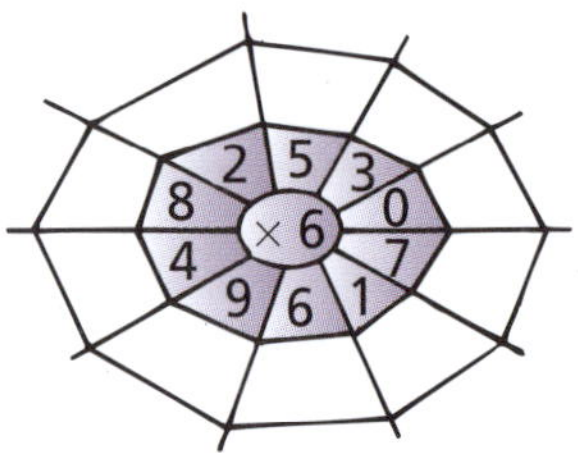

b

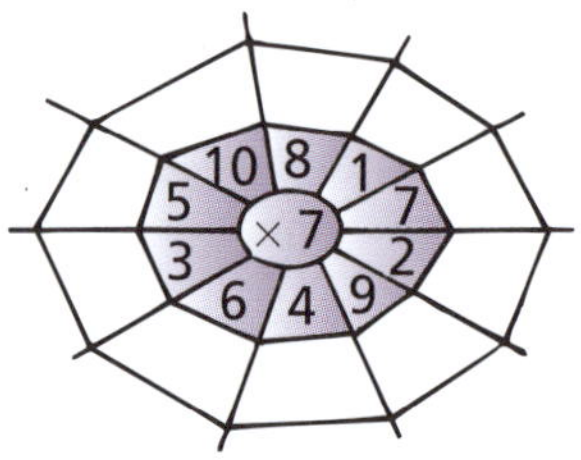

c

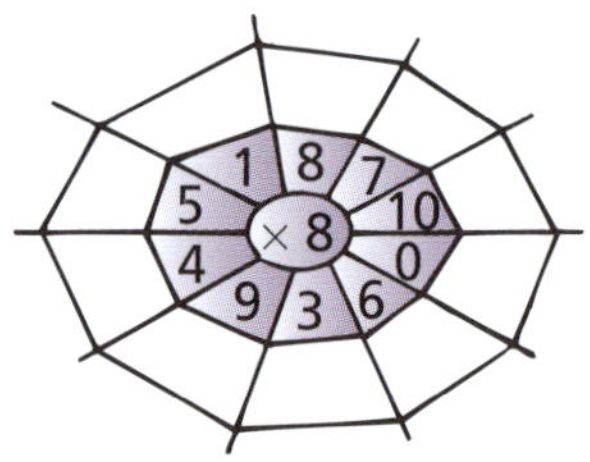

d

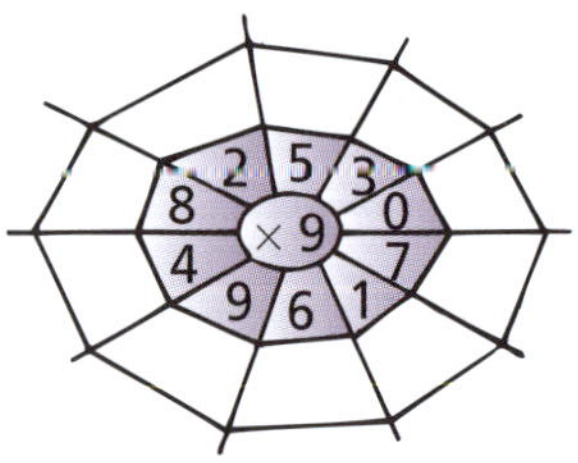

e

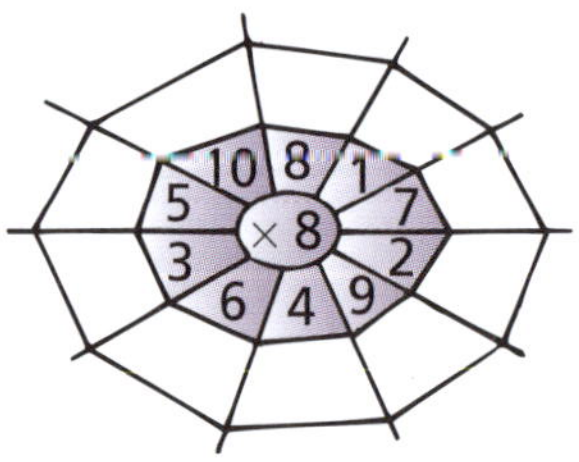

f

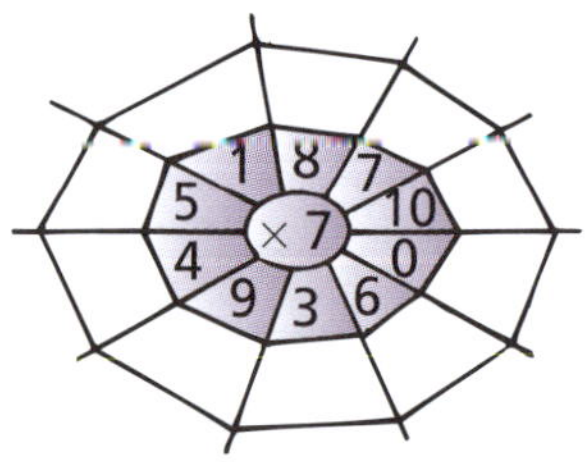

3 Write the first ten multiples of:

a 6 | 6 | | | | | | | | | |

b 7 | 7 | | | | | | | | | |

c 9 | 9 | | | | | | | | | |

 ISBN 978 1 4886 2190 1

2:03 Learning your Multiplication Tables

- Have someone test you.
- For each table you don't know, make a card with the question on one side and the answer on the other.
- Carry these cards with you, testing yourself until you know them.

CONCEPT

1 Try to do these without using the table below.

a 3 × 3 = ☐ **b** 6 × 2 = ☐ **c** 4 × 3 = ☐ **d** 9 × 2 = ☐

e 8 × 3 = ☐ **f** 5 × 5 = ☐ **g** 4 × 6 = ☐ **h** 6 × 6 = ☐

i 4 × 4 = ☐ **j** 7 × 2 = ☐ **k** 9 × 3 = ☐ **l** 8 × 2 = ☐

m 5 × 3 = ☐ **n** 5 × 6 = ☐ **o** 8 × 8 = ☐ **p** 6 × 3 = ☐

q 9 × 6 = ☐ **r** 7 × 3 = ☐ **s** 10 × 6 = ☐ **t** 8 × 6 = ☐

u 7 × 7 = ☐ **v** 10 × 8 = ☐ **w** 7 × 6 = ☐ **x** 9 × 9 = ☐

2 **a**

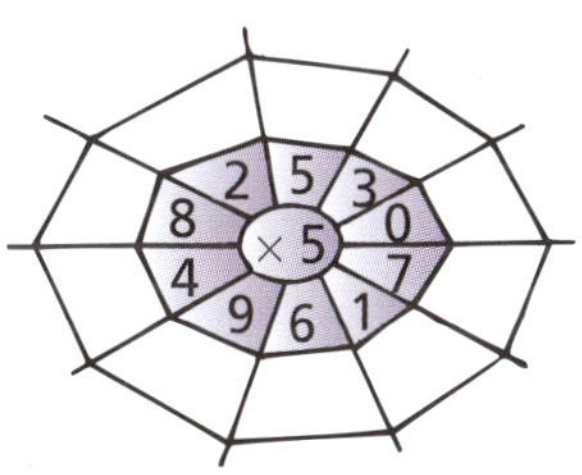

b

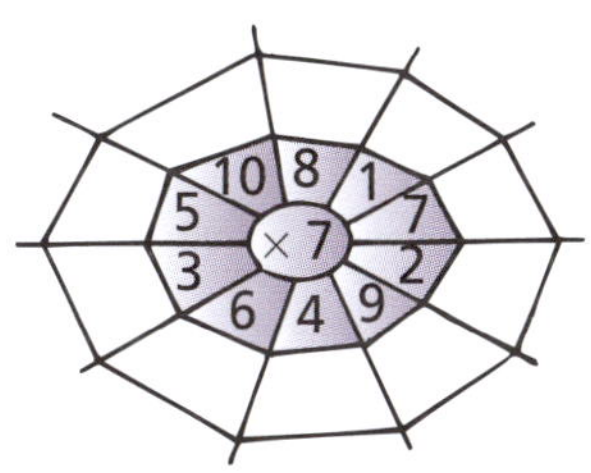

c

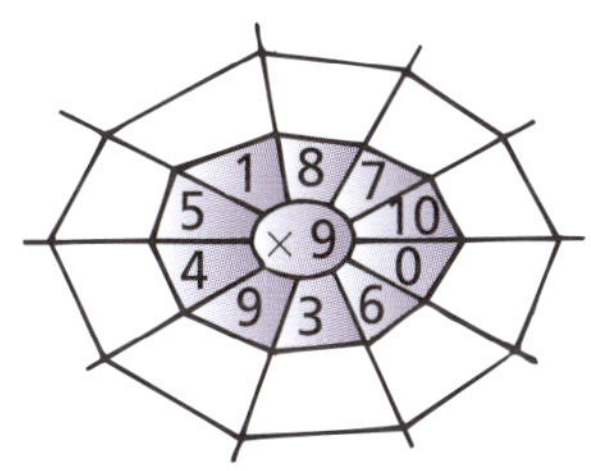

3 Write the first ten multiples of:

a 4 [4] ☐ ☐ ☐ ☐ ☐ ☐ ☐ ☐ ☐

b 8 [8] ☐ ☐ ☐ ☐ ☐ ☐ ☐ ☐ ☐

Can you see a connection between the multiples in parts **a** and **b**?

The multiples of 8 are ☐ the size of the multiples of 4.

0 × 1 = 0	0 × 2 = 0	0 × 3 = 0	0 × 4 = 0	0 × 5 = 0	0 × 6 = 0	0 × 7 = 0	0 × 8 = 0	0 × 9 = 0	0 × 10 = 0
1 × 1 = 1	1 × 2 = 2	1 × 3 = 3	1 × 4 = 4	1 × 5 = 5	1 × 6 = 6	1 × 7 = 7	1 × 8 = 8	1 × 9 = 9	1 × 10 = 10
2 × 1 = 2	2 × 2 = 4	2 × 3 = 6	2 × 4 = 8	2 × 5 = 10	2 × 6 = 12	2 × 7 = 14	2 × 8 = 16	2 × 9 = 18	2 × 10 = 20
3 × 1 = 3	3 × 2 = 6	3 × 3 = 9	3 × 4 = 12	3 × 5 = 15	3 × 6 = 18	3 × 7 = 21	3 × 8 = 24	3 × 9 = 27	3 × 10 = 30
4 × 1 = 4	4 × 2 = 8	4 × 3 = 12	4 × 4 = 16	4 × 5 = 20	4 × 6 = 24	4 × 7 = 28	4 × 8 = 32	4 × 9 = 36	4 × 10 = 40
5 × 1 = 5	5 × 2 = 10	5 × 3 = 15	5 × 4 = 20	5 × 5 = 25	5 × 6 = 30	5 × 7 = 35	5 × 8 = 40	5 × 9 = 45	5 × 10 = 50
6 × 1 = 6	6 × 2 = 12	6 × 3 = 18	6 × 4 = 24	6 × 5 = 30	6 × 6 = 36	6 × 7 = 42	6 × 8 = 48	6 × 9 = 54	6 × 10 = 60
7 × 1 = 7	7 × 2 = 14	7 × 3 = 21	7 × 4 = 28	7 × 5 = 35	7 × 6 = 42	7 × 7 = 49	7 × 8 = 56	7 × 9 = 63	7 × 10 = 70
8 × 1 = 8	8 × 2 = 16	8 × 3 = 24	8 × 4 = 32	8 × 5 = 40	8 × 6 = 48	8 × 7 = 56	8 × 8 = 64	8 × 9 = 72	8 × 10 = 80
9 × 1 = 9	9 × 2 = 18	9 × 3 = 27	9 × 4 = 36	9 × 5 = 45	9 × 6 = 54	9 × 7 = 63	9 × 8 = 72	9 × 9 = 81	9 × 10 = 90
10 × 1 = 10	10 × 2 = 20	10 × 3 = 30	10 × 4 = 40	10 × 5 = 50	10 × 6 = 60	10 × 7 = 70	10 × 8 = 80	10 × 9 = 90	10 × 10 = 100

2:04 Division Facts

30 ÷ 5
Ask ☐ × 5 = 30?

6 × 5 = 30

$6\overline{)30}$ = 5 $5\overline{)30}$ = 6

CONCEPT

1 Use the multiplication table to answer the division questions.

a 4 × 10 = 40, so, $4\overline{)40}$ ☐ $10\overline{)40}$ ☐
b 5 × 8 = 40, so, $8\overline{)40}$ ☐ $5\overline{)40}$ ☐
c 7 × 3 = 21, so, $7\overline{)21}$ ☐ $3\overline{)21}$ ☐
d 9 × 8 = 72, so, $8\overline{)72}$ ☐ $9\overline{)72}$ ☐
e 8 × 9 = 72, so, $9\overline{)72}$ ☐ $8\overline{)72}$ ☐
f 4 × 9 = 36, so, $9\overline{)36}$ ☐ $4\overline{)36}$ ☐

2
a ☐ × 2 = 22 **b** 4 × ☐ = 40 **c** ☐ × 8 = 56 **d** 5 × ☐ = 10
e ☐ × 3 = 18 **f** 9 × ☐ = 81 **g** ☐ × 9 = 9 **h** 7 × ☐ = 0
i ☐ × 7 = 49 **j** 4 × ☐ = 32 **k** ☐ × 5 = 35 **l** 3 × ☐ = 27

3
a 18 ÷ 3 = ☐
Ask 3 × ☐ = 18
b 28 ÷ 4 = ☐
Ask 4 × ☐ = 28
c 40 ÷ 5 = ☐
Ask 5 × ☐ = 40
d 30 ÷ 3 = ☐
Ask 3 × ☐ = 30
e 42 ÷ 6 = ☐
Ask 6 × ☐ = 42
f 24 ÷ 4 = ☐
Ask 4 × ☐ = 24

4 Use the method in Question 3 to answer these.

a 18 ÷ 2 = ☐ **b** 25 ÷ 5 = ☐ **c** 18 ÷ 6 = ☐ **d** 24 ÷ 6 = ☐
e 30 ÷ 3 = ☐ **f** 4 ÷ 4 = ☐ **g** 35 ÷ 7 = ☐ **h** 27 ÷ 9 = ☐
i 0 ÷ 8 = ☐ **j** 48 ÷ 6 = ☐ **k** 16 ÷ 8 = ☐ **l** 36 ÷ 6 = ☐
m 27 ÷ 3 = ☐ **n** 25 ÷ 5 = ☐ **o** 63 ÷ 9 = ☐ **p** 45 ÷ 5 = ☐

5
a 15 ÷ ☐ = 3
Ask 3 × ☐ = 15
b 54 ÷ ☐ = 6
Ask 6 × ☐ = 54
c 49 ÷ ☐ = 7
Ask 7 × ☐ = 49
d 90 ÷ ☐ = 10
Ask 10 × ☐ = 90
e 55 ÷ ☐ = 11
Ask 11 × ☐ = 55
f 12 ÷ ☐ = 3
Ask 3 × ☐ = 12
g 81 ÷ ☐ = 9 **h** 42 ÷ ☐ = 7 **i** 36 ÷ ☐ = 4

2:05 Rounding

That's 3900 to the nearest 100, or that's 3860 to the nearest 10.

Examples:
96 → 100
527 → 500
4510 → 5000

It says 3864 attended.

Rounding

- If the digit is 4 or less ... round down.
- If the digit is 5 or more ... round up, e.g. 5 → 10, 96 → 100.

CONCEPT

1 Round these numbers to the nearest ten.

a 37	**b** 82	**c** 111	**d** 126
e 305	**f** 927	**g** 792	**h** 329
i 1894	**j** 1895	**k** 1896	**l** 2067

2 Round these numbers to the nearest hundred.

a 742	**b** 381	**c** 151	**d** 853
e 937	**f** 679	**g** 405	**h** 249
i 3481	**j** 9048	**k** 4093	**l** 2145

3 Round these numbers to the nearest thousand.

a 2790	**b** 4281	**c** 8021	**d** 3724
e 7814	**f** 6930	**g** 1398	**h** 5316

4 a Circle the numbers that round to 3000.

3672	3196	2878
3945	2764	3285
3812	2246	3349

b Circle the numbers that round to 6000.

6249	5862	6591
5396	6478	6487
5723	6823	5734

5 a Circle the numbers that round to 3200.

3275	3227	3270
3249	3185	3175
3159	3129	3207

b Circle the numbers that round to 75000.

75247	75672	74503
74820	75981	75199
74498	74981	74500

6 Answer **true** or **false** for each statement.

a 7251 rounds to 7300.

b 4500 rounds to 5000.

c 6172 rounds to 6100.

d 3409 rounds to 4000.

 • *AUSTRALIAN SIGNPOST MATHS 5* • ISBN 978 1 4886 2190 1

2:06 Addition to 999

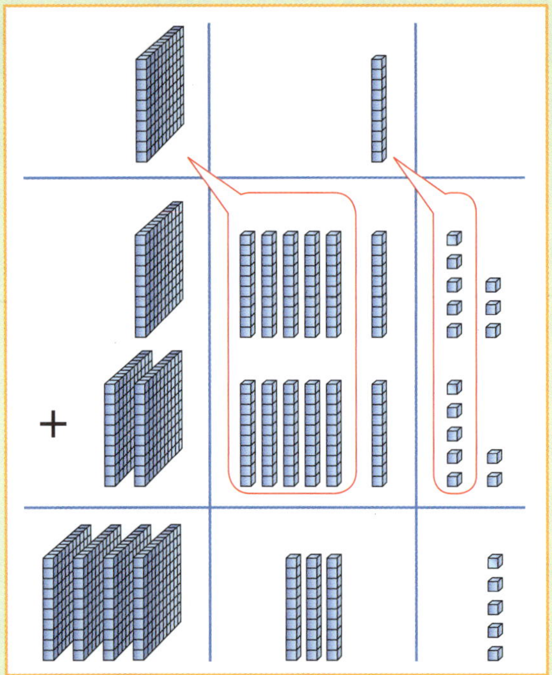

CONCEPT

168 sheep and 267 cattle are for sale.

How many animals are there altogether?

	Hund	Tens	Ones
	1	1	
	1	6	8
+	2	6	7
	4	3	5

There are 435 animals altogether.

We can check the answer by rounding to the nearest 100 (or 10).

200 + 300 = 500

435 is reasonably close to 500.

1 a

	Hund	Tens	Ones
		6	1
+		9	3

b

	Hund	Tens	Ones
		7	2
+		4	5

c

	Hund	Tens	Ones
		3	0
+		7	8

d

	Hund	Tens	Ones
	2	5	8
+		2	5

e

	Hund	Tens	Ones
		7	6
+	7	1	4

f

	Hund	Tens	Ones
	3	6	9
+		1	9

g

	Hund	Tens	Ones
	4	3	6
+		8	0

h

	Hund	Tens	Ones
	4	2	5
+		8	5

i

	Hund	Tens	Ones
		8	7
+	3	4	1

j

	Hund	Tens	Ones
	1	8	8
+	1	4	3

k

	Hund	Tens	Ones
	3	6	5
+	1	4	9

l

	Hund	Tens	Ones
	4	5	9
+	2	7	4

2 Check your answers by rounding each number to the nearest 100 (or 10), then adding.

 • *AUSTRALIAN SIGNPOST MATHS 5* • ISBN 978 1 4886 2190 1

2:07 Addition to 999

1

a

	H	T	U
	2	8	3
+		4	5

b

	H	T	U
	5	3	1
+		7	0

c

	H	T	U
		8	2
+	7	3	3

d

	H	T	U
	6	5	4
+		5	9

e

	H	T	U
		8	7
+	4	7	4

f

	H	T	U
	3	6	4
+		9	2

g

	H	T	U
	1	3	8
+	2	0	4

h

	H	T	U
	3	1	6
+	1	4	7

i

	H	T	U
	3	4	6
+	2	2	4

j

	H	T	U
	1	8	4
+	1	0	9

2

a

	H	T	U
	2	4	8
+	1	1	7

b

	H	T	U
		7	4
+	5	3	1

c

	H	T	U
	4	1	9
+	2	0	9

d

	H	T	U
	1	6	8
+	6	1	1

e

	H	T	U
	2	4	8
+	1	6	6

f

	H	T	U
	2	2	7
+	5	9	2

Rounding Rules

CONCEPT

We round down if the digit is 4 or less. We round up if the digit is 5 or more.

When both numbers are rounded down the answer is an underestimate.	When both numbers are rounded up the answer is an overestimate.
319 + 443 = 862	482 + 257 = 739
300 + 400 = 700	**500 + 300 = 800**
The estimate is less than the actual answer.	The estimate is greater than the actual answer.

 • *AUSTRALIAN SIGNPOST MATHS 5* • ISBN 978 1 4886 2190 1

2:08 Subtraction Without Trading to 999

438 students.
216 were in uniform.
How many were out of uniform?

I show

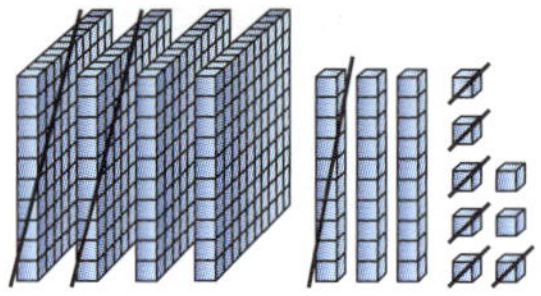

Take away 6 ones.
Take away 1 ten.
Take away 2 hundreds.

I write

	Hund	Tens	Ones
	4	3	8
−	2	1	6
	2	2	2

222 students out of uniform

CONCEPT

1 **a**

	4 hundreds	7 tens	6 ones
−	2 hundreds	3 tens	1 one

b

	8 hundreds	9 tens	7 ones
−	2 hundreds	5 tens	2 ones

c

	Hund	Tens	Ones
	8	5	2
−	7	2	1

d

	Hund	Tens	Ones
	9	6	7
−	2	3	4

e

	Hund	Tens	Ones
	9	9	8
−	4	6	5

f

	Hund	Tens	Ones
	3	4	2
−		1	1

g

	Hund	Tens	Ones
	6	9	8
−		2	4

h

	Hund	Tens	Ones
	9	4	5
−		1	1

i

	Hund	Tens	Ones
	6	8	6
−			4

j

	Hund	Tens	Ones
	7	4	7
−			4

k

	Hund	Tens	Ones
	3	6	9
−			7

l

	Hund	Tens	Ones
	5	5	9
−	4	3	0

m

	Hund	Tens	Ones
	1	7	3
−		5	0

n

	Hund	Tens	Ones
	6	4	6
−	2	0	3

o

	Hund	Tens	Ones
	7	7	8
−	6	0	0

p

	Hund	Tens	Ones
	5	2	9
−	1	0	0

q

	Hund	Tens	Ones
	1	8	3
−	1	1	3

r

	Hund	Tens	Ones
	5	8	7
−	3	8	2

2 Estimate to check each answer.

Make up a problem to match each of the sums in parts **c**, **d**, **e** and **f**.

I had 275 teeth but 122 fell out. How many are left?

FUN SPOT

2:09 Writing the Addition Algorithm

CONCEPT

	Hund	Tens	Ones
	1	1	
	5	8	3
+	1	7	9
	7	6	2

	Hund	Tens	Ones
	1	1	
	5	8	3
+	1	7	9
	7	6	2

	H	T	U
	1	1	
	5	8	3
+	1	7	9
	7	6	2

	1	1	
	5	8	3
+	1	7	9
	7	6	2

762 pieces of fruit were sold altogether.

1

a $250 + 84$

b $534 + 128$

c $154 + 207$

d $92 + 340$

e $245 + 370$

f $365 + 166$

g $209 + 583$

h $345 + 312$

i $507 + 220$

j $365 + 397$

k $188 + 133$

l $365 + 366$

m $509 + 296$

n $381 + 259$

o $777 + 185$

2 Estimate each answer and use the estimate to check your exact answer.

a Luke watched 287 minutes of television on Saturday and 145 minutes on Sunday. How much did he watch altogether? Give your answer in minutes. ☐ minutes

b Church attendances in the two services on Sunday were 156 and 218. How many people attended altogether? ☐ people

c This year farmer McDonald sold 180 bags of corn. Last year he sold 17 bags more than this. How many bags were sold in the two years? ☐ bags

d Rex and Lyn bought the house next door, adding 346 square metres to their original block of 575 square metres. How large is the block now? ☐ m^2

2:10 Subtraction with Trading to 999

CONCEPT

28 of my 384 stamps are Chinese. How many are not Chinese?

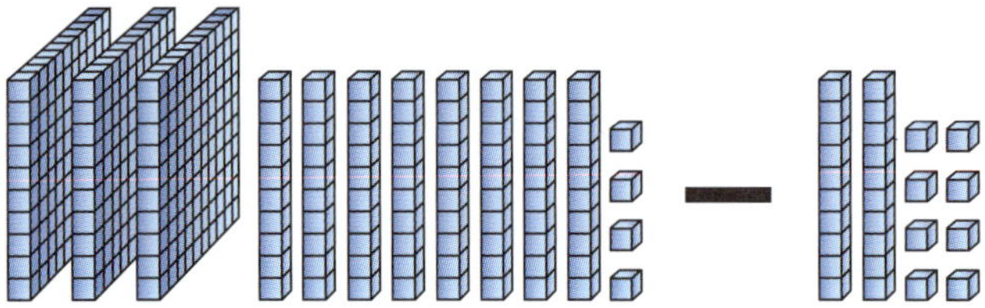

We can't take 8 ones from 4 ones, so trade 1 ten for 10 ones.

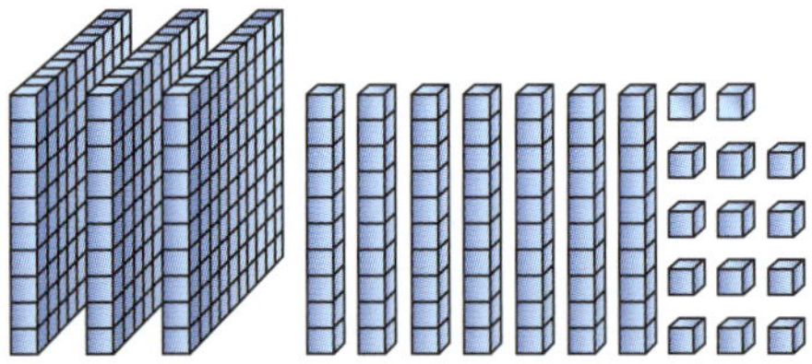

From this we take 2 tens and 8 ones.

	Hund	Tens	Ones
		7	14
	3	~~8~~	~~4~~
−		2	8
	3	5	6

1

a $581 - 53$

b $343 - 28$

c $296 - 67$

d $460 - 39$

e $273 - 9$

f $166 - 58$

g $844 - 8$

h $973 - 65$

i $750 - 39$

j $341 - 26$

k $280 - 32$

l $647 - 19$

m $835 - 108$

n $891 - 836$

o $520 - 415$

p $684 - 76$

q $214 - 208$

r $784 - 50$

s $391 - 375$

Use estimation to check your answers.

 • *AUSTRALIAN SIGNPOST MATHS 5* • ISBN 978 1 4886 2190 1

2:11 Subtraction with Trading to 999

CONCEPT

354 of the 718 birds in the zoo are parrots.
How many are not parrots?

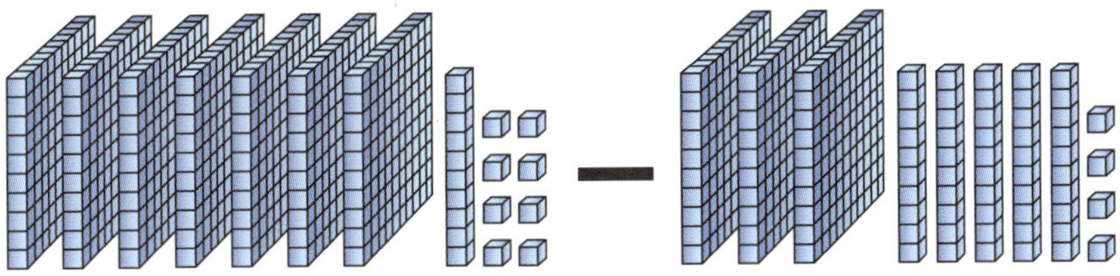

We can't take 5 tens from 1 ten, so trade 1 hundred for 10 tens.

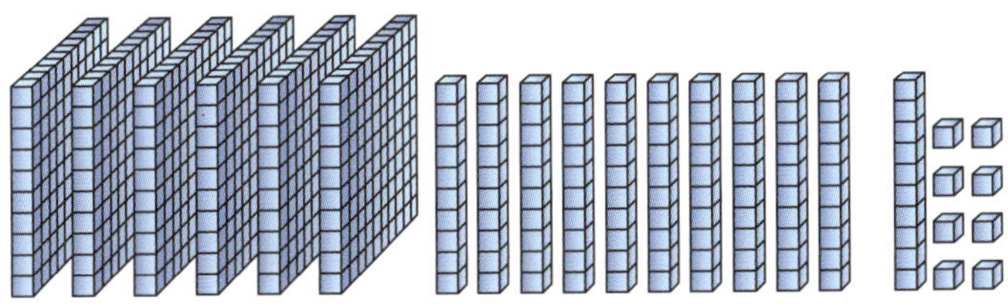

From this we take 3 hundreds, 5 tens and 4 ones.

	Hund	Tens	Ones
	6	11	
	~~7~~	~~1~~	8
−	3	5	4
	3	6	4

364 of the birds are not parrots.

1

a $617 - 153$ b $826 - 571$ c $434 - 240$ d $729 - 38$

e $528 - 152$ f $607 - 352$ g $685 - 91$ h $215 - 175$

i $450 - 139$ j $617 - 326$ k $504 - 192$ l $947 - 309$

m $733 - 108$ n $916 - 836$ o $608 - 415$ p $729 - 276$

q $414 - 208$ r $984 - 550$ s $866 - 375$

Use estimation to check your answers.

2:12 Multiples

CONCEPT

A multiple of 5 is found when you multiply it by a counting number.

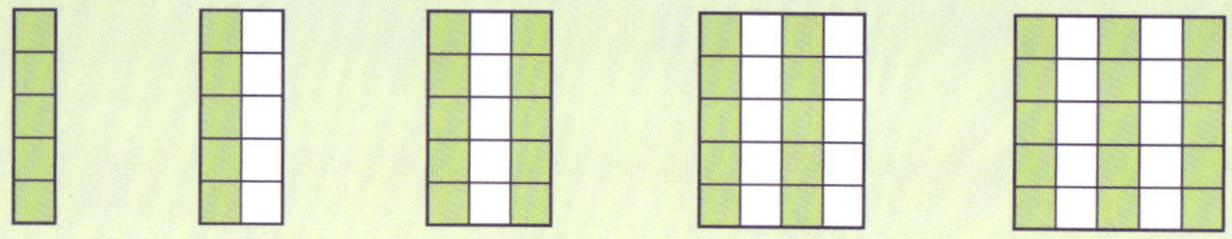

1 × 5 = 5   2 × 5 = 10   3 × 5 = 15   4 × 5 = 20   5 × 5 = 25

1 Complete these number sentences to find the first nine multiples of 5.

1 × 5 = ____   2 × 5 = ____   3 × 5 = ____
4 × 5 = ____   5 × 5 = ____   6 × 5 = ____
7 × 5 = ____   8 × 5 = ____   9 × 5 = ____

The first ten multiples of 5 are 5, 10, ____, ____, ____, ____, ____, ____, ____ and ____.

2 Use skip counting to write down the first ten multiples of:

a 2 2, ____, ____, ____, ____, ____, ____, ____, ____ and ____
b 4 4, ____, ____, ____, ____, ____, ____, ____, ____ and ____
c 5 ____, ____, ____, ____, ____, ____, ____, ____, ____ and ____
d 6 ____, ____, ____, ____, ____, ____, ____, ____, ____ and ____
e 8 ____, ____, ____, ____, ____, ____, ____, ____, ____ and ____
f 10 ____, ____, ____, ____, ____, ____, ____, ____, ____ and ____

You could use a calculator to do this.

3 By looking at the answers in Question 2, find the smallest number that is a multiple of both:

a 2 and 5 ____   **b** 5 and 6 ____   **c** 6 and 10 ____
d 5 and 8 ____   **e** 2 and 10 ____   **f** 4 and 10 ____
g 6 and 8 ____   **h** 4 and 5 ____   **i** 8 and 10 ____
j 2 and 8 ____   **k** 4 and 8 ____   **l** 5 and 10 ____

ACTIVITY

4 Use different coloured connecting cubes to find the common multiples of 2 and 5.

The first 10 common multiples are:

10, 20, ____, ____, ____, ____, ____, ____, ____, ____.

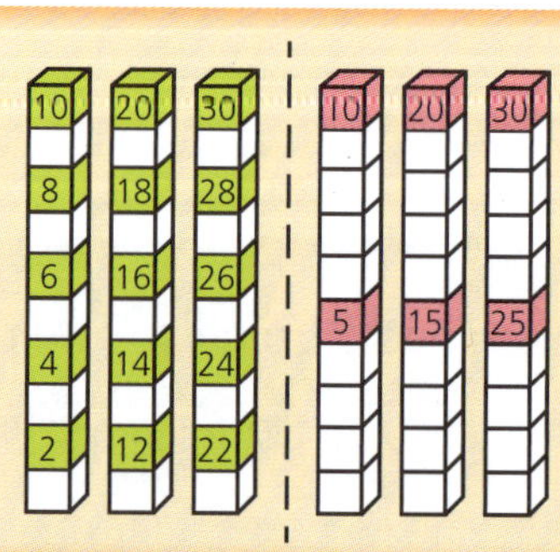

2:13 Factors

CONCEPT

The factors of 18:

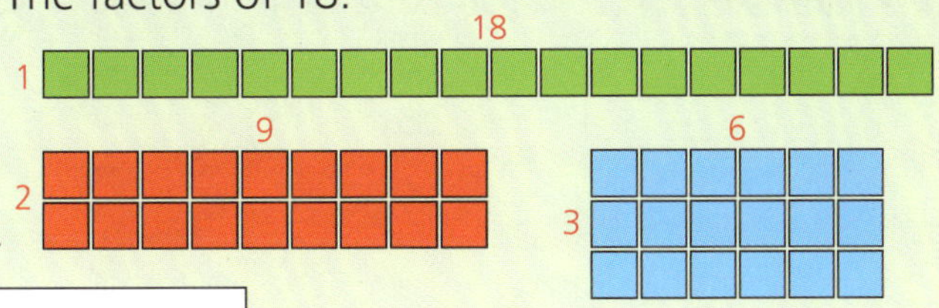

$18 = 1 \times 18$
$18 = 2 \times 9$
$18 = 3 \times 6$

The red numbers are factors of 18.
18 is a multiple of each factor.

1 Use these rectangles to complete different number sentences below.

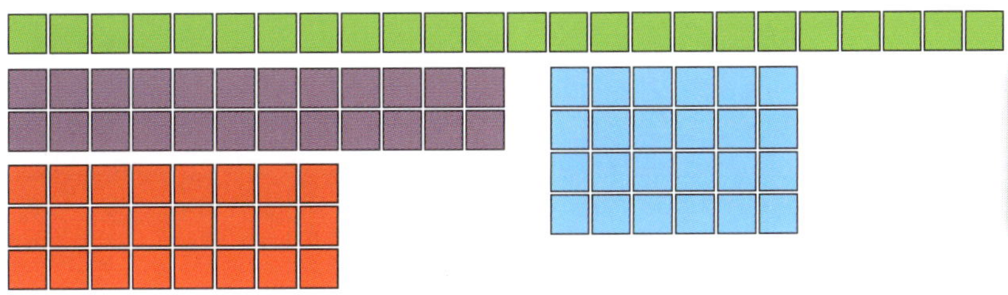

a 24 = ___ × ___ **b** 24 = ___ × ___ **c** 24 = ___ × ___

d 24 = ___ × ___ **e** The factors of 24 are ___, ___, ___, ___, ___, ___, ___ and ___.

2 Make rectangles using place-value blocks or square counters to find all factors of these numbers.

a 6: ___, ___, ___, ___

b 12: ___, ___, ___, ___, ___, ___

c 5: ___, ___

d 20: ___, ___, ___, ___, ___, ___

e 15: ___, ___, ___, ___

f 28: ___, ___, ___, ___, ___, ___

g 30: ___, ___, ___, ___, ___, ___, ___, ___

Counting numbers that multiply to give a whole number are its factors.

3 If $80 = 1 \times 80$, $80 = 2 \times 40$, $80 = 4 \times 20$, $80 = 5 \times 16$ and $80 = 8 \times 10$, then the factors of 80 are: ___, ___, ___, ___, ___, ___, ___, ___, ___ and ___.

4 **a** 16 soldiers are to march in equal rows. How wide could the rows be? ___

b Sam had 21 plants. If he planted them in rows, how many could be in each row? ___

c Debbie had 12 lollies. She wanted to give the same number to each child until she had none left. How many children might be given lollies? ___

2:14 Addition of Money

100 cents
10 cents
cents

```
   1  1
 $ 3. 4 6
+$ 4. 8 5
 $ 8. 3 1
```

CONCEPT

1

a $0.41 + $0.37
b $0.64 + $0.88
c $1.26 + $3.47
d $2.24 + $3.07
e $3.75 + $2.94

f $4.69 + $3.56
g $2.63 + $2.07
h $5.83 + $2.68
i $4.96 + $1.47
j $2.79 + $4.32

k $2.41 + $3.67
l $4.06 + $2.91
m $3.77 + $1.63
n $1.92 + $4.37
o $5.16 + $1.98

2 Write each as an algorithm or use a mental strategy to find your answer.

a $1.56 + $2.23 ______
b $4.01 + $1.76 ______
c $3.51 + $2.17 ______
d $5.03 + $3.65 ______
e $2.37 + $4.12 ______
f $6.45 + $2.03 ______

3 Find the total if I bought:

a a cake for $5.50 and a drink for $4.05. ______
b a dessert for $4.95 and a drink for $3.25. ______
c a sandwich for $7.30 and a drink for $1.55. ______
d a biscuit for $3.75 and a drink for $4.95. ______

Make $10

FUN SPOT

- Each player begins the game with $5 and takes turns to roll a dice.
- The number shown on each dice is multiplied by 10 and that number of cents is added to the player's total.
- The game continues until one player reaches $10.

 • *AUSTRALIAN SIGNPOST MATHS 5* • ISBN 978 1 4886 2190 1

2:15 Subtraction of Money

100 cents
10 cents
cents

	$7.39 (7 crossed to 6, 3 crossed to 13)
−	$5.63
	$1.76

CONCEPT

1

a $6.52 − $1.30	**b** $8.79 − $3.62	**c** $9.85 − $3.51	**d** $7.82 − $5.43	**e** $6.24 − $2.07
f $4.47 − $2.63	**g** $5.38 − $2.61	**h** $7.50 − $1.42	**i** $3.54 − $0.20	**j** $7.55 − $3.47
k $8.57 − $6.87	**l** $9.57 − $6.99	**m** $6.50 − $2.61	**n** $7.90 − $2.45	**o** $3.52 − $1.48

2 Write each as an algorithm or use a mental strategy to find your answer.

a $3.35 – $1.24 ______ **b** $8.98 – $4.00 ______

c $8.56 – $3.21 ______ **d** $4.54 – $4.49 ______

e $7.00 – $1.50 ______ **f** $3.50 – $1.99 ______

3 I started with $8.50. What is my change if I buy:

a a book for $7.45? ______ **b** an icy pole for $2.30? ______

c a drink for $3.05? ______ **d** a keyring for $7.30? ______

First to $5

FUN SPOT

- Each player begins the game with $10 and takes turns to roll a dice.
- The number shown on each dice is multiplied by 10 and that number of cents is subtracted from the player's total.
- The game continues until one player reaches $5.

2:16 Shopping

1 If you had $5, would you be able to buy:

a the hat? ☐ **b** the book? ☐ **c** the trumpet? ☐ **d** the bear? ☐

e the book and the bear? ☐ **f** the hat and the bear? ☐

g the hat and the trumpet? ☐ **h** the trumpet and the bear? ☐

2 Find the total of each bill.

a

hat	$2.90
bear	$1.70
total	

b

hat	$2.90
book	$4.50
total	

c

book	$4.50
trumpet	$3.20
total	

d

book	$4.50
hat	$2.90
bear	$1.70
total	

e

hat	$2.90
bear	$1.70
trumpet	$3.20
total	

f

book	$4.50
bear	$1.70
trumpet	$3.20
total	

3 Find the amount of money left.

a $9.00 − $4.60

b $9.00 − $6.20

c $9.00 − $7.80

d $9.00 − $4.90

e $9.00 − $7.40

4

a $3.78 + $5.93

b $2.75 + $6.27

c $1.82 + $7.36

d $4.24 + $3.57

e $5.51 + $1.98

f $7.63 − $4.36

g $9.74 − $3.29

h $8.11 − $4.72

i $9.80 − $3.47

j $8.25 − $4.99

5 Round the answers from Question 4 to the nearest 5 cents.

FUN SPOT

a ☐ **b** ☐ **c** ☐ **d** ☐ **e** ☐

f ☐ **g** ☐ **h** ☐ **i** ☐ **j** ☐

2:17 Using Strategies to Solve Problems

INVESTIGATION

Use this diagram of 5 dots to find how many intervals can be drawn that join pairs of dots.

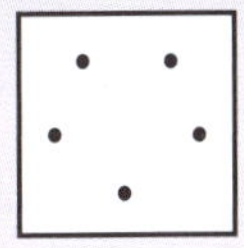

Solution and Reasoning

- Start at one point. Draw intervals to each of the other points.
- From the next point, draw any additional intervals possible.
- Do this at all other points, counting the intervals as you draw them.

$4 + 3 + 2 + 1 = 10$ Ten intervals can be drawn.

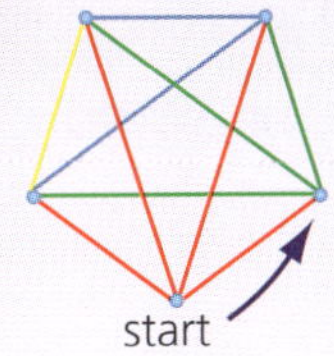

1 Use this diagram of 6 dots to find how many intervals can be drawn that join pairs of dots. ______

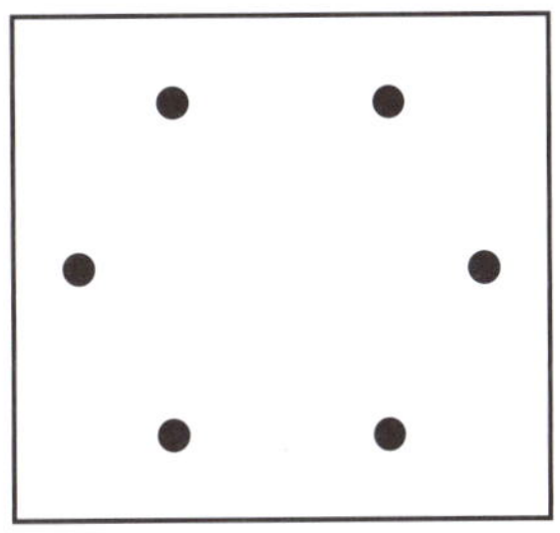

2 Use this diagram of 8 dots to find how many intervals can be drawn that join pairs of dots. ______

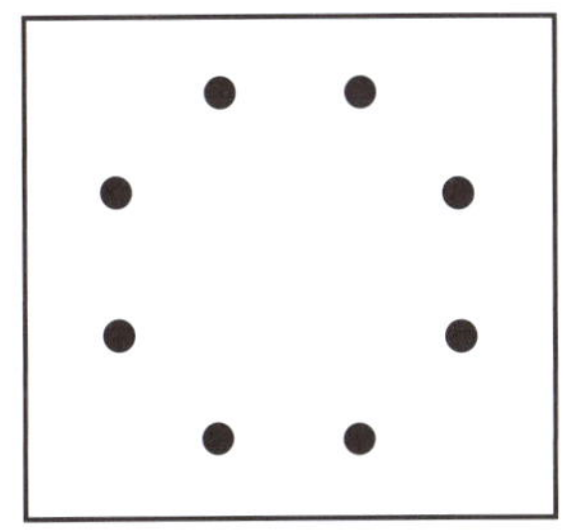

3 **a** Use the diagram of 9 dots to find the number of intervals that can be drawn from the centre to the other dots. ______

b How many intervals can be drawn from just one corner dot to any other dot, if the interval does not pass through a third dot? ______

c How many intervals can be drawn if each interval joins two dots and does not pass through a third dot? ______

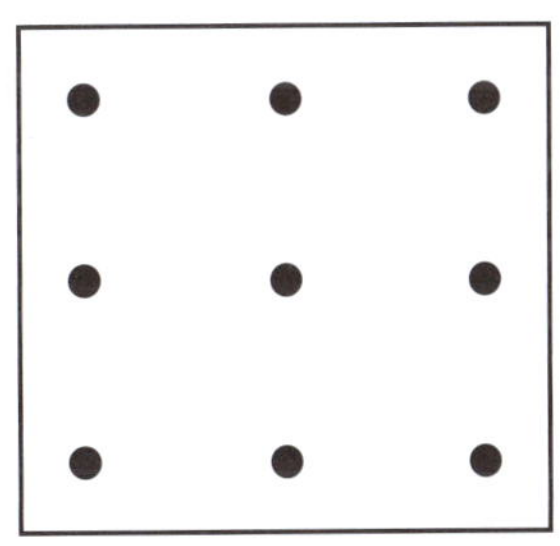

4 Use the diagram of 9 dots to find the number of squares that can be drawn using four of the dots as corners. ______

5 Write each of the numbers 1 to 24 in its proper place on this Venn diagram.

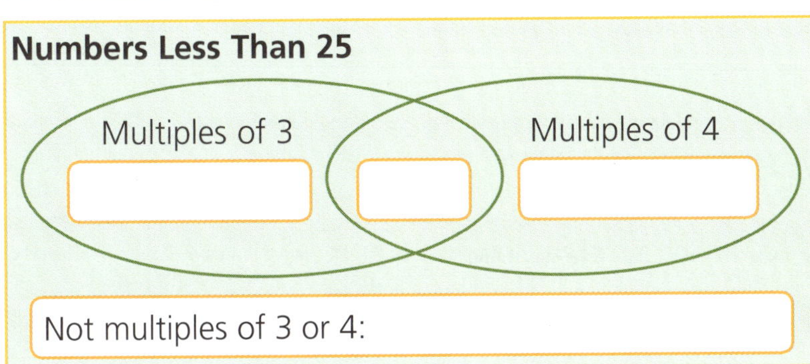

6 Does the height of a person depend on age? Explain your answer.

2:18 Division with Remainders

CONCEPT

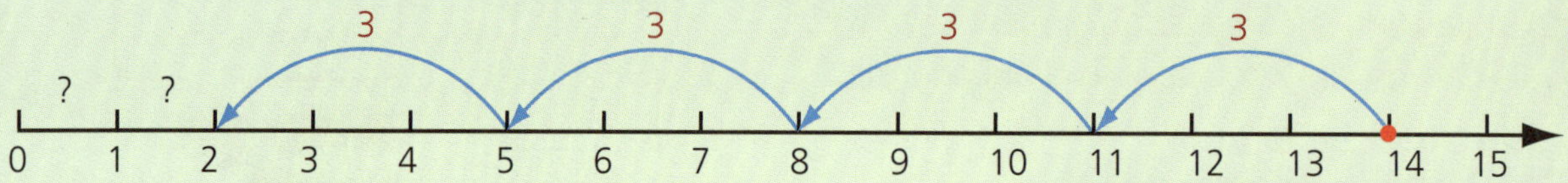

This number line shows 14 ÷ 3.

Only 4 groups of 3 can be taken from 14 and 2 are left over.

14 ÷ 3 = 4 remainder 2
14 ÷ 3 = 4 r 2

1 Use a number line to find the answers.

a 8 ÷ 3 = ☐ r ☐ **b** 9 ÷ 4 = ☐ r ☐ **c** 12 ÷ 5 = ☐ r ☐ **d** 14 ÷ 5 = ☐ r ☐

e 13 ÷ 5 = ☐ r ☐ **f** 11 ÷ 4 = ☐ r ☐ **g** 15 ÷ 4 = ☐ r ☐ **h** 15 ÷ 6 = ☐ r ☐

2 Use the pictures to answer these division questions.

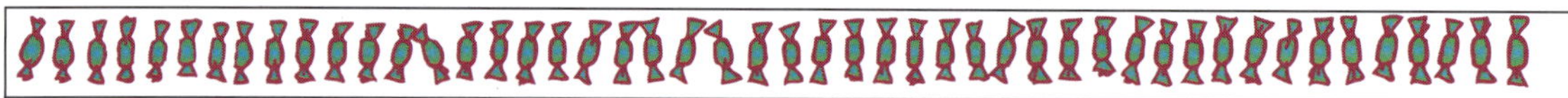	48
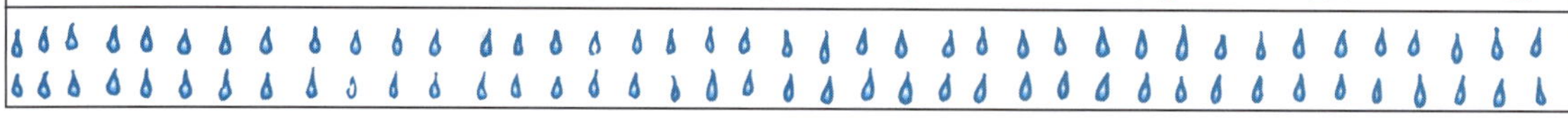	80

a How many packets of 7 lollies in 48? 48 ÷ 7 = ☐ packets and ☐ remaining

b How many groups of 10 lollies in 48? 48 ÷ 10 = ☐ groups and ☐ remaining

c How many rows of 8 lollies in 48? 48 ÷ 8 = ☐ rows and ☐ remaining

d How many lots of 8 drops in 80? 80 ÷ 8 = ☐ lots and ☐ remaining

e How many lots of 10 drops in 80? 80 ÷ 10 = ☐ lots and ☐ remaining

f How many groups of 9 drops in 80? 80 ÷ 9 = ☐ groups and ☐ remaining

3 **a** How many groups of 3 hats can be made from 17 hats?

17 ÷ 3 = ☐ groups and ☐ remaining

b 17 ÷ 6 = ☐ groups of 6 and ☐ remaining

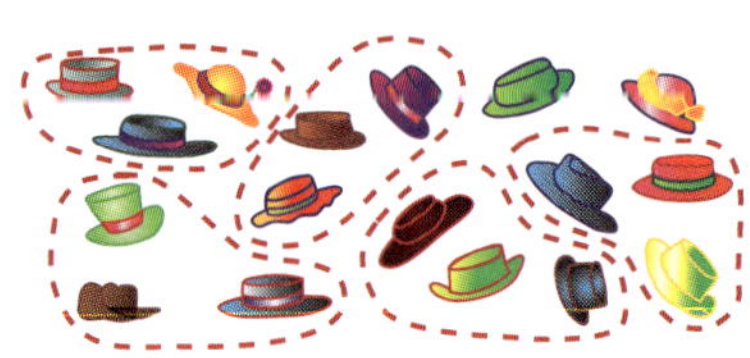

4 Use counters or place-value blocks to answer these.

a 10 pencils fill one tin. How many tins will 35 pencils fill? ☐ tins, 35 ÷ 10 = ☐ ,(r ☐)

b 33 books are in sets of 4. How many sets are there? ☐ sets, 33 ÷ 4 = ☐ ,(r ☐)

c 8 stamps fill a page. 42 stamps fill … ☐ pages, 42 ÷ 8 = ☐ ,(r ☐)

d 6 pens fill a box. 50 pens fill … ☐ boxes, 50 ÷ 6 = ☐ ,(r ☐)

2:19 Division of 2-Digit Numbers

CONCEPT

Alana puts 21 plates into four equal piles. How many are in each pile? How many are left over?

That's 21 divided by 4.

4)21 = 5 r 1

That's 5 in each pile and 1 left over.

1 You could use place-value blocks to answer these.

a 2)9 = ☐ r	**b** 3)7 = ☐ r	**c** 2)15 = ☐ r	**d** 4)7 = ☐ r	**e** 5)13 = ☐ r
f 4)22 = r	**g** 5)8 = r	**h** 6)16 = r	**i** 3)26 = r	**j** 10)78 = r
k 4)34 = r	**l** 10)65 = r	**m** 7)25 = r	**n** 10)92 = r	**o** 8)20 = r
p 9)19 = r	**q** 5)44 = r	**r** 6)32 = r	**s** 8)35 = r	**t** 7)40 = r

Share 60 sweets among 3 boys. How many sweets each? That's 60 divided by 3.

3)60 = 20

Each boy has 20 sweets.

2

a 4)40	**b** 2)60	**c** 3)30	**d** 5)50	**e** 2)40
f 3)60	**g** 2)20	**h** 3)90	**i** 8)80	**j** 6)60

3 **a** We have 60 kg of rice to give away. How many people can be given 8 kg?

8)60 = r ☐ people

b We wrapped 25 exercise books in packs of 4. How many packs did we wrap?

4)25 = r ☐ packs

2:20 Using Division Facts

CONCEPT

How many teams of eight can be made with 52 students?

52 ÷ 8 = 6 r 4

There are 6 teams of eight (and 4 students remaining).

1 Use the first number sentence to fill in the other one.

a (6 × 3) + 2 = 20 20 ÷ 3 = ___ r ___

b (4 × 10) + 1 = 41 41 ÷ 10 = ___ r ___

c (7 × 9) + 2 = 65 65 ÷ 9 = ___ r ___

2

a 10 ÷ 4 = ___ r ___ **b** 13 ÷ 5 = ___ r ___ **c** 28 ÷ 6 = ___ r ___

d 40 ÷ 7 = ___ r ___ **e** 21 ÷ 2 = ___ r ___ **f** 17 ÷ 10 = ___ r ___

g 66 ÷ 8 = ___ r ___ **h** 43 ÷ 6 = ___ r ___ **i** 39 ÷ 5 = ___ r ___

j 14 ÷ 4 = ___ r ___ **k** 76 ÷ 9 = ___ r ___ **l** 53 ÷ 7 = ___ r ___

3 Write as a number sentence, showing remainders:

a 17 shared among 3.
___ ÷ ___ = ___ r ___

b 27 divided by 7.
___ ÷ ___ = ___ r ___

c 30 in groups of eight.
___ ÷ ___ = ___ r ___

d 25 oranges put into groups of 6. How many groups?
___ ÷ ___ = ___ r ___

e 53 cows put into herds of 10. How many herds?
___ ÷ ___ = ___ r ___

f 33 desks put into groups of 5. How many groups?
___ ÷ ___ = ___ r ___

g 60 wombats shared among 9 zoos. How many in each zoo?
___ ___ ___ = ___ r ___

h 36 biscuits put into rows of 8. How many rows?
___ ___ ___ = ___ r ___

i 47 galahs living in groups of 6. How many groups?
___ ___ ___ = ___ r ___

j 32 bottles put into 8 rows. How many in each row?
___ ___ ___ = ___ r ___

k 29 birds flying in a flock. How many pairs?
___ ___ ___ = ___ r ___

l 53 days, 7 days in a week. How many weeks?
___ ___ ___ = ___ r ___

m 25 divided by 4.
___ ___ ___ = ___ r ___

n 27 divided by 2.
___ ___ ___ = ___ r ___

o 46 divided by 5.
___ ___ ___ = ___ r ___

4 How many teams of 5 can be made with 47 students? ______

 • *AUSTRALIAN SIGNPOST MATHS 5* • ISBN 978 1 4886 2190 1

2:21 Number Patterns

1 Complete each pattern and write the rule.

a 7, 10, 13, ___, ___, ___, ___ The rule is: ___

b 7, 19, 31, ___, ___, ___, ___ The rule is: ___

c 87, 78, 69, ___, ___, ___, ___ The rule is: ___

d 53, 46, 39, ___, ___, ___, ___ The rule is: ___

2 Write and continue the pattern made by the jumps on the number line. Write the rule for each.

a

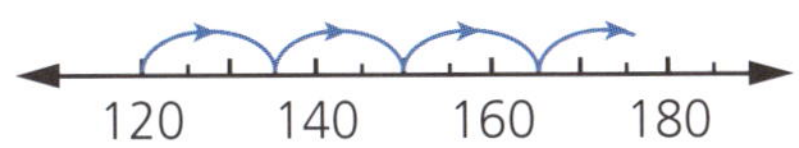

120, ___, ___, ___, ___

The rule is: ___

b

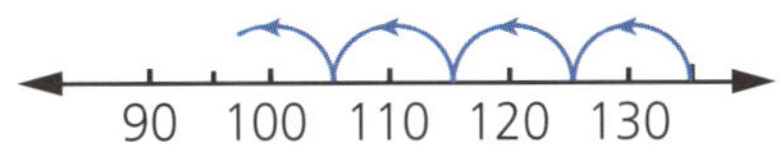

135, ___, ___, ___, ___

The rule is: ___

3 Create your own number pattern using jumps on the number line. Write the rule.

a

___, ___, ___, ___, ___, ___ The rule is: ___

4 Write and continue the pattern shown in these diagrams. Write the rule for each.

a The number of small squares

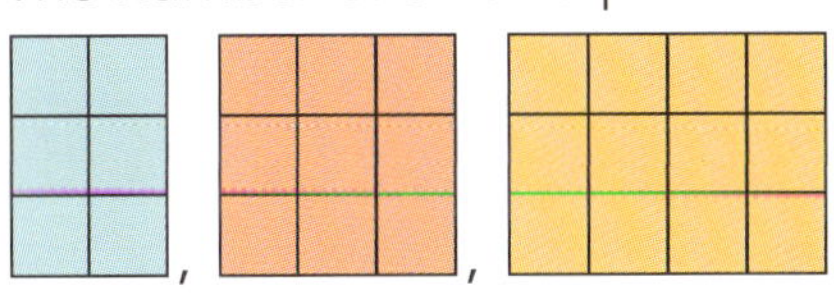

, , …

6, ___, ___, ___, ___, ___, ___

The rule is: ___

b The number of star points

, , …

5, ___, ___, ___, ___, ___, ___

The rule is: ___

c The number of small triangles

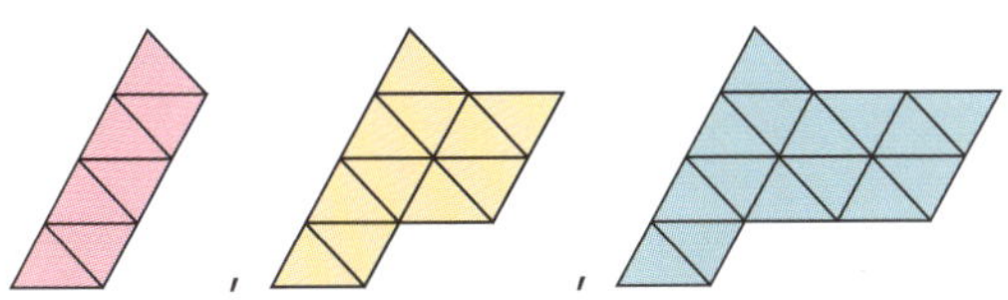

7, ___, ___, ___, ___, ___, ___

The rule is: ___

d The number of sides drawn

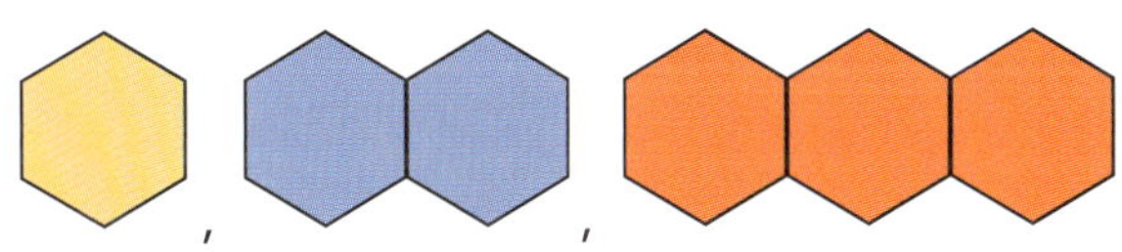

6, ___, ___, ___, ___, ___, ___

The rule is: ___

 • *AUSTRALIAN SIGNPOST MATHS 5* • ISBN 978 1 4886 2190 1

2:22 Subtraction with Trading to 999

CONCEPT

I had $524 and spent $245. How much do I have left?

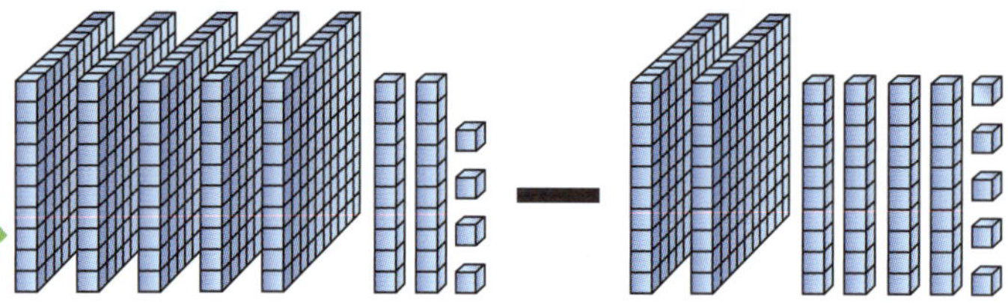

I trade 1 ten for 10 ones, and 1 hundred for 10 tens.

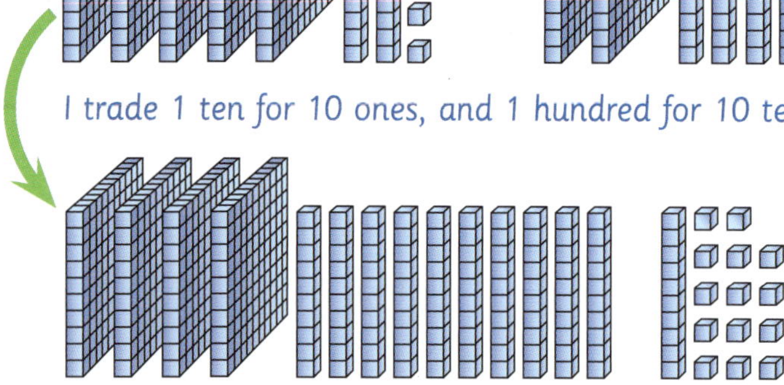

From this we take 2 hundreds, 4 tens and 5 ones.

	Hund	Tens	Ones
		11	
	4	~~1~~	14
	5	2	4
−	2	4	5
	2	7	9

1

	H	T	U
a	3	1	4
−		5	6

	H	T	U
b	2	4	6
−		6	7

	H	T	U
c	1	3	2
−		9	8

	H	T	U
d	7	3	5
−		8	9

	H	T	U
e	4	3	0
−	1	5	6

	H	T	U
f	3	4	1
−	2	7	8

	H	T	U
g	6	2	4
−	1	3	5

	H	T	U
h	9	2	2
−	3	3	3

	H	T	U
i	7	1	5
−	5	7	7

	H	T	U
j	4	7	0
−	2	9	3

	H	T	U
k	8	4	2
−	1	8	3

	H	T	U
l	9	1	1
−	3	4	4

2 Use this method to subtract money.

Use estimation to check your answers.

	Hundreds cents	Tens cents	Ones cent
		10	
	3	~~0~~	15
$	~~4~~ .	~~1~~	~~5~~
− $	3 .	8	7
$	0 .	2	8

a $6.04 − $5.75

b $5.28 − $4.29

c $7.55 − $6.68

d $7.46 − $2.49

e $6.81 − $1.84

f $5.76 − $2.78

 • *AUSTRALIAN SIGNPOST MATHS 5* • ISBN 978 1 4886 2190 1

2:23 Subtraction from Hundreds

CONCEPT

Of $600 Kim earned, he gave $155 to relieve suffering due to famine in Africa. How much money did he have left?

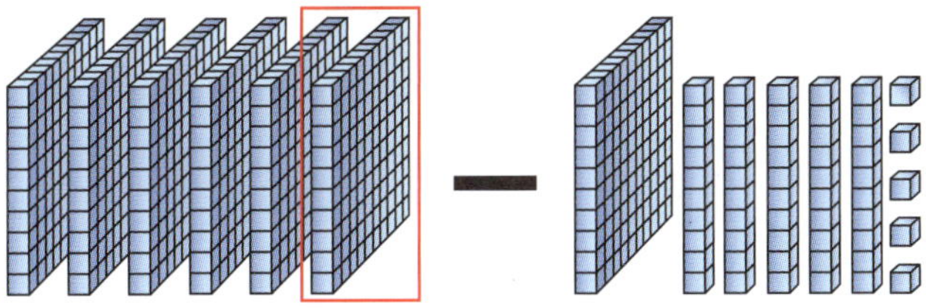

We can't take 5 ones from 0 ones, so we need to trade 1 ten for 10 ones. Since there are no tens we must first trade 1 hundred for 10 tens. Then trade 1 of the tens for 10 ones.

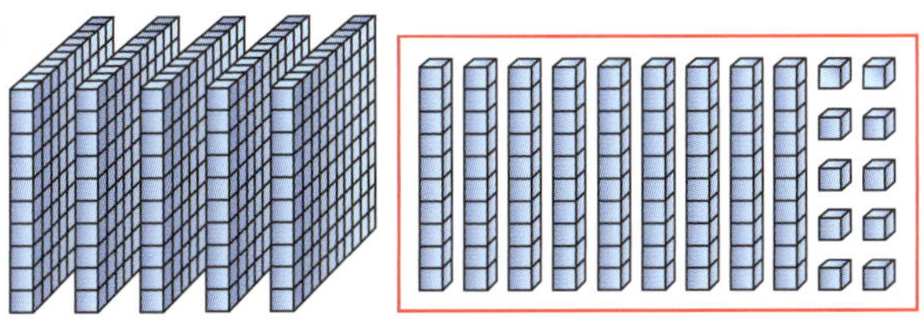

From this we take 1 hundred, 5 tens and 5 ones.

Hund	Tens	Ones
	9	
5	~~10~~	10
~~6~~	~~0~~	~~0~~
− 1	5	5
4	4	5

1

a (H T U)
```
  1 0 0
−   2 7
```

b (H T U)
```
  2 0 0
−   3 8
```

c (H T U)
```
  9 0 0
−   6 1
```

d (H T U)
```
  6 0 0
−   5 5
```

e
```
  7 0 0
− 3 0 2
```

f
```
  9 0 0
− 7 9 3
```

g
```
  4 0 0
− 1 0 6
```

h
```
  3 0 0
− 2 5 3
```

2

a
```
  $6.00
− $5.34
```

b
```
  $5.00
− $3.99
```

c
```
  $8.00
− $6.74
```

d
```
  $7.00
− $6.68
```

3 Estimate each answer and use the estimates to check your answers.

a Chloe needed 600 points to qualify for a state team. She earned only 487 points. How far short of her target was she?

b Blackburn Post Office was sent 900 stamp albums to sell. If 481 albums were sold in the first month, how many were left?

c In our class library we have 400 books. I have read 147. How many books have I not read?

d What is left if I had $10 and spent $3.65?

We could write 600 as 599 + 1.

2:24 Addition to 9999

CONCEPT

Rebecca had 1 559 stamps. Jake had 756. How many stamps altogether?

Thous	Hund	Tens	Ones
1	1	1	
1	5	5	9
	7	5	6
2	3	1	5

2315 stamps altogether.

I show

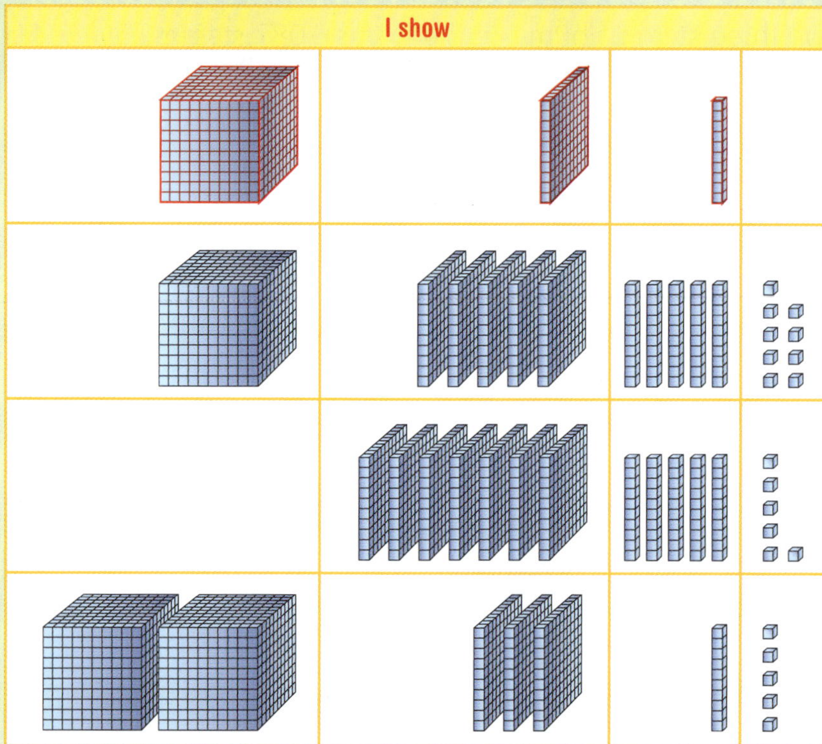

1

a	b	c	d	e
3416 + 2052	8104 + 871	6318 + 1270	392 + 3406	35 + 7932

f	g	h	i	j
3418 + 2906	7950 + 650	6087 + 1164	946 + 2627	6492 + 1748

2

a	b	c	d	e
5071 + 2124	3831 + 1054	2516 + 1827	3195 + 5730	7041 + 978

f	g	h	i	j
4692 + 2813	7472 + 1809	479 + 3860	1793 + 5824	6097 + 1835

3

a	b	c	d	e
\$7542 + \$1607	\$6549 + \$1742	\$3581 + \$4328	\$2591 + \$3706	\$7059 + \$2371

 • *AUSTRALIAN SIGNPOST MATHS 5* • ISBN 978 1 4886 2190 1

2:25 Addition to 9999

CONCEPT

I have three stamp albums. In the first is 3087 stamps, in the second 1872 stamps and in the third 4378. How many stamps have I altogether?

How many stamps?

3087 + 1872 + 4378 = ______

Answer

I have 9337 stamps.

	Thous	Hund	Tens	Ones
	1	2	1	
	3	0	8	7
	1	8	7	2
+	4	3	7	8
	9	3	3	7

1

a

	Th	H	T	U
	1	8	3	5
		4	0	8
+	5	3	1	8

b

	Th	H	T	U
		9	9	9
	4	2	5	6
+	1	6	8	0

c

	Th	H	T	U
	8	4	1	8
		6	4	7
+		7	5	3

d

	Th	H	T	U
	3	8	4	2
		7	0	6
+	2	9	3	4

2

a

```
   386
  2514
   977
+   27
```

b

```
   183
   964
  1863
+ 2098
```

c

```
  5142
   863
   744
+ 1088
```

d

```
  2099
  1847
  1935
+ 1207
```

3

a

```
  $61.29
  $13.91
+ $10.94
```

b

```
  $28.18
  $14.63
+ $ 9.72
```

c

```
  $ 8.47
  $60.08
+ $21.55
```

d

```
  $28.47
  $29.66
+ $28.67
```

4 Estimate then calculate. (E = estimate, A = answer)

a Residents were phoned at 3187 homes in Bendigo, 1394 in Ballarat and 914 in Geelong. How many residents were phoned? E = ______ A = ______

b A bookshop kept 3096 books in room A, 2515 in room B and 2845 in room C. How many books were in the three rooms? E = ______ A = ______

c Sandy spent $3245 renovating her laundry, $2188 on landscaping and $2840 on furniture. How much did she spend? E = ______ A = ______

 • *AUSTRALIAN SIGNPOST MATHS 5* • ISBN 978 1 4886 2190 1

2:26 Subtraction with Trading to 9999

CONCEPT

2216 people started in a *Town To Surf* race, but 865 people dropped out before the finish. How many people finished the race?

- We can show 2216 with place-value blocks.
- Trade 1 hundred for 10 tens.
- Trade 1 thousand for 10 hundreds.
- Now we can subtract 865 by taking away 8 hundreds, 6 tens and 5 ones.

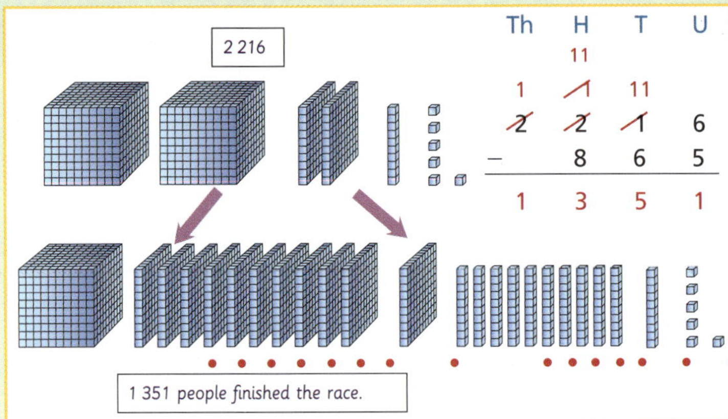

1

	Th	H	T	U
a	5	6	9	5
	− 1	3	4	2
b	7	8	5	0
	−	7	5	0
c	6	9	2	5
	− 1	5	0	5
d	8	0	4	6
	−		4	6
e	9	3	8	8
	− 3	1	7	5
f	6	4	5	0
	−	3	8	0
g	2	3	9	2
	− 1	9	5	0
h	7	5	0	0
	− 2	1	8	0
i	6	4	3	2
	−	1	2	9
j	5	2	9	9
	− 3	4	7	5

2

	Th	H	T	U
a	4	3	8	1
	− 1	6	0	5
b	2	1	8	6
	−	9	9	3
c	4	5	0	6
	− 3	8	0	0
d	7	8	6	0
	−	8	7	0
e	9	9	4	5
	− 6	8	7	7
f	2	3	1	5
	− 1	9	4	6
g	6	1	0	5
	− 3	8	2	7
h	9	0	3	1
	− 7	4	2	9
i	7	6	3	7
	−	8	8	8
j	5	4	5	5
	− 1	9	6	6

3 a Stephanie used her phone 9690 times last year. If 7815 were related to her business, how many were not business calls?

b 3495 calls were on her mobile phone. How many were not on her mobile phone?

c During June her business received $8475 and paid out $4816. How much money did her business make in June?

4 Tran bought 8095 bricks to pave part of his backyard. He used only 7186 bricks. How many were not used?

 • *AUSTRALIAN SIGNPOST MATHS 5* • ISBN 978 1 4886 2190 1

2:27 Four-Digit Subtraction from Thousands

CONCEPT

We received 6 000 books from the printers.
We have sold 3 721. How many have we left?

	Th	H	T	U
		9	9	
	5	~~10~~	~~10~~	10
	~~6~~	~~0~~	~~0~~	~~0~~
−	3	7	2	1
	2	2	7	9

We trade 1 thousand for 10 hundreds, then trade 1 hundred for 10 tens, then trade 1 ten for 10 ones, and then subtract.

2 279 books are left.

1

a 1 0 0 0 − 4 3 6
b 2 0 0 0 − 3 0 1
c 1 0 0 0 − 8 9 7
d 3 0 0 0 − 7 2 5
e 7 0 0 0 − 1 8 3 2
f 4 0 0 0 − 2 8 6 0
g 5 0 0 0 − 3 9 0 0
h 9 0 0 0 − 4 4 2 7

2

a 8 0 0 0 − 3 4 5 0
b 6 0 0 0 − 2 9 8 8
c 2 0 0 0 − 1 2 2 5
d 7 0 0 0 − 5 3 5 5
e 7 0 0 0 − 4 1 8 6
f 4 0 0 0 − 2 6 9 5
g 9 0 0 0 − 8 9 7 2
h 8 0 0 0 − 1 3 6 7

3

a The longest river is the Nile in North Africa (6 690 km). Australia's longest river system is the Murray–Darling which is 3 370 km. How much shorter is this than the Nile?

b How much deeper is the Indian Ocean (7 125 m at the Java Trench) than the Arctic Ocean (5 450 m at the Eurasia Basin)?

c Mount Everest, the highest mountain on earth, has an altitude of 8 850 m. We needed to use a supply of oxygen as we climbed the last 5 192 m of our ascent to the top. At what altitude did we begin to use oxygen?

d Tom McSeveny was born on 24 March 1916. How old was he on 24 March 2004?

e Chub Travers was born on 27 March 1913. How old was he on 27 March 2004?

● Use estimation to check your answers.

2:28 Subtraction from Thousands Strategy

CONCEPT

$5000 - 1278$
$= 4999 + 1 - 1278$
$= 4999 - 1278 + 1$
$= 3721 + 1$
$= 3722$

	5	0	0	0
−	1	2	7	8
	3	7	2	2

	Th	H	T	U	
	4	9	9	9	+1
−	1	2	7	8	
	3	7	2	1	+1
	3	7	2	2	

1 Use the strategy above to complete these.

a Th H T U
  9 9 9 +1
~~1 0 0 0~~
− 3 8 2
+1
= ☐

b Th H T U
3 9 9 9 +1
~~4 0 0 0~~
− 2 6 9 1
+1
= ☐

c Th H T U
1 9 9 9 +1
~~2 0 0 0~~
− 1 9 8 3
+1
= ☐

d Th H T U
2 9 9 9 +1
~~3 0 0 0~~
− 1 9 2 4
+1
= ☐

e Th H T U
6 9 9 9 +1
~~7 0 0 0~~
− 3 4 8 1
+1
= ☐

f Th H T U
5 9 9 9 +1
~~6 0 0 0~~
− 4 9 7 5
+1
= ☐

g Th H T U
+1
~~$9 0 0 0~~
− $8 2 0 7
+1
= ☐

h Th H T U
+1
~~$5 0 0 0~~
− $3 0 8 2
+1
= ☐

i Th H T U
+1
~~$8 0 0 0~~
− $3 7 9 1
+1
= ☐

2 Complete these on your own paper.

a 3000 – 1973 = ☐ **b** 5000 – 1903 = ☐

c 9000 – 2752 = ☐ **d** 6000 – 2246 = ☐

e I had $4000. I spent $2731. My change is ☐.

2:29 Mental Strategies

The Jump Strategy: Addition

487 + 237

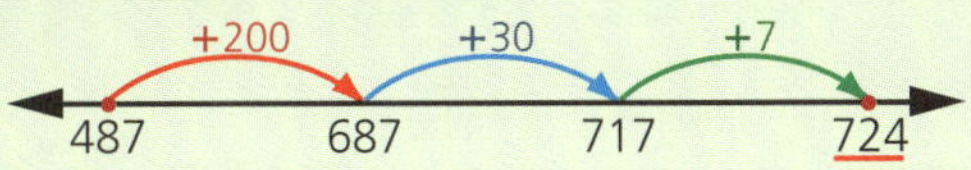

487 + 237 = 724

CONCEPT

1 Use the addition jump strategy to solve these.

a 347 + 518 = ☐

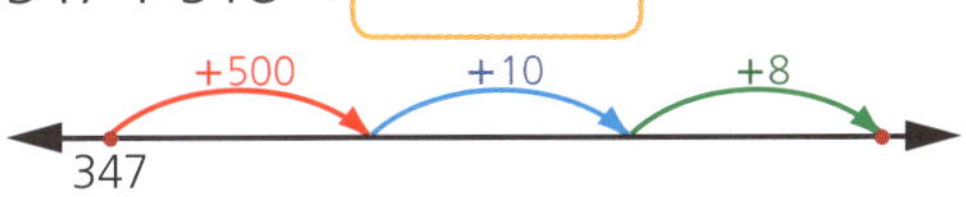

b 416 + 342 = ☐

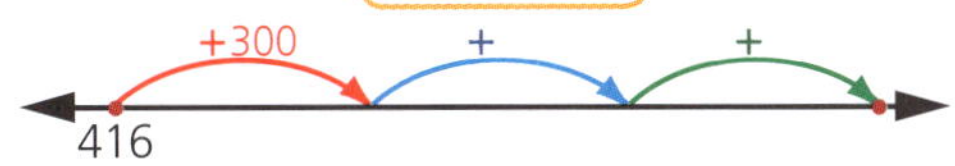

c 682 + 247 = ☐

d 236 + 497 = ☐

e 437 + 416 = ☐

f 365 + 366 = ☐

The Jump Strategy: Subtraction

625 – 237

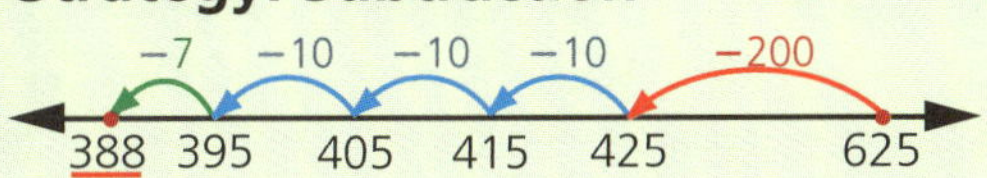

625 – 237 = 388

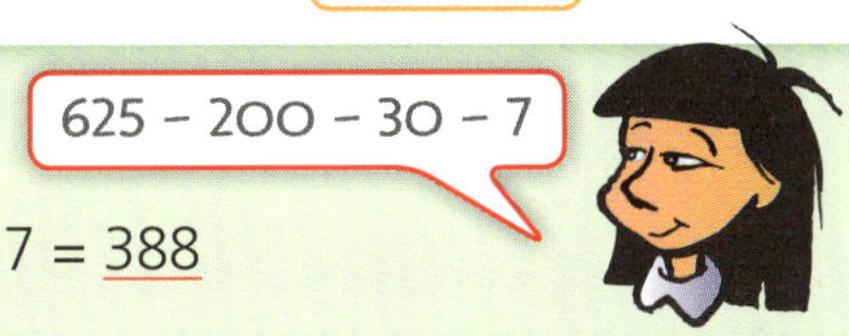

CONCEPT

2 Use the subtraction jump strategy to solve these.

a 856 – 231 = ☐

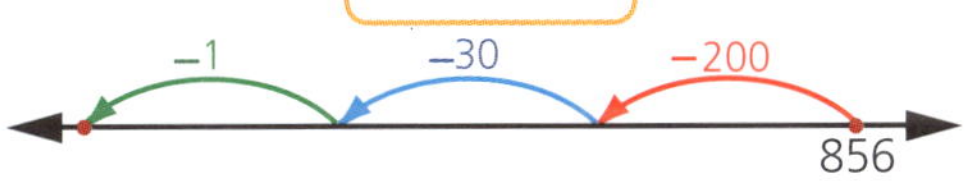

b 742 – 337 = ☐

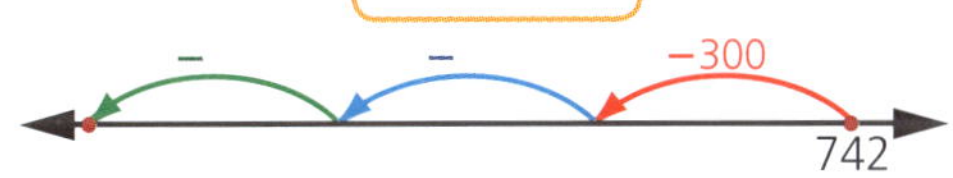

c 538 – 184 = ☐

d 961 – 327 = ☐

e 635 – 417 = ☐

f 471 – 122 = ☐

Subtracting from Numbers Ending in Zeros

8000 – 732 = 7999 + 1 – 732 = 7267 + 1 = 7268

CONCEPT

3 Use the strategy for subtracting numbers ending in zeros to solve these.

a 700 – 93 = 699 + 1 – 93 = ☐ = ☐

b 4000 – 346 = 3999 + 1 – 346 = ☐ = ☐

c 400 – 82 = ☐

d 800 – 165 = ☐

e 900 – 418 = ☐

f 3000 – 172 = ☐

g 5000 – 436 = ☐

h 9000 – 136 = ☐

i 8000 – 1888 = ☐

j 4000 – 2914 = ☐

k 7000 – 3415 = ☐

2:30 Factors and Multiples

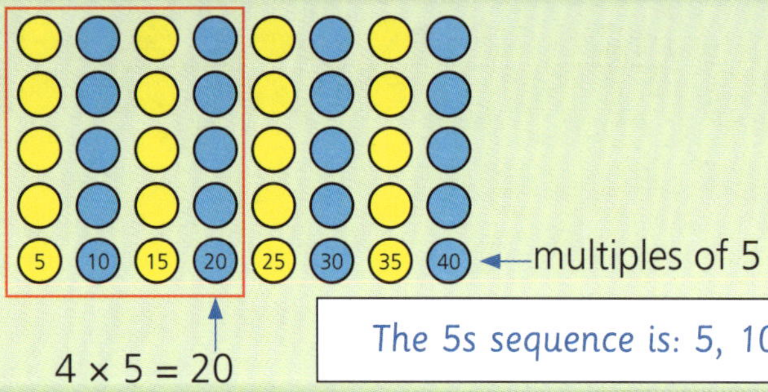

The 5s sequence is: 5, 10, 15, 20, 25, 30, 35, 40, 45, 50, …

CONCEPT

1 Complete the following.

- **a** The 7th number in the 5s sequence above is ____ . 7 × 5 = ____
- **b** The 3rd number in the 5s sequence above is ____ . 3 × 5 = ____
- **c** The 5th number in the 5s sequence above is ____ . 5 × 5 = ____
- **d** The 9th number in the 5s sequence above is ____ . 9 × 5 = ____
- **e** The 6th number in the 5s sequence above is ____ . 6 × 5 = ____
- **f** The 8th number in the 5s sequence above is ____ . 8 × 5 = ____

2 Continue the patterns.

- **a** 30, 35, 40, 45, ____, ____, ____, ____, ____, ____.
- **b** 14, 16, 18, 20, ____, ____, ____, ____, ____, ____.
- **c** 70, 80, 90, 100, ____, ____, ____, ____, ____, ____.
- **d** 58, 60, 62, 64, ____, ____, ____, ____, ____, ____.

3 From the numbers 12, 14, 21, 24, 28, 35, 42 and 70, find two numbers that are multiples of both:

- **a** 2 and 7 ____ and ____
- **b** 5 and 7 ____ and ____
- **c** 2 and 3 ____ and ____
- **d** 3 and 7 ____ and ____

4

- **a** 10 will be a factor of any number ending in ____.
- **b** 5 will be a factor of any number ending in ____ or ____.
- **c** 2 will be a factor of any number ending in ____, ____, ____, ____ or ____.

5 Using a calculator, fill in the next nine multiples.

ICT

- **a** 36, 42, 48, ____, ____, ____, ____, ____, ____, ____, ____, ____
- **b** 63, 72, 81, ____, ____, ____, ____, ____, ____, ____, ____, ____

2:31 Factors and Multiples

- **Factors** come in pairs. Any number of a product is a factor, e.g. 10 is equal to 1 × 10 and 2 × 5. Factors of 10 are 1, 10, 2 and 5.
- A **multiple** of a counting number is found when you multiply it by a counting number. The first 5 multiples of 7 are 7, 14, 21, 28 and 35.

CONCEPT

1. Circle the factors of 21. 3 7 8 13 20 21
2. List the multiples of 2 that are less than 16. ☐
3. What are the factors of 32? ☐, ☐, ☐, ☐, ☐ and ☐
4. What are the factors of 24? ☐, ☐, ☐, ☐, ☐, ☐, ☐ and ☐
5. What are the factors of 100? ☐, ☐, ☐, ☐, ☐, ☐, ☐, ☐, and ☐

6. Write down the first ten multiples of:
   a 2 ☐ ☐ ☐ ☐ ☐ ☐ ☐ ☐ ☐ ☐
   b 4 ☐ ☐ ☐ ☐ ☐ ☐ ☐ ☐ ☐ ☐
   c 8 ☐ ☐ ☐ ☐ ☐ ☐ ☐ ☐ ☐ ☐
   d 10 ☐ ☐ ☐ ☐ ☐ ☐ ☐ ☐ ☐ ☐

7. Write the answer. Is it odd or even?
   a 4 × 4 = ☐ which is ☐
   b 4 × 3 = ☐ which is ☐
   c 4 + 3 = ☐ which is ☐
   d 4 + 6 = ☐ which is ☐
   e 7 × 2 = ☐ which is ☐
   f 5 × 7 = ☐ which is ☐

8. Complete this for counting numbers.
   a even × even = ☐
   b even × odd = ☐
   c odd × odd = ☐

9. Give two examples to show that, for counting numbers:
   a even × even = even ☐
   b even × odd = even ☐
   c odd × odd = odd ☐

   Discuss with others why this is so.

INVESTIGATION

2:32 Dividing 2-Digit Numbers

CONCEPT

Share 44 cups among 4 cupboards.	Put 10 cups in each cupboard.	4 remain, so put 1 cup in each cupboard.
$4\overline{)44}$	$\overset{1}{4\overline{)44}}$ One ten Write 1 in the tens column.	$\overset{11}{4\overline{)44}}$

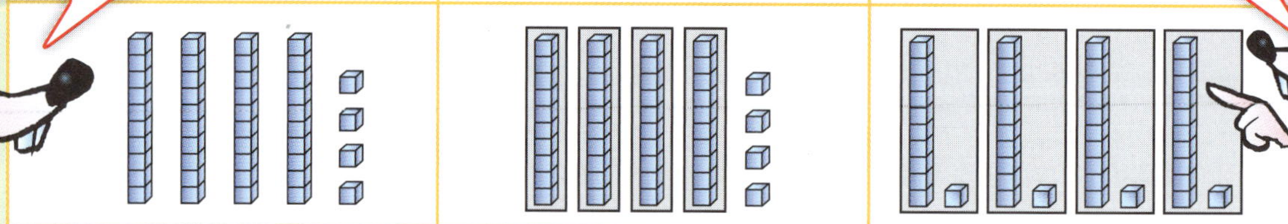

1 Here you will need to trade tens for ones. Use a calculator to check your answers.

a $2\overline{)44}$ **b** $2\overline{)68}$ **c** $2\overline{)88}$ **d** $2\overline{)24}$ **e** $3\overline{)33}$ **f** $3\overline{)66}$

g $3\overline{)39}$ **h** $3\overline{)36}$ **i** $4\overline{)44}$ **j** $4\overline{)48}$ **k** $4\overline{)84}$ **l** $4\overline{)88}$

CONCEPT

Share 52 pens among 4 boxes.	Put 10 in each box and trade 1 ten for 10 ones.	12 remain, so put 3 in each box.

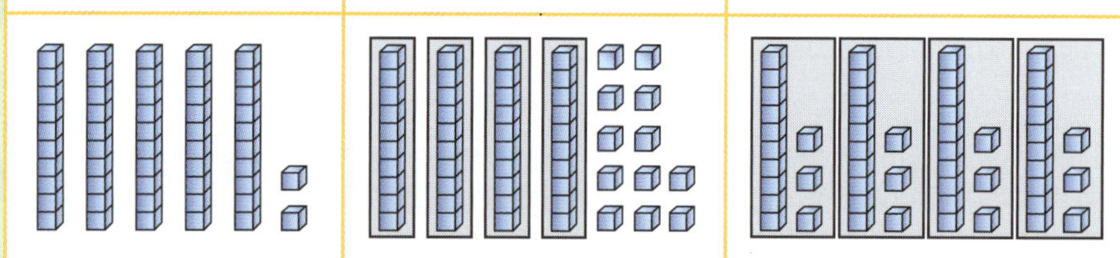

We can use place-value blocks to divide.

2 Use place-value blocks or mental strategies to do these. Use a calculator to check your answers.

a 34 ÷ 2 = ☐ **b** 45 ÷ 3 = ☐ **c** 92 ÷ 4 = ☐

d 54 ÷ 3 = ☐ **e** 65 ÷ 5 = ☐ **f** 68 ÷ 4 = ☐

FUN SPOT

Snakes and Ladders

- Use the board on the right to play the game.
- Take turns to roll two dice. If one number can be divided exactly by the other, your counter can move one place forwards on the board.
- The first to land on **Finish** wins.

 • *AUSTRALIAN SIGNPOST MATHS 5* • ISBN 978 1 4886 2190 1

2:33 Dividing 2-Digit Numbers

CONCEPT

$84 \div 3 =$ ☐

8 tens ÷ 3 = 2 tens

and 2 tens left over.

3)8²4 — 2 — 3 into 8 goes twice.

Trade two tens for 20 ones.

24 ones ÷ 3 = 8 ones

3)8²4 — 28 — Trade 2 tens. 24 ÷ 3 gives 8.

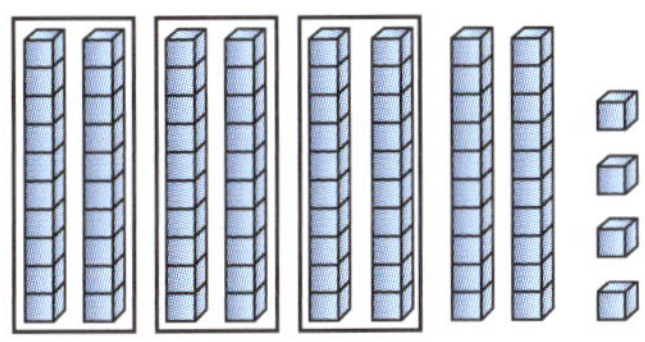

1
a 2)46 b 5)55 c 4)84 d 2)82
e 3)90 f 2)24 g 2)68 h 6)66

18
4)7³2

2 In these you need to trade 1 ten for 10 ones.

a 3)75 b 2)96 c 5)65 d 4)96
e 6)78 f 3)45 g 7)84 h 8)96

3 In these, you need to trade tens for ones.

a 3)54 b 7)98 c 5)70 d 4)68
e 5)85 f 5)95 g 6)96 h 3)87

280 ÷ 14
(divide both by 7)
40 ÷ 2 = 20

4 In each case, find one share.

a 52 brushes shared by 4.
4)52
One share = ☐

b 85 ladders shared by 5.
5)85
One share = ☐

c 91 globes shared by 7.
7)91
One share = ☐

d 90 plants shared by 5.
5)90
One share = ☐

e 66 caps shared by 3.
3)66
One share = ☐

f 90 spades shared by 6.
6)90
One share = ☐

2:34 Dividing 2-Digit Numbers

1 Show all remainders as fractions.

a 2)25 **b** 3)35 **c** 4)89

d 5)53 **e** 6)69 **f** 3)37

2 In these, you will need to trade 1 ten for 10 ones.

a 2)35 **b** 3)74 **c** 2)33

d 8)99 **e** 6)79 **f** 4)59

CONCEPT

Share 23 pieces of toast between four people.

$$4\overline{)23}\quad 5 \text{ r } 3$$

In this case the remainder can be shared also.

$\frac{1}{4}$	$\frac{1}{4}$	$\frac{1}{4}$	$\frac{1}{4}$	$\frac{1}{4}$	$\frac{1}{4}$
$\frac{1}{4}$	$\frac{1}{4}$	$\frac{1}{4}$	$\frac{1}{4}$	$\frac{1}{4}$	$\frac{1}{4}$

Each person would be given an extra $\frac{3}{4}$ of a piece of toast.

$$4\overline{)23}\quad 5\tfrac{3}{4}$$

Each person receives $5\frac{3}{4}$ pieces of toast.

3 In these, you will need to trade 2 tens for 20 ones.

a 5)75 **b** 3)51 **c** 7)98 **d** 3)57

e 4)64 **f** 3)84 **g** 6)84 **h** 7)91

$$3\overline{)8^{2}1}\quad 2\ 7$$

4 Here you will need to trade tens for ones.

a 5)85 **b** 6)96 **c** 4)76 **d** 5)90 **e** 2)96

f 2)74 **g** 5)80 **h** 3)81 **i** 4)72 **j** 6)90

5 **a** In two days I walked 36 km. What was the average distance walked each day? ☐)____

b Seven people contributed a total of $84. What was the average amount given? ☐)____

c Three towns in Australia recorded temperatures of 37°C, 15°C and 23°C. What was the average temperature for these towns? ☐)____

d Five Goliath spiders were caught. The spans of their legs were 28 cm, 28 cm, 25 cm, 18 cm and 16 cm. What was the average span of their legs? ☐)____

To find the average, find the total and divide by the number sharing.

 ISBN 978 1 4886 2190 1

2:35 Dividing 3-Digit Numbers

CONCEPT

Amy was told to send the same number of glasses to six restaurants. What is the greatest number she can send to each if she has 897 glasses?

Estimate to check the answer.

149 r3
6)8²9⁵7

6 into **8** goes 1 time with 2 left over. Trade the 2.
6 into **29** goes 4 times with 5 left. Trade the 5.
6 into **57** goes 9 times with a remainder of 3.

Amy can send 149 glasses to each restaurant. She would then have 3 glasses left.

1 There are no remainders in these.

a 3)402	**b** 2)570	**c** 4)576	**d** 6)732	**e** 5)665
f 4)696	**g** 7)945	**h** 6)870	**i** 8)952	**j** 7)889

2 These have remainders.

a 6)887	**b** 5)706	**c** 4)937	**d** 7)824	**e** 3)772
f 8)939	**g** 4)786	**h** 6)814	**i** 5)918	**j** 8)951

3

a 4)324	**b** 5)250	**c** 3)216
d 6)378	**e** 8)264	**f** 9)657

6 into 3? No.
6 into 30? 5.

4

a 4)169	**b** 5)155	**c** 8)649	**d** 7)800	
e 4)300	**f** 8)900	**g** 9)400	**h** 7)200	
i 9)342	**j** 4)196	**k** 5)335	**l** 3)282	**m** 7)588

5 **a** The cheetah is the fastest land animal. If it can travel 105 metres in 5 seconds, how far could it travel in 1 second?

b The fastest animal is the peregrine falcon. It can fly 776 metres in 8 seconds. How far can it travel in 1 second?

 • *AUSTRALIAN SIGNPOST MATHS 5* • ISBN 978 1 4886 2190 1

2:36 Multiplying Tens

10 tens = 100

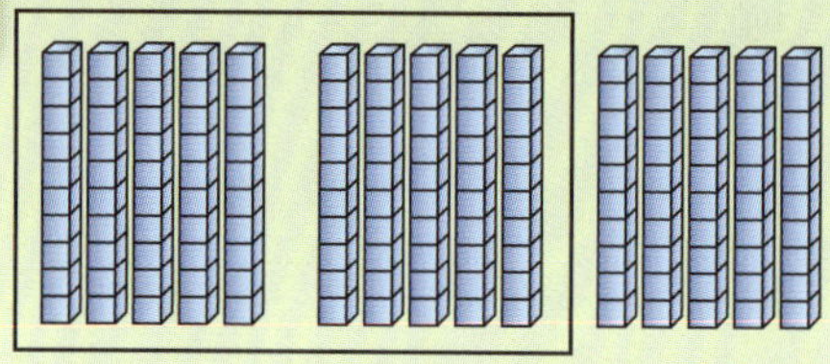

15 tens = 150   15 × 10 = 150

To multiply by 10, put a zero at the end.

CONCEPT

$8 \times 10 = 80$
21 tens = 210
$47 \times 10 = 470$

1
a 8 tens = ☐ b 9 tens = ☐ c 6 tens = ☐ d 4 tens = ☐
e 11 tens = ☐ f 14 tens = ☐ g 12 tens = ☐ h 13 tens = ☐
i 32 tens = ☐ j 21 tens = ☐ k 48 tens = ☐ l 85 tens = ☐
m 90 tens = ☐ n 77 tens = ☐ o 60 tens = ☐ p 99 tens = ☐

2
a $8 \times 10 =$ ☐ b $7 \times 10 =$ ☐ c $10 \times 10 =$ ☐ d $9 \times 10 =$ ☐
e $13 \times 10 =$ ☐ f $18 \times 10 =$ ☐ g $27 \times 10 =$ ☐ h $22 \times 10 =$ ☐
i $35 \times 10 =$ ☐ j $61 \times 10 =$ ☐ k $84 \times 10 =$ ☐ l $75 \times 10 =$ ☐

3
a 6 × 3 tens = ☐ tens = ☐
b 7 × 3 tens = ☐ tens = ☐
c 4 × 5 tens = ☐ tens = ☐
d 2 × 9 tens = ☐ tens = ☐
e 8 × 4 tens = ☐
f 5 × 8 tens = ☐
g 4 × 9 tens = ☐
h 7 × 6 tens = ☐

4
a $6 \times 30 = 6 \times 3 \times 10 =$ ☐
b $8 \times 50 = 8 \times 5 \times 10 =$ ☐
c $7 \times 40 = 7 \times 4 \times 10 =$ ☐
d $5 \times 40 = 5 \times 4 \times 10 =$ ☐

5
a $4 \times 30 =$ ☐
b $3 \times 60 =$ ☐
c $4 \times 80 =$ ☐
d $7 \times 50 =$ ☐
e 7 × $20 = ☐
f 9 × $60 = ☐
g 8 × $80 = ☐
h 6 × $60 = ☐

 ISBN 978 1 4886 2190 1

2:37 Multiplying Tens or Hundreds

CONCEPT

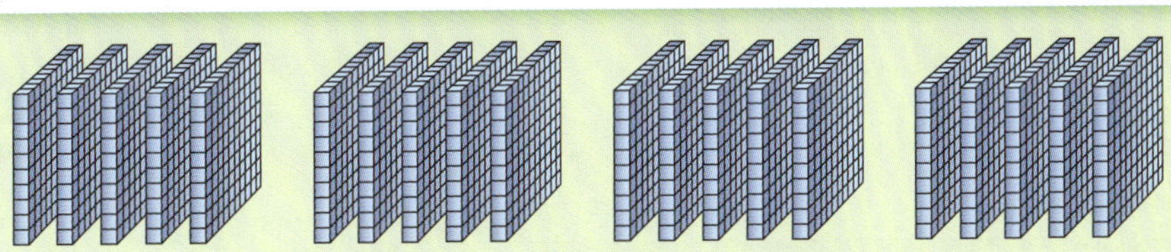

$4 \times 500 = 4 \times 5 \times 100$
$= 20 \times 100$
$= 2000$

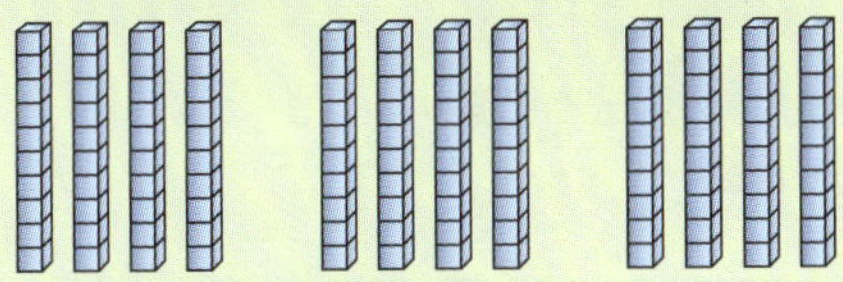

$3 \times 40 = 3 \times 4 \times 10$
$= 12 \times 10$
$= 120$

500
× 4
2000

To multiply by 10, put a zero at the end of the number.

To multiply by 100, put 2 zeros at the end of the number.

1 a 7 tens = ☐ b 21 tens = ☐ c 9 hundreds = ☐
d 12 hundreds = ☐ e 10 hundreds = ☐ f 11 hundreds = ☐
g 25 hundreds = ☐ h 30 hundreds = ☐ i 49 hundreds = ☐

2 a 9 × 10 = ☐ b 27 × 10 = ☐ c 56 × 10 = ☐
d 6 × 100 = ☐ e 13 × 100 = ☐ f 25 × 100 = ☐

3 a 6 × 4 tens
= ☐ tens
= ☐

b 7 × 3 hundreds
= ☐ hundreds
= ☐

c 4 × 8 hundreds
= ☐ hundreds
= ☐

d 7 × 5 tens
= ☐ tens
= ☐

e 3 × 5 hundreds
= ☐ hundreds
= ☐

f 9 × 4 hundreds
= ☐ hundreds
= ☐

4 a 3 × 80 = ☐ b 7 × 50 = ☐ c 9 × 60 = ☐
d 4 × 200 = ☐ e 7 × 300 = ☐ f 5 × 600 = ☐

5 a

	H	T	U
		4	0
×			3

b

	H	T	U
		8	0
×			4

c

	H	T	U
		6	0
×			5

d

	Th	H	T	U
		3	0	0
×				6

e

	Th	H	T	U
		7	0	0
×				4

f

	Th	H	T	U
		9	0	0
×				8

You could use a calculator to check your answers.

2:38 Dividing 3-Digit Numbers by 10

Ten eggs fill a carton. How many cartons can be filled with 593 eggs?

10)5 9⁹3 = 5 9 r3

59 cartons can be filled. 3 eggs remain.

Ten desks fill our truck. How many trips would I need to make to transport 700 desks?

10)7 0 0 = 7 0

I would need to make 70 trips.

CONCEPT

1 These have no remainder.

a 10)8 0 0	**b** 10)6 0 0	**c** 10)5 0 0	**d** 10)3 0 0	**e** 10)9 0 0
f 10)3 7 0	**g** 10)8 6 0	**h** 10)5 4 0	**i** 10)290	**j** 10)1 8 0
k 10)1 1 0	**l** 10)2 0 0	**m** 10)9 9 0	**n** 10)730	**o** 10)5 5 0

2 Most of these have remainders.

a 10)1 5 6	**b** 10)3 7 5	**c** 10)5 2 1	**d** 10)2 1 8	**e** 10)8 6 6
f 10)6 8 4	**g** 10)5 6 0	**h** 10)7 3 7	**i** 10)9 0 8	**j** 10)3 6 7
k 10)7 1 1	**l** 10)4 0 7	**m** 10)8 9 9	**n** 10)4 0 0	**o** 10)6 0 1

3 **a** Ashley has $950. This is 10 times as much as Blake. How much has Blake? $ ______

b The total number of runs I scored in 10 games is 510 runs. What is my average number of runs scored per game? ______ runs

c Mia used 470 mL of oil to fill ten identical lamps. How much was in each? ______ mL

d Sarah has 200 g of sugar. Into how many cups could she place 10 g of it? ______ cups

4 Find the missing digits.

a 10)☐☐☐ = 75

b 10)☐☐☐ = 8 3 r1

c 10)☐☐☐ = 60

d 10)☐☐☐ = 9 0 r9

FUN SPOT

2:39 Division Involving Zeros in Answers

CONCEPT

Six boys shared 612 toy soldiers. How many was each boy given?

$$\begin{array}{r} 102 \\ 6\overline{)61^{1}2} \end{array}$$

Each boy was given 102 soldiers.

Each girl was given 4 of 403 stickers. How many girls could be given stickers?

$$\begin{array}{r} 100r3 \\ 4\overline{)403} \end{array}$$

100 girls could each be given 4 stickers. Three stickers were left over.

1 Don't forget the zeros as you do these.

a	b	c	d	e
$3\overline{)612}$	$4\overline{)832}$	$2\overline{)816}$	$3\overline{)921}$	$4\overline{)820}$
f	**g**	**h**	**i**	**j**
$3\overline{)312}$	$8\overline{)840}$	$6\overline{)612}$	$4\overline{)816}$	$3\overline{)627}$

2 Some of these will have remainders.

a	b	c	d	e
$5\overline{)505}$	$3\overline{)306}$	$4\overline{)408}$	$2\overline{)806}$	$3\overline{)609}$
f	**g**	**h**	**i**	**j**
$8\overline{)802}$	$5\overline{)508}$	$2\overline{)609}$	$7\overline{)708}$	$6\overline{)605}$

3
- **a** 864 sheep are in 8 paddocks. What is the average number per paddock? ☐
- **b** How many groups of 6 tins are in 612 tins? ☐
- **c** I caught 812 fish, 4 times as many as Peter. How many did Peter catch? ☐
- **d** How many people can be given 4 sandwiches each if we have 416? ☐
- **e** 6 bananas are needed to fill a bucket. How many buckets can be filled with 639 bananas? ☐

FUN SPOT

4 Find the missing digits.

a	b	c	d
$\begin{array}{r} 210r3 \\ 4\overline{)\square\square\square} \end{array}$	$\begin{array}{r} 304 \\ 3\overline{)\square\square\square} \end{array}$	$\begin{array}{r} 407r1 \\ 2\overline{)\square\square\square} \end{array}$	$\begin{array}{r} 100r3 \\ 4\overline{)\square\square\square} \end{array}$

2:40 Divisibility

These are quick checks for divisibility.

- Any even number is exactly divisible by 2.

Example
36 ÷ 2 = 18, (36 is even.)

- A number is exactly divisible by 3 if the sum of the digits is divisible by 3.

Example
243 ÷ 3 = 81, (2 + 4 + 3 = 9)

- Any number ending with a 5 or 0 is exactly divisible by 5.

Example
60 ÷ 5 = 12,
(60 ends with 0.)

- A number is exactly divisible by 9 if the sum of the digits is divisible by 9.

Example
567 ÷ 9 = 63, (5 + 6 + 7 = 18)

CONCEPT

1 Circle the numbers which are:

a divisible by 2:	24	29	30	46	55	76	81	88	93	100
b divisible by 5:	30	44	45	60	73	75	86	95	120	135
c divisible by 3:	21	34	63	74	76	81	97	102	124	162
d divisible by 9:	24	36	56	72	684	693	115	126	154	234

2 Write the numbers below in the columns of the table.

Some numbers can by placed in more than one column.

64 96 80 54 75 92 134 145 168
135 190 162 348 460 645 378 385 649

If a number is divisible by 9, then 9 is a ***factor*** *of the number.*

Divisible by 2	Divisible by 5	Divisible by 3	Divisible by 9

3 Complete these sets of related number facts.

a 14 × 3 = 42, 3 × 14 = ☐, ☐ ÷ 3 = 14, ☐ ÷ 14 = 3

b 16 × 5 = 80, 5 × ☐ = ☐, ☐ ÷ 5 = ☐, ☐ ÷ 16 = ☐

c 13 × 9 = 117, 9 × ☐ = ☐, ☐ ÷ 9 = ☐, ☐ ÷ 13 = ☐

4 Write all the factors of each number.

a 8 ☐ **b** 16 ☐ **c** 35 ☐

d 28 ☐ **e** 49 ☐ **f** 99 ☐

5 Do larger numbers have more factors? ☐ Discuss.

2:41 Factors and Multiples

1 Write down all factors of:

a 15 ☐ ☐ ☐ ☐ b 6 ☐ ☐ ☐ ☐

c 36 ☐ ☐ ☐ ☐ ☐ ☐ ☐ ☐ ☐

2 Write down the first ten multiples of:

a 6 ☐ ☐ ☐ ☐ ☐ ☐ ☐ ☐ ☐ ☐

b 9 ☐ ☐ ☐ ☐ ☐ ☐ ☐ ☐ ☐ ☐

c 7 ☐ ☐ ☐ ☐ ☐ ☐ ☐ ☐ ☐ ☐

Since 10 = 1 × 10 and 10 = 2 × 5, the factors of ten are 1, 10, 2 and 5.

ACTIVITY

- Some of the multiples of 6 are: 6, 12, 18, 24, 30, 36, 42, 48, 54, 60.
- The unit digits are: 6, 2, 8, 4, 0.
- We can draw the unit digit pattern like this. Begin and end at 6.

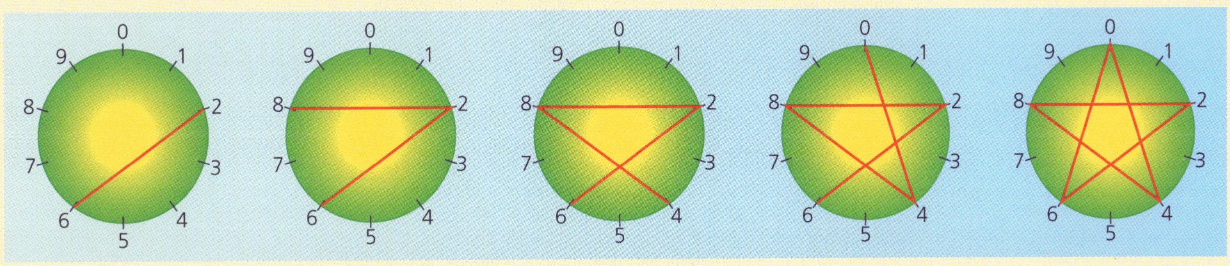

- Investigate the unit digit pattern for other multiples.

INVESTIGATION

Problem

Find all the factors of 64.

Solution

Use a calculator to divide 64 by 1, 2, 3, 4, etc.

Whole number answers, together with their divisors, are factors.

The factors of 64 are 1, 64, 2, 32, 4, 16 and 8.

3 Use this method to find all the factors of:

a 80 ☐ b 92 ☐ c 63 ☐

d 56 ☐ e 160 ☐ f 108 ☐

g 48 ☐ h 60 ☐ i 144 ☐

2:42 Averages

To find the average, add the scores then divide by the number of scores.

An average is a fair share.

CONCEPT

1 Measure the length of each nail.

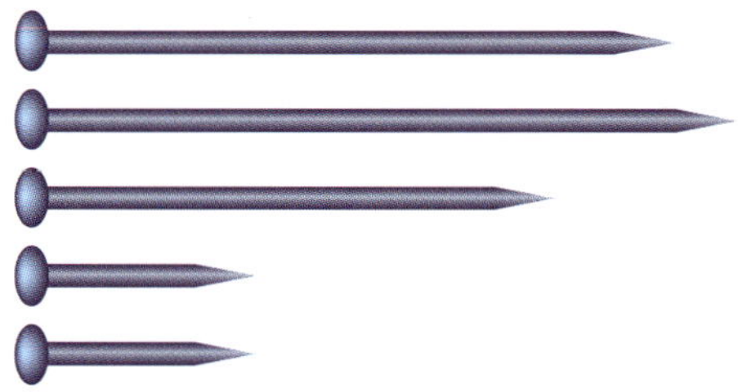

a What is the average length of the nails?

b Tom needed two 4 cm long nails. He bought a packet of nails which said 'average length 4 cm'. These are the nails from that packet. Was he disappointed? Why or why not?

The average does not tell us about single cases, it gives the amount when shared.

2 A B C D

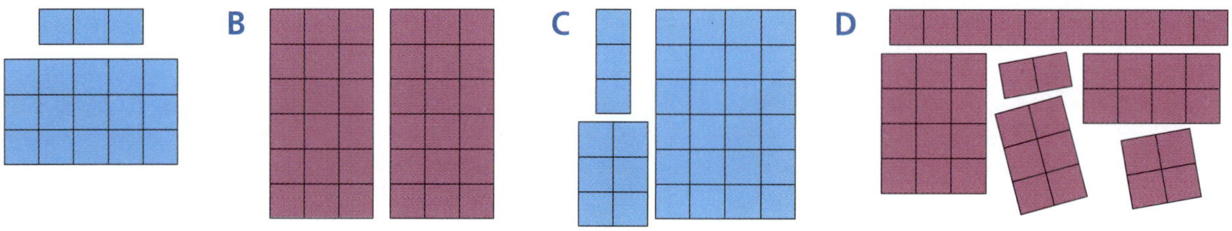

a Find the average number of small squares in each group of rectangles.

Average **A** = Average **B** = Average **C** = Average **D** =

b What is the average height of the rectangles in **C**?

c What is the average width of the rectangles in **A**?

3 a Sam has $40. I have $28. What is the average of these amounts?

b In three days, I earned $45, $28 and $17. What were my average daily earnings?

c For our school play, 34 tickets were sold on the first day, 17 on the second, 25 on the third and 12 on the fourth. What was the average number of tickets sold per day?

An average can be written using a fraction or a decimal. Examples: Kim averaged $2\frac{1}{2}$ goals per game, I averaged 1·2 fish each hour.

4 Ben bought 2 tickets to the play, Sarah bought 4, Kuan bought 3 and Anna bought 1.

a What was the average number of tickets bought by these people?

b What was the average number of tickets bought by the two boys?

© PEARSON AUSTRALIA 2018 • *AUSTRALIAN SIGNPOST MATHS 5* • ISBN 978 1 4886 2190 1

2:43 Averages

CONCEPT

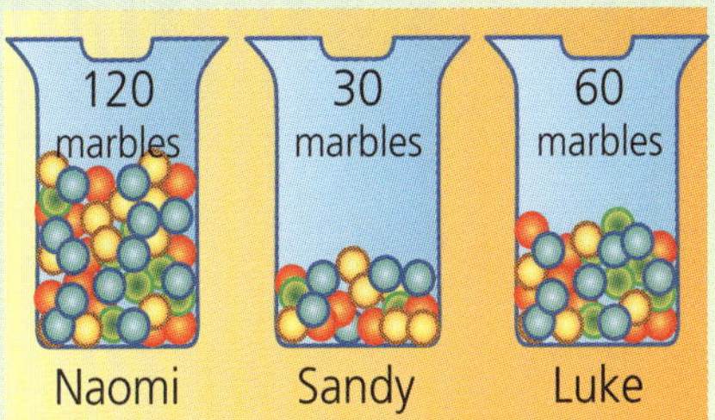

Find the average number of marbles in the three containers.

- Let's share the marbles.
  Step 1: Naomi gives Sandy 40.
  Step 2: Naomi gives Luke 10.

  The average is 70.

N	S	L
80	70	60
70	70	70

- Another way to share is to divide the total number of marbles by 3.
  The average = (120 + 30 + 60) ÷ 3

  The average is 70.

$$3\overline{)210} = 70$$

The average (or mean) is an **equal share**.

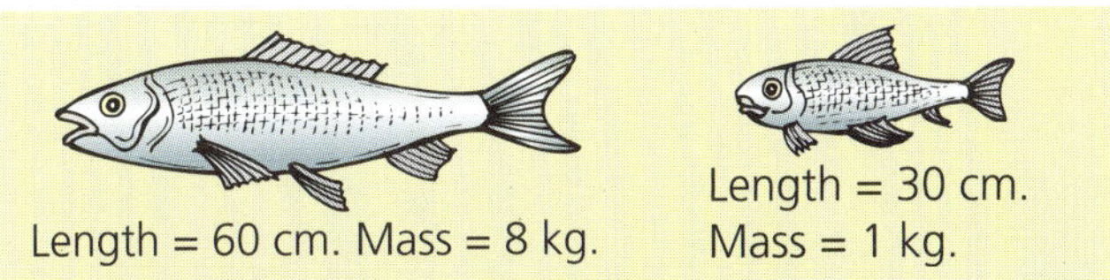

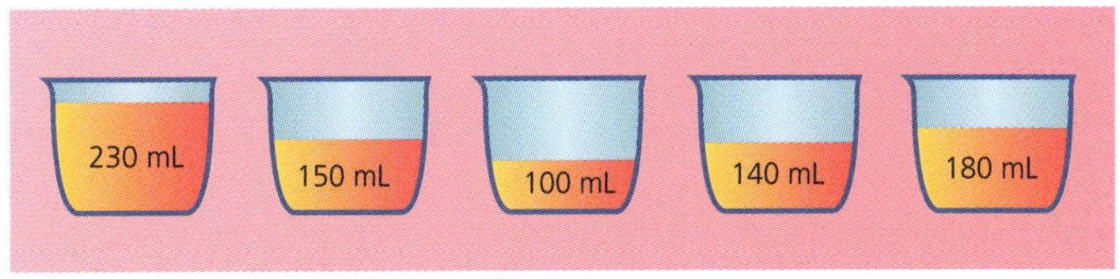

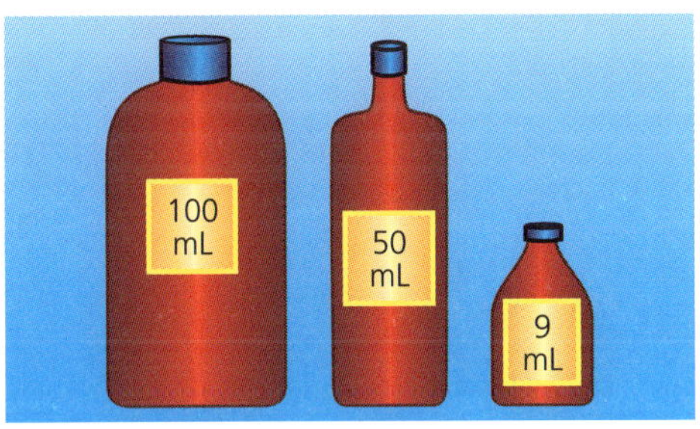

1 a Find the average length of the two fish.

b Find the average mass of the two fish.

c Find the average height of the kangaroo, the boy and the crocodile.

d What is the average contents of the bottles?

e What is the average contents of the cups?

f The contents of the cups are shared among five. How much does each receive?

2 At Heather's birthday party, the children played 'find the chocolates'. Alana found 12, Rachel 15, Naomi 10, Luke 6 and Miranda 12.

a If they shared all fairly, how many would each child receive?

b What was the average number of chocolates each found?

Always estimate first.

3 Within these rectangles, each small square has a length of 3 mm.

a What is the average width of these rectangles?

b What is the average height of these rectangles?

c What is the average number of small squares?

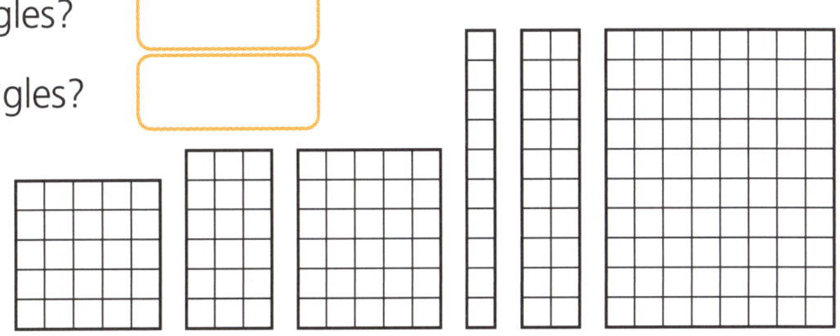

2:44 Using Factors in Multiplication

Factorising a number can make a question easier.

18 × 5
= 9 × 2 × 5
= 9 × 10
= 90

24 × 25
= 6 × 4 × 25
= 6 × 100
= 600

CONCEPT

1 Complete each question to show how factors can be used.

a 16 × 5
= ☐ × 2 × 5
= ☐

b 12 × 5
= ☐ × 2 × 5
= ☐

c 18 × 5
= ☐ × 2 × 5
= ☐

d 14 × 5
= ☐ × 2 × 5
= ☐

e 8 × 25
= ☐ × 4 × 25
= ☐

f 16 × 25
= ☐ × 4 × 25
= ☐

g 28 × 25
= ☐ × 4 × 25
= ☐

h 36 × 25
= ☐ × 4 × 25
= ☐

i 20 × 5
= ☐ × 2 × 5
= ☐

j 12 × 25
= ☐ × 4 × 25
= ☐

k 20 × 25
= ☐ × 4 × 25
= ☐

l 32 × 25
= ☐ × 4 × 25
= ☐

To multiply by 12, multiply by 6 and then double.

12 × 8
= 2 × 6 × 8
= 2 × 48
= 96

8 × 12
= 8 × 6 × 2
= 48 × 2
= 96

	4 (carry 1)	8
+	4	8
	9	6

CONCEPT

2

a 12 × 6
= 2 × 6 × 6
= 2 × ☐
= ☐

b 14 × 4
= 2 × ☐ × ☐
= 2 × ☐
= ☐

c 8 × 14
= 8 × 7 × ☐
= 56 × ☐
= ☐

d 6 × 16
= 6 × 8 × ☐
= ☐
= ☐

e 4 × 18
= 2 × 2 × 18
= 2 × ☐
= ☐

f 20 × 9
= 2 × 10 × 9
= 2 × ☐
= ☐

g 16 × 4
= 2 × ☐ × ☐
= ☐
= ☐

h 14 × 6
= 2 × ☐ × ☐
= ☐
= ☐

 • *AUSTRALIAN SIGNPOST MATHS 5* • ISBN 978 1 4886 2190 1

2:45 Mental Strategies for Multiplication

CONCEPT

Each of 13 students carried 9 bricks. How many bricks were carried altogether?

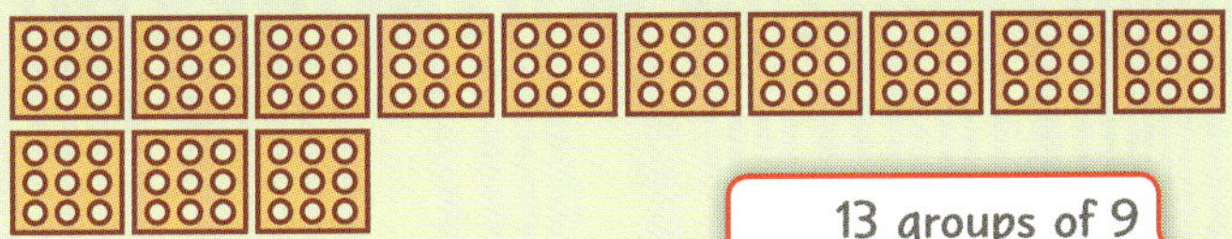

Here we use known facts to find the answer.

$13 \times 9 = (10 \times 9) + (3 \times 9)$
$= 90 + 27$
$= 117$

1 Use the mental strategy above to answer these questions.

a 14×8
$= (10 \times 8) + (4 \times 8)$
= ___ + ___
= ___

b 12×4
$= (10 \times 4) + (2 \times 4)$
= ___ + ___
= ___

c 15×6
$= (10 \times 6) + (5 \times 6)$
= ___ + ___
= ___

d 19×5
$= (10 \times 5) + (9 \times 5)$
= ___ + ___
= ___

e 17×9
$= (10 \times 9) + (7 \times 9)$
= ___ + ___
= ___

f 16×8
$= (10 \times 8) + (6 \times 8)$
= ___ + ___
= ___

4 groups of 16 is the same as 4 groups of 10 + 4 groups of 6.

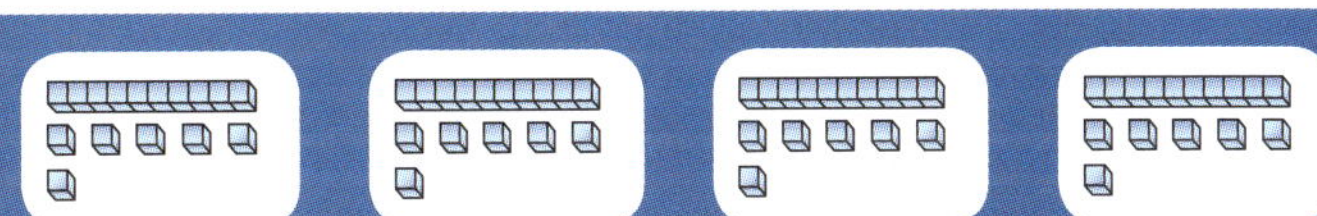

2 Here we change the second number.

a 4×16
$= (4 \times 10) + (4 \times 6)$
= ___ + ___
= ___

b 8×23
$= (8 \times 20) + (8 \times 3)$
= ___ + ___
= ___

c 5×43
$= (5 \times 40) + (5 \times 3)$
= ___ + ___
= ___

We can also use relationships between facts.
$23 \times 4 = 4 \times 23$, so double 23 and then double again.
(To double a number, we can add it to itself.)

$23 \times 4 = 2 \times (2 \times 23)$
$= 2 \times 46$
$= 92$

3 Use doubling to calculate these.

a $2 \times 23 =$ 46
$4 \times 23 = 2 \times$ 46
= ___

b $51 \times 2 =$ ___
$51 \times 4 = 2 \times$ ___
= ___

c $2 \times 21 =$ ___
$4 \times 21 = 2 \times$ ___
= ___

 • *AUSTRALIAN SIGNPOST MATHS 5* • ISBN 978 1 4886 2190 1

2:46 Number Patterns

The rule is: 'add 10 to the first number'.

Add 10 to the first number to find the second one.

First number	1	2	3	4	5	6
Second number	11	12	13	14	15	16

CONCEPT

1 **a** Complete the table.

First number	1	2	3	4	5	6
Second number	15	16	17	18		

b Write a rule to describe the pattern.

[ ] to the [1st] number.

c What would be the 7th term in the second number pattern? [ ]

2 **a** Complete the table.

First number	1	2	3	4	5	6
Second number	4	8	12	16		

b Write a rule to describe the pattern.

[ ] the [ ] number by [ ].

c What would be the 10th term in the second number pattern? [ ]

3 **a** Complete the table.

First number	90	80	70	60	50	40
Second number	82	72	62	52		

b Write a rule to describe the pattern.

[ ] from the [ ] number.

c What would be the 7th term in the second number pattern? [ ]

4 **a** Complete the table.

Input number	36	46	56	66	76	86
Output number	43	53	63	73		

b Write a rule to describe the pattern.

[ ] to the input number.

c What would be the 10th term in the output pattern? [ ]

5 **a** Complete the table.

Input number	29	39	49	59	69	79
Output number	12	22	32			

b Write a rule to describe the pattern.

[ ] from the input number.

c What would be the 10th term in the output pattern? [ ]

6 Make your own number patterns using multiplication and addition.

ACTIVITY

a Multiply the first number by [ ].

First number	21	22	23	24	25	26
Second number						

b Add [ ] to the first number.

First number	1	2	3	4	5	6
Second number						

 ISBN 978 1 4886 2190 1

2:47 Number Patterns

CONCEPT

First number	1	2	3	4	5	6
Second number	16	17	18	19	20	21

1 Complete the tables and write a rule for each number pattern.

a

First number	1	2	3	4	5	6
Second number	7	14	21	28		

Rule: ______ the first number by ______.

b

First number	1	2	3	4	5	6
Second number	9	18	27	36		

Rule: ______ the first number by ______.

c

First number	1	2	3	4	5	6
Second number	24	25	26			

Rule: ______ to the first number.

d

First number	1	2	3	4	5	6
Second number	36	37	38			

Rule: ______ to the first number.

e

First number	35	45	55	65	75	85
Second number	27	37	47	57		

Rule: ______

f

First number	92	82	72	62	52	42
Second number	104	94	84	74		

Rule: ______

g

First number	37	47	57	67	77	87
Second number	24	34	44			

Rule: ______

h

First number	78	68	58	48	38	28
Second number	53	43	33			

Rule: ______

ICT

2 **a** Enter the following into a calculator. Under each [=], write the answer shown.

[1] [0] [+] [+] [5] [=] [=] [=] [=] [=] [=] [=] [=]

b What is the difference between each pair of numbers in the number pattern? ______

c Write a rule for the number pattern. ______

2:48 Multiplying 2-Digit Numbers

CONCEPT

3 × 72
= 3 groups of 72
= (3 × 7 tens) + (3 × 2 ones)
= 210 + 6
= 216
There are 216 pegs.

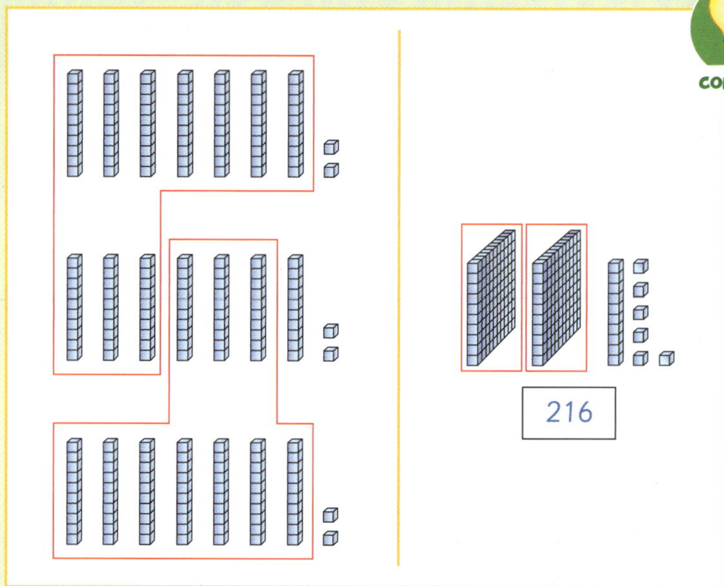

1 **a** 2 × 43 = (2 × 4 tens) + (2 × 3 ones)
= ___ + ___
= ___

b 5 × 15 = (5 × 1 ten) + (5 × 5 ones)
= ___ + ___
= ___

c 6 × 38 = (6 × 3 tens) + (6 × 8 ones)
= ___ + ___
= ___

d 4 × 29 = (4 × 2 tens) + (4 × 9 ones)
= ___ + ___
= ___

2 **a** 2 × 85
= (2 × 80) + (2 × 5)
= ___ + ___
= ___

b 4 × 28
= (4 × 20) + (4 × 8)
= ___ + ___
= ___

c 6 × 32
= (6 × 30) + (6 × 2)
= ___ + ___
= ___

d 7 × 16
= (7 × 10) + (7 × 6)
= ___ + ___
= ___

e 9 × 31
= (9 × 30) + (9 × 1)
= ___ + ___
= ___

f 8 × 23
= (8 × 20) + (8 × 3)
= ___ + ___
= ___

g 5 × 44
= (5 × 40) + (5 × 4)
= ___ + ___
= ___

h 3 × 82
= (3 × 80) + (3 × 2)
= ___ + ___
= ___

i 4 × 93
= (4 × 90) + (4 × 3)
= ___ + ___
= ___

2:49 Introducing Extended Multiplication

$6 \times 32 = (6 \times 30) + (6 \times 2)$

H	T	U
	3	0
×		6
1	8	0

+

H	T	U
		2
×		6
	1	2

=

H	T	U
1	8	0
+	1	2
1	9	2

CONCEPT

1. One tin holds 25 blocks. How much would six tins hold?

6×25 blocks
$= (6 \times 20) + (6 \times 5)$
= ____ + ____
= ____ blocks

H	T	U
	2	0
×		6

+

H	T	U
		5
×		6

=

H	T	U
+		

2. I bought 7 tickets for $38 each. How much did it cost?

7 × $38
$= (7 \times 30) + (7 \times 8)$
= ____ + ____
= ____ dollars

H	T	U
	3	0
×		7

+

H	T	U
		8
×		7

=

H	T	U
+		

3. There were 16 people in 9 boats. How many people were there?

9×16 people
$= (9 \times 10) + (9 \times 6)$
= ____ + ____
= ____ people

H	T	U
	1	0
×		9

+

H	T	U
		6
×		9

=

H	T	U
+		

4. 73 students were each given $8. How much was given altogether?

73 × $8
$= (70 \times 8) + (3 \times 8)$
= ____ + ____
= ____ dollars

H	T	U
	7	0
×		8

+

H	T	U
		3
×		8

=

H	T	U
+		

2:50 The Extended Form of Multiplication

CONCEPT

We have 8 tennis balls. Each has a mass of 57 g. What is their total mass?

$8 \times 57 = (8 \times 50) + (8 \times 7)$

	H	T	U	
		5	7	
×			8	
		5	6	← (8 × 7)
	4	0	0	← (8 × 50)
	4	5	6	

1

a

	H	T	U
		4	3
×			5

b

	H	T	U
		7	3
×			4

c

	H	T	U
		2	7
×			8

d

	H	T	U
		7	7
×			3

e

	H	T	U
		7	9
×			6

f

	H	T	U
		9	1
×			2

g

	H	T	U
		3	8
×			7

h

	H	T	U
		1	9
×			8

2

a

	H	T	U
		4	9
×			3

b

	H	T	U
		3	2
×			9

c

	H	T	U
		5	4
×			6

d

	H	T	U
		7	5
×			5

e

	H	T	U
		9	6
×			4

f

	H	T	U
		4	4
×			7

g

	H	T	U
		3	7
×			8

h

	H	T	U
		2	1
×			9

3 Use rounding to check your answers in Questions 1 and 2.

2:51 The Extended Form of Multiplication

Gabriela was given $175 but her mother was given 5 times as much. How much was her mother given?

	H	T	U	
	$ 1	7	5	
×			5	
		2	5	← (5 × 5)
	3	5	0	← (5 × 70)
	5	0	0	← (5 × 100)
	$ 8	7	5	

Her mother was given $875.

CONCEPT

1

a

	H	T	U
	4	2	1
×			2

b

	H	T	U
	3	1	9
×			3

c

	H	T	U
	1	2	9
×			7

d

	H	T	U
	3	9	5
×			2

e

	H	T	U
	1	3	4
×			6

f

	H	T	U
	1	1	5
×			8

g

	H	T	U
	2	1	7
×			4

h

	H	T	U
	1	7	5
×			5

i

	H	T	U
	2	3	6
×			4

j

	H	T	U
	3	0	8
×			3

k

	H	T	U
	1	4	1
×			7

l

	H	T	U
	1	7	9
×			7

2

a Jordan needed to buy new horseshoes for her 48 horses. How many shoes must she buy if she needs one for each hoof? ☐ horseshoes

b Alana bought six 128-page books. How many pages did she buy? ☐ pages

c Rhonda carried four boxes of eggs to the car. In each box there were 144 eggs. How many eggs did she carry? ☐ eggs

d A father gave his younger son three times as much money as his older son. If he gave his older son $297, how much did he give the younger? ☐ dollars

3 Check your answers to Question 2 by using rounding.

 • *AUSTRALIAN SIGNPOST MATHS 5* • ISBN 978 1 4886 2190 1

2:52 Estimating by Rounding

Round to the nearest 100 to estimate 339 + 850.

Estimate = 300 + 900
= 1 200

The answer is 1189 (by calculator).

An **estimate** should be near the real answer. It is a good guess.

CONCEPT

1. Round each number to the nearest 10 then work out the estimate.
   - a 157 + 31
     E = 160 + 30
     E = ______
   - b 84 – 77
     E = 80 – 80
     E = ______
   - c 63 + 88
     E = ______ + ______
     E = ______
   - d 174 – 52
     E = ______ – ______
     E = ______

2. Use the jump strategy or a calculator to write the answers to Question 1.
   a ______ b ______ c ______ d ______

3. Round each number to the nearest 100 then work out the estimate.
   - a 325 + 189
     E = 300 + 200
     = ______
   - b 608 – 518
     E = ______ – ______
     = ______
   - c 751 + 427
     E = ______ + ______
     = ______
   - d 846 – 749
     E = ______ – ______
     = ______

4. Use the jump strategy or a calculator to write the answers to Question 3.
   a ______ b ______ c ______ d ______

5. Round each number to the nearest 1 000 then work out the estimate.
   - a 3 804 + 621
     E = 4 000 + 1 000
     = ______
   - b 7 299 – 4 106
     E = ______ – ______
     = ______
   - c 6 337 + 3 904
     E = ______ + ______
     = ______

6. Use the jump strategy or a calculator to write the answers to Question 5.
   a ______ b ______ c ______

7. 693 2 199 4 904 7 057 9 251

   ACTIVITY

   Estimate which **two numbers** above should be added to give the answer closest to:
   - a 3 000 ______
   - b 7 000 ______
   - c 6 000 ______
   - d 11 000 ______
   - e 12 000 ______
   - f 16 000 ______
   - g Was it necessary to add all of the pairs of numbers? ______

© PEARSON AUSTRALIA 2018 • *AUSTRALIAN SIGNPOST MATHS 5* • ISBN 978 1 4886 2190 1

2:53 Estimating Products

Estimate 76 × 8.

76 × 8 < 76 × 10 but 76 × 8 > 76 × 5,
so 76 × 8 is between 380 and 760.
(76 × 5 is half of 76 × 10.)

76 × 8 is between 76 × 5 and 76 × 10.

CONCEPT

1 Circle the correct answers in each case.

a	47 × 7 is less than:	47 × 10	47 × 5	470	235
	47 × 7 is greater than:	47 × 10	47 × 5	470	235
b	58 × 8 is less than:	58 × 10	58 × 5	290	580
	58 × 8 is greater than:	58 × 10	58 × 5	290	580

2 Use the method in the concept box to find an upper and lower estimate.

a Estimate 73 × 9. 73 × 9 < ____ but > ____,
so 73 × 9 is between ____ and ____.

b Estimate 62 × 6. ____ < ____ but > ____,
so 62 × 6 is between ____ and ____.

c Estimate 95 × 7. ____ < ____ but > ____,
so 95 × 7 is between ____ and ____.

d Estimate 81 × 8. ____ < ____ but > ____,
so 81 × 8 is between ____ and ____.

e Estimate 79 × 9 ____ < ____ but > ____,
so 79 × 9 is between ____ and ____.

f Estimate 86 × 6 ____ < ____ but > ____,
so 86 × 6 is between ____ and ____.

To find a ×10 answer, add a zero.

To find a × 5 answer, halve the × 10 answer.

Numbers in History

ACTIVITY

Roman numerals are used in some books and on clocks.

- LX means 50 and 10.
- XL means 10 less than 50.

Number	1	5	10	50	100	500	1 000
Symbol	I	V	X	L	C	D	M

Examples: CCCXX = 320 CD = 400 DCC = 700 CM = 900 MMM = 3 000 MDCXCI = 1 691

3 Write our numeral for:

a CCCXL ____ **b** CDLXIV ____ **c** DCLXXV ____ **d** MMCDIX ____

4 Write the Roman numeral for:

a 346 ____ **b** 862 ____ **c** 928 ____ **d** 1 263 ____

2:54 The Contracted Form of Multiplication

CONCEPT

Flood lights cost $142 each. I bought seven. How much did I pay?

Extended form

1	4	2
×		7
	1	4
2	8	0
7	0	0
9	9	4

Contracted form

H	T	U
1	4	2
× 2	1	7
9	9	4

What is traded goes here.

- 7 × 2 ones is 14 ones. Write 4 in the units column and put a small 1 in the tens column.
- 7 × 4 tens plus the 1 traded is 29 tens. Write 9 in the tens column and a small 2 in the hundreds column.
- 7 × 1 hundred plus the 2 traded is 9 hundreds. Write the 9 in the hundreds column.

Use the contracted form to answer these and check your answers with a calculator.

$\begin{array}{r} 19 \\ \times\ {}_3 4 \\ \hline 76 \end{array}$

1
a 20 × 8 b 31 × 7 c 83 × 3
d 39 × 2 e 18 × 5 f 27 × 3
g 63 × 5 h 57 × 6 i 46 × 10 j 75 × 8 k 57 × 4
l 81 × 10 m 40 × 3 n 79 × 2 o 14 × 4 p 98 × 6

2
a 128 × 3 b 346 × 2 c 117 × 5 d 209 × 4 e 325 × 3
f 130 × 6 g 121 × 8 h 293 × 3 i 484 × 2 j 170 × 4
k $1.98 × 5 l $1.27 × 7 m $1.53 × 6 n $1.06 × 8 o $1.35 × 7

© PEARSON AUSTRALIA 2018 • *AUSTRALIAN SIGNPOST MATHS 5* • ISBN 978 1 4886 2190 1

2:55 The Contracted Form of Multiplication

CONCEPT

	Th	H	T	U
		3	5	2
×		2		4
	1	4	0	8

Use the contracted form to answer these.

1

a 630 × 3 **b** 811 × 6 **c** 700 × 8 **d** 824 × 2 **e** 933 × 3

f 583 × 2 **g** 419 × 4 **h** 407 × 9 **i** 312 × 8 **j** 971 × 6

k 546 × 5 **l** 768 × 10 **m** 369 × 7 **n** 257 × 9 **o** 498 × 10

p 609 × 7 **q** 850 × 9 **r** 999 × 6

2

a 3021 × 3 **b** 2112 × 4 **c** 3033 × 3

d 1602 × 4 **e** 501 × 7 **f** 1810 × 5 **g** 2186 × 3

h 4974 × 2 **i** 2324 × 4 **j** 1650 × 6 **k** 3258 × 2

l $13.85 × 7 **m** $17.99 × 5 **n** $12.47 × 6 **o** $11.66 × 8 **p** $41.35 × 10

2513
× (1) 3
7539

2:56 Using Algorithms to Solve Problems

Deaths from Traffic Accidents, 2003							
Type of road user	NSW	Vic	Qld	SA	WA	Tas	Total
Drivers of motor vehicles	284	188	138	71	78	16	
Motor cyclists	55	56	53	22	23	9	
Pedal cyclists	14	7	5	3	6	0	
Passengers	124	89	88	40	48	4	
Pedestrians	95	58	37	18	24	6	
Other	–	–	–	–	1	–	
Total							

Source: Australian Transport Safety Bureau

1 Use the table above to answer the following.
  a Fill in all the totals.
  b Which state had the smallest number of deaths?
  c Which state had the most deaths?
  d What was the total number of passengers killed in the six states?
  e What was the total number of pedal cyclists killed in the six states?

2 What was the difference between the total number of road deaths in New South Wales and Tasmania?

3 In Victoria, how many more drivers of motor vehicles died than passengers?

4 What was the difference between the total number of motor cyclists and pedal cyclists who died?

5 What was the total number of cyclists who died from traffic accidents in New South Wales?

6 Which two states had a combined total number of deaths of 501?

7 Which two states had a difference in the total number of deaths of 119?

8 How can you check that your totals in the table are correct?

ACTIVITY

- Collect newspaper reports of road accidents.
- Investigate safety measures which could be taken to reduce the number of traffic deaths. Discuss at least two safety measures for:
  - drivers
  - motor cyclists
  - pedal cyclists
  - passengers
  - pedestrians

2:57 Problems Involving Change of Units

CONCEPT

Anna bought 2 m of wire and used 135 cm of it to repair her aviary.
How much wire was left?
We must first change 2 m to cm.
2 m – 135 cm → 200 cm – 135 cm

		9	
	1	~~10~~	10
	~~2~~	~~0~~	~~0~~
–	1	3	5
		6	5

Length	Mass	Capacity	Time
1 m = 100 cm	1 kg = 1 000 g	1 L = 1 000 mL	1 min = 60 s
1 km = 1 000 m		1 kL = 1 000 L	1 h = 60 min
			1 day = 24 h

kilo (k) means 1000

1 a I bought 2 kg of seed and mice ate all but 750 g. How much was eaten?

b A jug holds 750 mL of milk. How many litres are needed to fill the jug four times?

c Eight children shared 2 kg of clay. How much clay was each child given?

d A movie on TV runs for 2 hours. Every 10 minutes, six advertisements are shown. How many ads are shown altogether?

2 a Nasha needed rice to prepare three meals. She needed 830 g for the first, 2 480 g for the second and 1 320 g for the third meal. How much rice must she buy if it is sold only in one kilogram packets?

b Every minute 875 L of water flows into our dam. How many litres would flow into it in 8 minutes?

c How many 200 g packets of rice can we make if we have a 1 kg bag of rice?

How many 200 g packets of rice can we make if we have a 1 850 kg sack of rice?

d Lee had four lengths of wire: 65 cm, 1 m 20 cm, 86 cm and 3 m 39 cm. How many metres of wire did she have altogether?

3 a The distance around a circular island is 5 km, while the distance across the middle is 1 600 m. Robinson Crusoe and Man Friday wanted to get from one side (**A**), to the opposite (**B**). If Robinson Crusoe walked around the island and Man Friday walked across the middle, how much further would Robinson Crusoe walk?

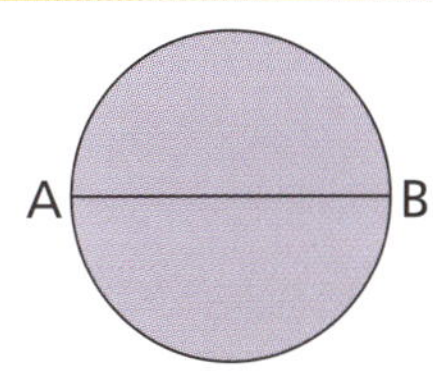

b Indu had to wait 7 days to have a leaking pipe repaired. How much water was lost if 6 L of water was lost each hour?

2:58 Estimation by Rounding

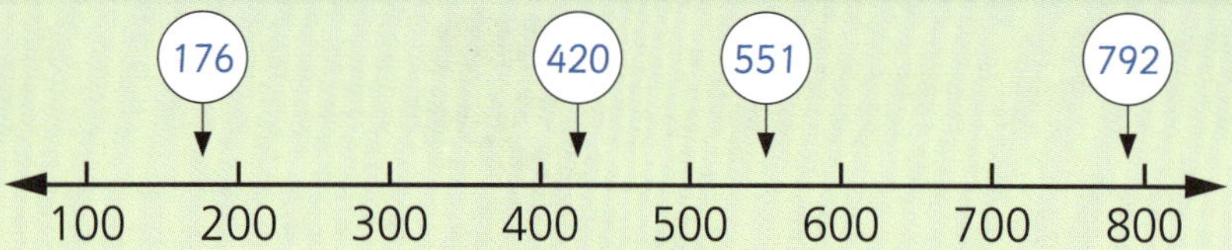

To round a number to the nearest hundred, write the 100 closest to it.

Estimating is useful for checking answers or for obtaining approximate answers.

Digits 5 and above are rounded up.

CONCEPT

1 Round these numbers to the nearest hundred.

a 297 ☐ **b** 306 ☐ **c** 541 ☐ **d** 183 ☐ **e** 748 ☐

f 418 ☐ **g** 853 ☐ **h** 660 ☐ **i** 928 ☐ **j** 573 ☐

2 **a** Is 48 × 6 closer to 50 × 5, 50 × 10 or 40 × 5? ☐

b Is 94 × 7 closer to 90 × 5, 90 × 10 or 100 × 10? ☐

c Is 351 × 8 closer to 400 × 5, 400 × 10 or 300 × 10? ☐

3 Circle the best answer from those given. Use your calculator to find the actual answer and compare this with your estimate.

	Question	Estimate
a	35 + 74 + 68 + 27	190, 210, 250
b	579 – 228	300, 400, 500
c	832 – 196	500, 600, 700
d	95 + 127 + 261 + 301	750, 790, 860

	Question	Estimate
e	28 × 6	100, 180, 210
f	156 ÷ 5	30, 50, 90
g	41 × 9	300, 400
h	401 ÷ 10	40, 60, 80

4 **a** Estimate the cost of 12 meals if they cost $9.50 each. ☐

b Tennis lessons cost $9.80 each. Estimate how much it costs for 21 lessons. ☐

c Estimate the cost of 33 litres of petrol at $1.47 per litre. ☐

INVESTIGATION

Estimate the average price per item at a grocery store. ☐

Estimate the number of grocery items needed to fill a shopping trolley. ☐

Circle the estimate that is closest to yours.

$150–$200	$200–$250	$250–$300	$300–$350	$350–$400

2:59 Estimating Products

78 × 34

Estimate
80 × 30
= 2 400

CONCEPT

1 Round each of these to the nearest ten.

a 27 ☐ b 49 ☐ c 83 ☐ d 47 ☐ e 25 ☐

f 91 ☐ g 65 ☐ h 48 ☐ i 71 ☐ j 62 ☐

2 Round each of these to the nearest hundred.

a 680 ☐ b 340 ☐ c 295 ☐ d 864 ☐ e 922 ☐

f 505 ☐ g 526 ☐ h 716 ☐ i 333 ☐ j 367 ☐

3 Find an estimate for each product by rounding each number first.

a 82 × 11
E = 80 × 10
= ☐

b 63 × 27
E = ☐ × ☐
= ☐

c 54 × 65
E = ☐ × ☐
= ☐

d 91 × 33
E = ☐ × ☐
= ☐

e 38 × 12
E = ☐ × ☐
= ☐

f 72 × 46
E = ☐ × ☐
= ☐

g 47 × 19
E = ☐ × ☐
= ☐

h 63 × 43
E = ☐ × ☐
= ☐

The **product** is the answer when you multiply.

Ask: Is the answer reasonable?

4 Round each number to its first digit and multiply to find an estimate.

a 43 × 631
E = 40 × 600
= ☐

b 36 × 194
E = ☐ × ☐
= ☐

c 82 × 666
E = ☐ × ☐
= ☐

d 58 × 749
E = ☐ × ☐
= ☐

43 × 461
Estimate
= 40 × 500
= 20 000

5 **True** or **false**?

a Approximating 27 × 7 with 30 × 10 will result in an estimate that is **larger** than the actual result. ☐

b Approximating 23 × 5 with 20 × 5 will result in an estimate that is **smaller** than the actual result. ☐

6 Estimate the answer, then check your estimate with a calculator. (E = estimate; A = answer)

ICT

a 94 × 37
E = ☐
A = ☐

b 65 × 31
E = ☐
A = ☐

c 57 × 347
E = ☐
A = ☐

d 625 × 47
E = ☐
A = ☐

 • *AUSTRALIAN SIGNPOST MATHS 5* • ISBN 978 1 4886 2190 1

2:60 Making a Budget

A budget is a way of working out how much a plan will cost. Work out what you need and the cost of each item. Add them to find the total cost.

"inc. GST" means "including Goods and Services Tax". This money, usually 10% of the cost, goes to the government.

CONCEPT

1. You plan to set up a class fish tank.
Decide what you want to buy.
Complete the table to find your budget.

Item	Number	Cost
tank		
filter		
stones		
food		
fish		
Total budget		

What you need:

tank	\$35 inc GST
filter	\$16.30 inc GST
stones	\$7.20 + GST = \$7.92
fish food	\$10.80 inc GST
fish	\$3.50 each inc GST

GST (10%)
\$7.20
+ .72
\$7.92

Optional extras:

cave	\$10.20 inc GST
plant	\$3.40 each inc GST

2. You plan to organise a class party.
Decide how much you want to buy.
Complete the table to find your budget.
Consider the number of people coming.

Item	Number	Cost
Total budget		

Food:

softdrink	\$3 per litre inc GST
juice	\$3.50 per litre inc GST
chips	\$4 a packet inc GST
donuts	\$6 a dozen + GST = \$ ____
lollies	\$4.30 a packet inc GST
popcorn	\$5.25 a packet inc GST

GST (10%)
\$6.00
+ .60
____

Equipment:

serviettes	\$2 for 20 inc GST
plates	\$4.40 for 40 inc GST
party hats	\$2 for 10 inc GST

Other items:

 ISBN 978 1 4886 2190 1

2:61 Shopping

Hannah and Yuri are working out their household accounts. At the end of each week any money left over is banked. Each person's payslip, expenses and bank balance are shown.

Hannah's Accounts

PAYSLIP No. 2364187

...$94.70

Charity $3.00

Guitar lessons $7.40

Clothes $22.00

Hannah Smith

Deposit	Withdraw	Balance
		$74.20
	$2.60	$71.60

1 **a** Estimate of Hannah's expenses: $ ______ **b** Hannah's actual expenses: $ ______

c At the end of the week, what must she deposit or withdraw from her bank account? $ ______

d The new bank balance is $ ______.

Yuri's Accounts

PAYSLIP No. 304295

...$88.20

Bus fares $7.60

Basketball fees $9.00

Yuri Stamos

Deposit	Withdraw	Balance
C.F		$66.90
$3.20		$70.10

2 **a** Estimate of Yuri's expenses: $ ______ **b** Yuri's actual expenses: $ ______

c At the end of the week, what must Yuri deposit or withdraw from his bank account? ______

d The new bank balance is $ ______.

 • *AUSTRALIAN SIGNPOST MATHS 5* • ISBN 978 1 4886 2190 1

2:62 Using Operations to Solve Problems

1. Ten people booked on a trip missed the bus. As they had already paid, they were given a $2330 refund to share. How much did each receive? ☐ dollars

2. Caroline is paid $153 per week. From this amount $19 is deducted for tax and other items. How much money does she receive in four weeks? ☐ dollars

3. The cost of hiring a hall for a Boys' Brigade display was shared by six leaders. If the cost was $774, how much did each leader pay? ☐ dollars

4. A truck has to move 400 tonnes of soil. On each trip it carries a load of 18 tonnes. It makes 8 trips before it breaks down. How much soil was moved before it broke down? ☐ tonnes

   How much soil was not moved? ☐ tonnes

5. Charity workers expected to raise $1 000 from three functions. The first raised $271, the second $406 and the third $590. How much more money did they raise than was expected? ☐ dollars

6. A person laying bricks is paid to lay 960 bricks. Each hour 185 bricks are laid. How many bricks will be laid in 3 hours? ☐ bricks

7. A DVD shop sold 576 DVDs in 8 hours. On average, how many DVDs were sold each hour? ☐ DVDs

8. I can travel 9 km on a litre of fuel. How much would I use to travel 72 km? ☐ litres

   What would be the cost of that fuel if I pay 160 cents per litre for it? ☐ dollars

9. After travelling for $6\frac{1}{2}$ hours, I had travelled 495 km of the 875 km to my home. If the trip takes 12 hours, how far did I still have to travel? ☐ km

   How many more hours should it take me? ☐ hours

10. A farmer sold some cattle for $2 360. He paid an agent $236 and a carrier $179. How much did he have left from the sale? ☐ dollars

The unit pattern for multiples of 2 is 2, 4, 6, 8, 0; 2, 4, 6, 8, 0.

11. Draw the unit digit patterns for the numbers below.

a multiples of 4

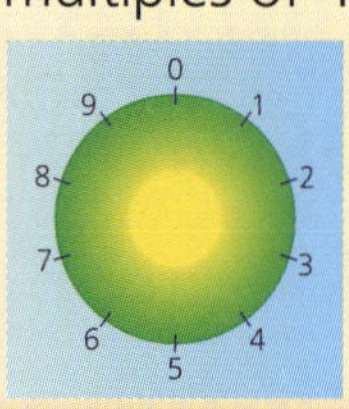

b multiples of 3

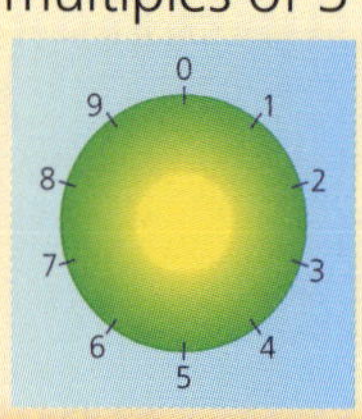

c multiples of 7

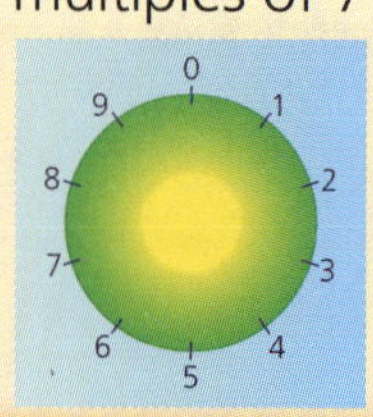

ACTIVITY

 • *AUSTRALIAN SIGNPOST MATHS 5* • ISBN 978 1 4886 2190 1

2:63 Strategies for Multiplication

The Area Model for 25 × 28

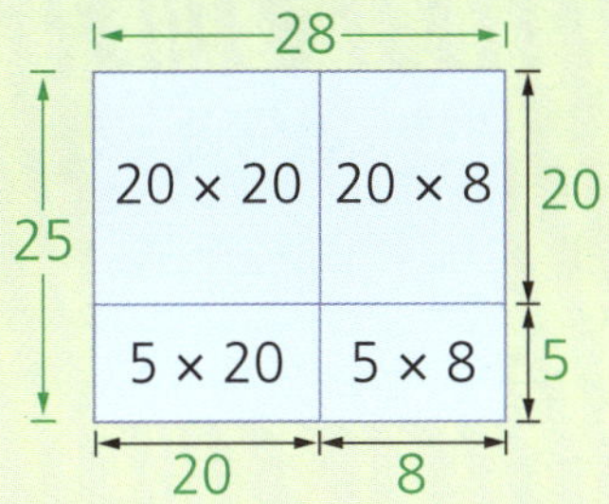

Break the whole into parts. Find each area, then add.

$25 \times 28 = (20 \times 20) + (20 \times 8) + (5 \times 20) + (5 \times 8)$
$= 400 + 160 + 100 + 40$
$= 700$

The Italian Lattice Method for 25 × 28

CONCEPT

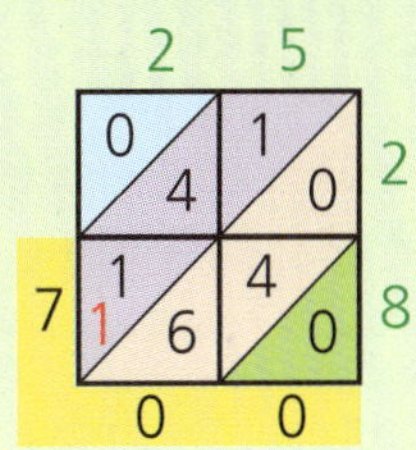

The answer is 700.

Multiply each column by each row. Write the tens and ones in separate triangles. Add digits diagonally to find your answer. Trade any tens to the sloping section to the left.

1 Use the area model to find:

a 34 × 57

= ( ) + ( ) + ( ) + ( )
= + + +
=

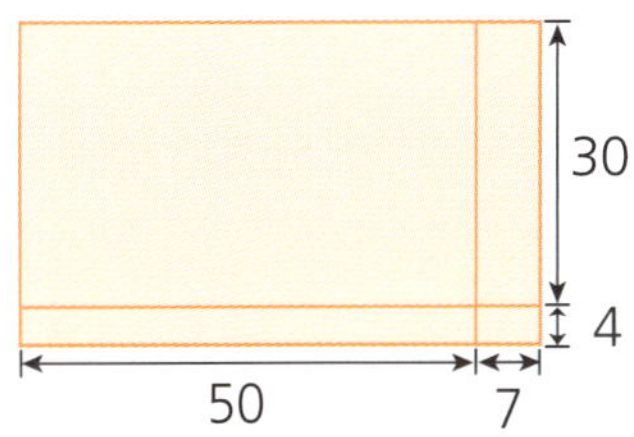

b 62 × 28

= ( ) + ( ) + ( ) + ( )
= + + +
=

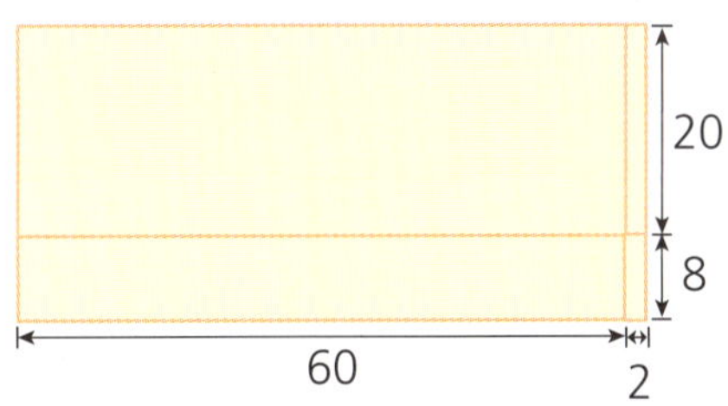

c 57 × 16

= ( ) + ( ) + ( ) + ( )
= + + +
=

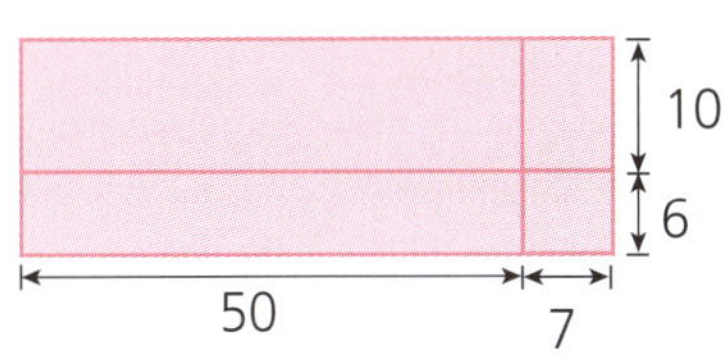

2 Use the Italian lattice method to find:

a 61 × 36 =

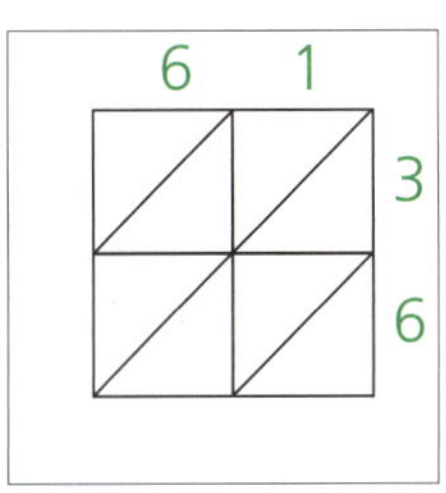

b 62 × 28 =

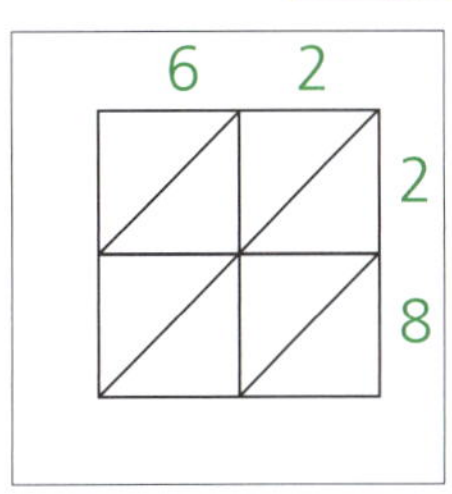

c 18 × 73 =

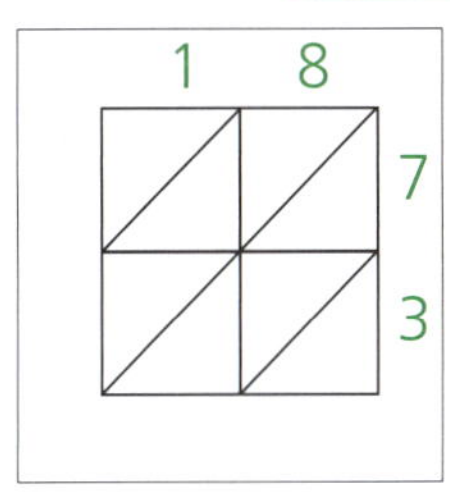

 • *AUSTRALIAN SIGNPOST MATHS 5* • ISBN 978 1 4886 2190 1

2:64 Finding Missing Numbers

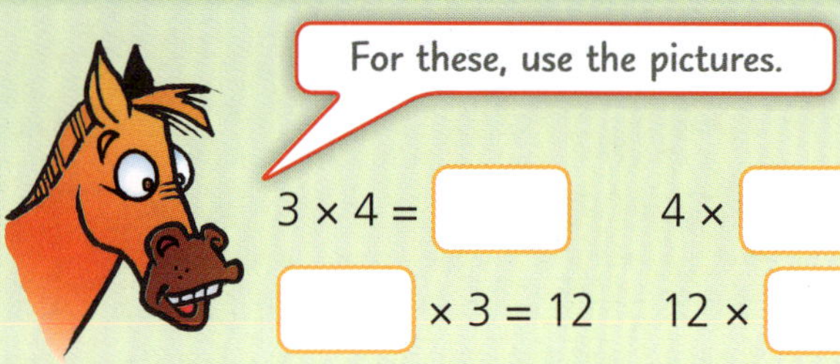

3 × 4 = ☐ 4 × ☐ = 12 12 ÷ ☐ = 3

☐ × 3 = 12 12 × ☐ = 4 12 – ☐ = 3

12 = 7 + ☐

CONCEPT

1 Find the missing numbers. Use a calculator to check your answers.

a ☐ × 8 = 24 **b** ☐ × 7 = 42 **c** ☐ × 6 = 36

d 1 × ☐ = 87 **e** 11 × ☐ = 99 **f** 12 × ☐ = 72

g 9 × ☐ = 63 **h** 6 × ☐ = 48 **i** 5 × ☐ = 55

2 **a** Michael had 93 coins in a bag. He drew out 15. How many were left? ☐

b There are 45 jars in 9 equal rows. How many are in each row? ☐

c 1 000 soldiers stood in 8 equal columns. How many were in one column? ☐

d In the game of Pludo there are 28 counters in each of four colours. After dropping them all in the car, I found 86. How many were still missing? ☐

Find the missing numbers.

3 **a** ☐ + 37 = 40 **b** ☐ + 63 = 100 **c** ☐ + 70 = 180

d ☐ + 140 = 360 **e** ☐ + 156 = 365 **f** ☐ + 200 = 471

g 311 + ☐ = 490 **h** 517 + ☐ = 995 **i** 840 + ☐ = 965

4 **a** ☐ – 8 = 7 **b** ☐ – 9 = 10 **c** ☐ – 6 = 5

d 16 – ☐ = 7 **e** 9 – ☐ = 4 **f** 15 – ☐ = 7

g 90 – ☐ = 30 **h** 180 – ☐ = 60 **i** 100 – ☐ = 31

5 **a** ☐ ÷ 6 = 5 **b** ☐ ÷ 4 = 8 **c** ☐ ÷ 7 = 6

d ☐ ÷ 9 = 9 **e** ☐ ÷ 3 = 36 **f** ☐ ÷ 2 = 45

g 100 ÷ ☐ – 20 **h** 60 ÷ ☐ = 10 **i** 54 : ☐ = 6

6 **a** ☐ = 436 ÷ 4 **b** ☐ + 14 = 36 **c** 17 = ☐ – 14

d 56 = ☐ – 37 **e** 9 × ☐ = 72 **f** 8 = 32 ÷ ☐

g ☐ = 50 × 7 **h** ☐ – 27 = 63 **i** 110 = ☐ + 36

 • *AUSTRALIAN SIGNPOST MATHS 5* • ISBN 978 1 4886 2190 1

2:65 Finding Missing Numbers

Write the missing number in each sentence.

1 a $\square + 8 = 10$ b $\square + 3 = 7$ c $\square + 8 = 15$

d $9 + \square = 12$ e $8 + \square = 17$ f $10 + \square = 10$

g $\square + 11 = 15$ h $\square + 9 = 14$ i $\square + 12 = 12$

j $14 + \square = 20$ k $18 + \square = 23$ l $17 + \square = 19$

m $\square + 9 = 20$ n $\square + 15 = 25$ o $\square + 11 = 40$

2 a $8 - \square = 1$ b $9 - \square = 0$ c $15 - \square = 10$ d $10 - \square = 10$

e $\square - 3 = 7$ f $\square - 10 = 10$ g $\square - 7 = 8$ h $\square - 5 = 4$

i $11 - \square = 4$ j $11 - \square = 7$ k $14 - \square = 6$ l $14 - \square = 8$

m $\square - 8 = 4$ n $\square - 3 = 9$ o $\square - 1 = 19$ p $\square - 11 = 20$

3 a $\square \times 6 = 18$ b $\square \times 5 = 35$ c $\square \times 4 = 24$ d $\square \times 3 = 27$

e $4 \times \square = 20$ f $7 \times \square = 21$ g $9 \times \square = 90$ h $8 \times \square = 40$

i $\square \times 10 = 80$ j $\square \times 6 = 36$ k $\square \times 4 = 28$ l $\square \times 5 = 35$

m $7 \times \square = 56$ n $8 \times \square = 72$ o $5 \times \square = 45$ p $9 \times \square = 63$

4 a $\square \div 2 = 8$ b $\square \div 4 = 3$ c $\square \div 5 = 8$

d $24 \div \square = 3$ e $24 \div \square = 4$ f $24 \div \square = 8$

g $\square$ squared = 9 h $\square$ squared = 64

i $\square$ squared = 100 j $\square$ squared = 49

To square a number, multiply it by itself.

5 a $6 + 7 + \square = 15$ b $3 - 1 + \square = 12$ c $8 + 7 + \square = 20$

d $4 \times 2 + \square = 10$ e $6 \times 5 - \square = 26$ f $21 \div 3 + \square = 13$

g $2 \times \square + 1 = 9$ h $3 \times \square + 2 = 17$ i $2 \times \square + 1 = 7$

6 a $5 \times (4 + 6) = (5 \times \square) + (5 \times 6)$ b $3 \times (9 + \square) = (3 \times 9) + (3 \times 2)$

c $7 \times (3 + 4) = (7 \times 3) + (7 \times \square)$ d $4 \times (\square + 2) = (4 \times 8) + (4 \times 2)$

2:66 Using Strategies to Solve Problems

How many faces would be showing if 22 cubes were placed on a table, end to end in a straight line?

Solution and reasoning

- Consider how many faces would show using 4 cubes. The two end cubes each have 4 faces showing. The two inside cubes each have 3 faces showing.
- Consider how many faces would show using 22 cubes. Each end cube would show 4 faces. Each inside cube would show 3 faces. The number of faces showing using 22 cubes: (2 × 4) + (20 × 3) = 68.

Altogether, 68 faces would be showing.

CONCEPT

1 Matches are used to form squares end to end as shown.

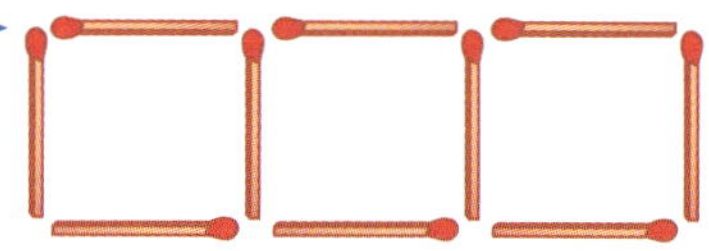

a How many matches would be needed to form 70 squares? ☐

b How many matches would be used in the perimeter of this line of 70 squares? ☐ (Hint: Consider what is needed for each additional square.)

2 Matches are used to form triangles in a line as shown.

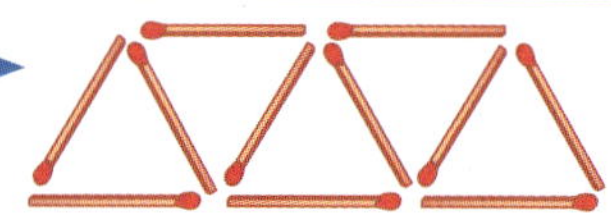

a How many would be needed to form 45 triangles? ☐

b How many would be used in the perimeter of this line of 45 triangles? ☐

3 Children's blocks (triangular prisms) are placed on a desk as shown.

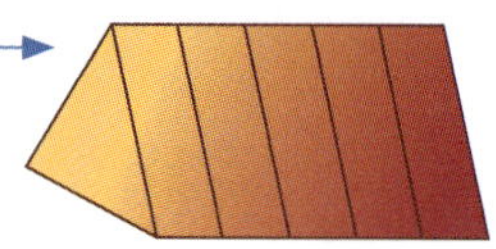

How many faces would be visible if there are:

a 5 blocks? ☐ b 50 blocks? ☐ c 132 blocks? ☐

4 Hexagonal paving bricks are stacked as shown.

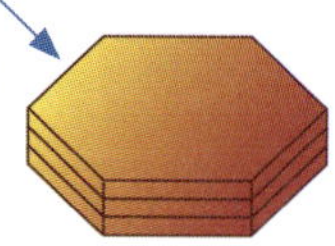

How many faces would be visible if we stacked:

a 3 bricks? ☐ b 20 bricks? ☐

c 37 bricks? ☐

How many 2-digit numbers can be made using the digits 2, 3, 4 and 5 if each digit is used only once in any one number?

Solution and reasoning

- List numbers starting with 2, 3, 4 then 5.

23	32	42	52
24	34	43	53
25	35	45	54

- No digit has been used twice in the same number.

Twelve 2-digit numbers can be made.

CONCEPT

5 How many 2-digit numbers can be made using the digits:

a 5, 6, 7, 8 and 9 if each digit is used only once in any one number? ☐

b 0, 1, 2 and 3 if each digit is used only once in any one number? ☐

c 0, 1, 2, 3 and 4 if each digit can be used more than once in any one number? ☐

2:67 Problem Solving

CONCEPT

Three different coloured cubes are used to make a tower. How many differently coloured towers is it possible to make?

Solution

These are the towers that can be built.

Answer

There are 6 different coloured towers.

1. Two different coloured cubes are used to make a tower. How many towers is it possible to make? ☐

2. Three different coloured cubes are used to make a tower which is two blocks high. How many towers is it possible to make? ☐

3. Four different coloured cubes are used to make a tower which is three blocks high. How many towers is it possible to make? ☐

4. Using only the coins shown above, list all the different ways you can make:
   - **a** 25 cents ☐
   - **b** 40 cents ☐
   - **c** 65 cents ☐

5. Using only the coins shown above, what is the least number of coins needed to make:
   a 35 cents? ☐ **b** 60 cents? ☐ **c** 75 cents? ☐ **d** $1.25? ☐

FUN SPOT

6. A bee wants to fly without stopping over these 9 flowers.

   Show how it can do this by flying in only 4 straight lines.

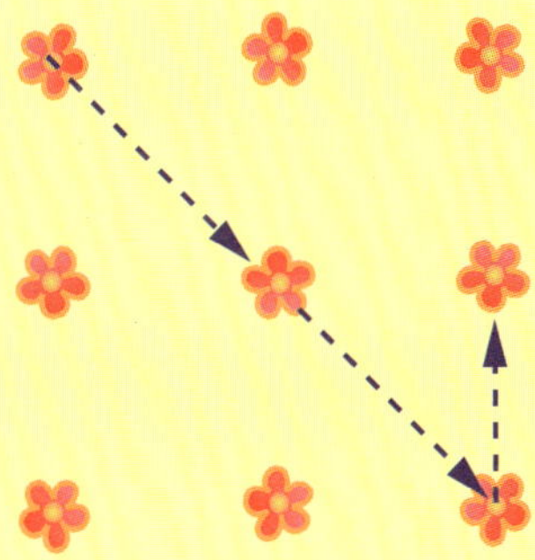

2:68 Number Machines

CONCEPT

Number Machines

A number goes in on the left.
The **Rule** is used and the answer comes out on the right.

If 4 goes in, 4 × 3 + 10 = 22, so 22 comes out.

In	4	5	6	7	8	9
Out	22	25	28	31	34	37

We can say that **Out = In × 3 + 10**

1 Use the rule for each number machine to complete the table. Rewrite the rule underneath.

a

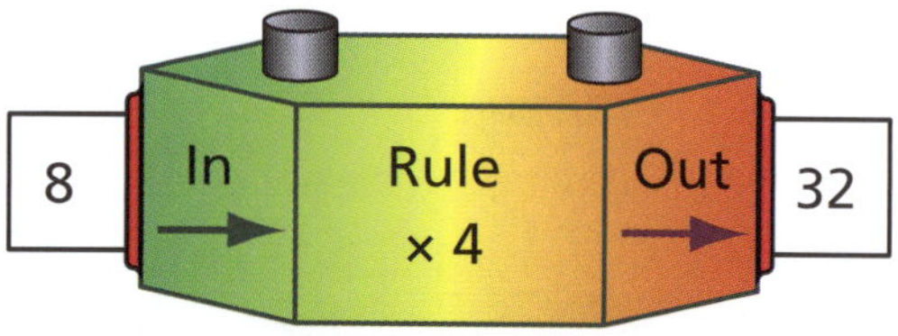

In	1	2	3	4	5	6
Out						

Rule: Out =

b

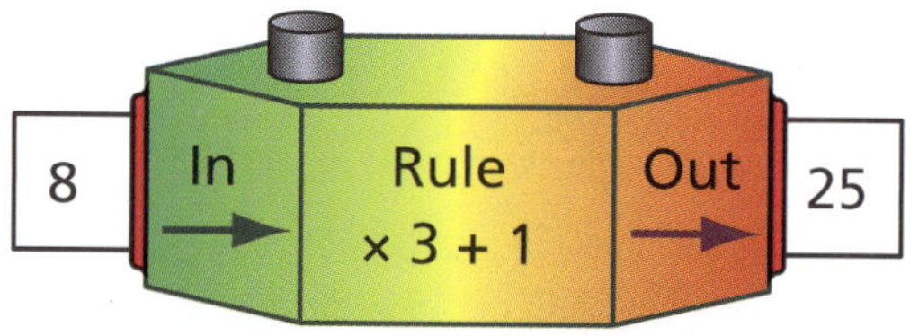

In	1	2	3	4	5	6
Out						

Rule: Out =

c

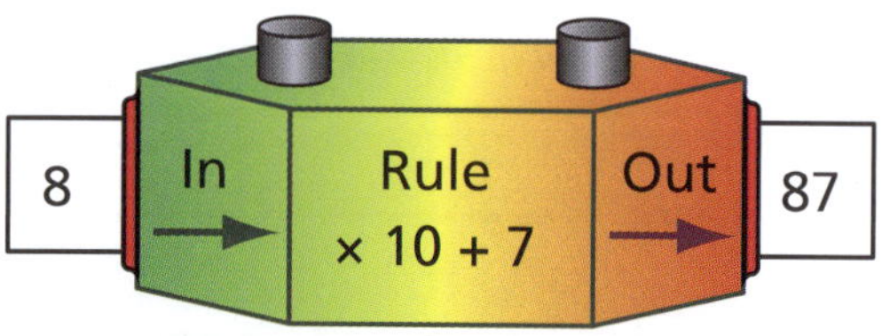

In	1	2	3	4	5	6
Out						

Rule: Out =

d

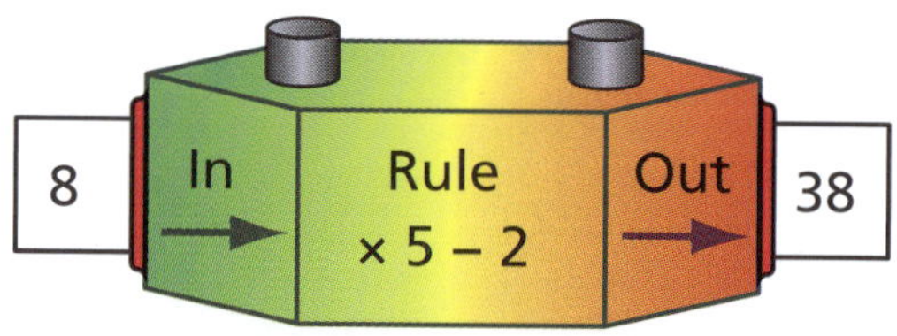

In	1	2	3	4	5	6
Out						

Rule: Out =

e

In	1	2	3	4	5	6
Out						

Rule: Out =

f

In	1	2	3	4	5	6
Out						

Rule: Out =

3:01 Time Units

This year is: ______

A millennium is 1 000 years.

We live in the 3rd millennium.

A century is 100 years.

We live in the 21st century.

CONCEPT

A decade is 10 years.
A year has 12 months.
A year is 52 weeks.
A fortnight is 2 weeks.
A week is 7 days.
A day is 24 hours.
An hour is 60 minutes.
A minute is 60 seconds.

1 Complete the following.

a 3 weeks = ☐ days	**b** 2 days = ☐ hours	**c** 7 minutes = ☐ seconds
d 180 minutes = ☐ hours	**e** 4 hours = ☐ minutes	**f** 6 fortnights = ☐ weeks
g 4 years = ☐ months	**h** 50 years = ☐ decades	**i** 120 minutes = ☐ hours
j 56 days = ☐ weeks	**k** 700 days = ☐ weeks	**l** 28 weeks = ☐ fortnights
m 84 days = ☐ fortnights	**n** 3 years = ☐ weeks	**o** 600 seconds = ☐ minutes

2 How much time has passed from:

a 8 am to 8:35 am? ☐

b 4:10 pm to 4:36 pm? ☐

c 5:13 am to 6:19 am? ☐

d 9 am to 10:49 am? ☐

e 7:08 pm to 9:52 pm? ☐

f 6:30 pm to 8:36 pm? ☐

3 If it is 8:17 am, how long will Graham have to wait until:

a school starts at 9 am? ☐

b sport starts at 10:30 am? ☐

c lunch starts at 1:05 pm? ☐

d school ends at 3:05 pm? ☐

e tennis coaching at 4:15 pm? ☐

f his dad gets home at 7:20 pm? ☐

4 At 8 am Jessica was looking forward to her party at 8:30 pm that night. Write how long she has to wait in:

a hours and minutes ☐

b hours ☐

c minutes ☐

d seconds ☐

Indigenous Methods for Identifying Time and Season

INVESTIGATION

Australia's Indigenous peoples have always used nature to identify time. Use the internet to investigate Indigenous methods of telling time. Search the internet for Australian Indigenous astronomy, seasons and moon cycles.

 • *AUSTRALIAN SIGNPOST MATHS 5* • ISBN 978 1 4886 2190 1

3:02 Kilometres

Practise estimating 1 kilometre.

km means kilometre.
1 km = 1 000 metres

CONCEPT

1 To measure these distances, would you use metres or kilometres?

- **a** the length of a skipping rope
- **b** the height of a chimney
- **c** the distance to the next town
- **d** the distance to Fiji
- **e** the distance you would walk in one day
- **f** the distance around a house
- **g** the perimeter of the school fence
- **h** the length of the classroom

2 How many metres in:

a 2 km? **b** 5 km? **c** 4 km? **d** 10 km? **e** 3 km?

3 How many kilometres in:

a 6000 m? **b** 10 000 m? **c** 7000 m? **d** 9000 m? **e** 8000 m?

4 Use the scale on the map to find the shortest distance by road (in **km**) from:

- **a** the police station to the school
- **b** the police station to the hospital
- **c** the school to the post office
- **d** the school to the shop
- **e** the post office to the shop
- **f** the church to the service station

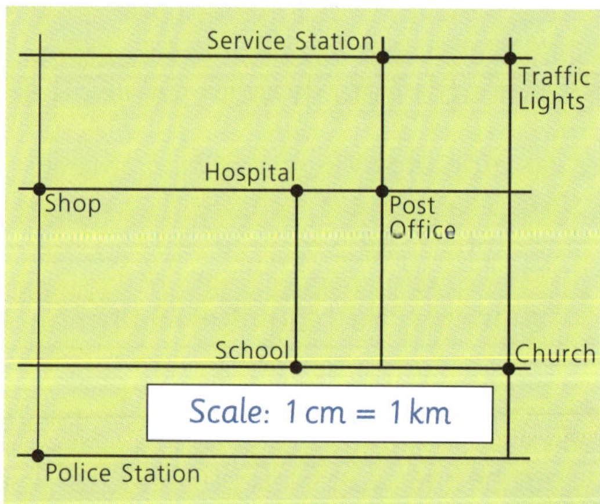

Researching Distance

- Make a chart to show the distances to nearby towns. Draw a map to show these distances.
- Make a chart to show the distances between capital cities in Australia.

ACTIVITY

3:03 Kilometres and Metres

1 Write as metres.

a 4 km ☐ b 7 km ☐ c 9 km ☐
d 6 km ☐ e 2 km ☐ f 3 km ☐
g 5 km ☐ h 10 km ☐ i 14 km ☐

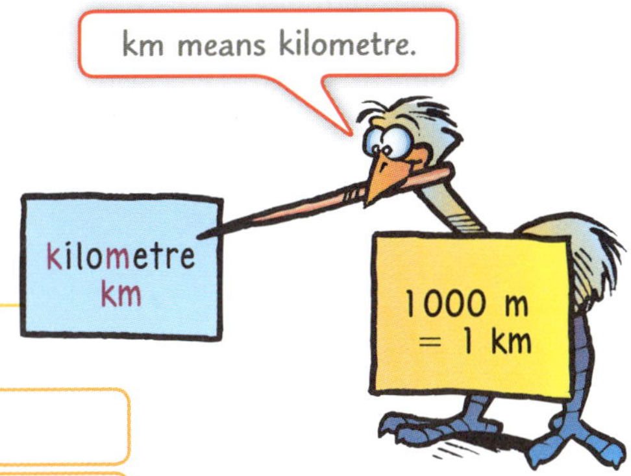

2 Find how many kilometres.

a 3000 m ☐ b 9000 m ☐ c 5000 m ☐
d 7000 m ☐ e 12 000 m ☐ f 16 000 m ☐

3 Choose the most suitable unit (**km**, **m**, **cm**, **mm**) to measure:

a the length of a road ☐
b the height of a house ☐
c the width of your hand ☐
d the length of an ant ☐
e the depth of a lake ☐
f the distance to Melbourne ☐
g the height of a flagpole ☐
h the width of Torres Strait ☐

4 Write each speed as kilometres per hour (**km/h**).

a 60 km travelled in one hour ☐
b 95 km travelled in one hour ☐
c 140 km travelled in two hours ☐
d 200 km travelled in four hours ☐
e 950 km travelled in five hours ☐

5 Write each length as kilometres and metres.

a 3850 m = ☐ km ☐ m
b 1056 m = ☐ km ☐ m
c 9005 m = ☐ km ☐ m
d 5600 m = ☐ km ☐ m
e 3275 m = ☐ km ☐ m
f 8241 m = ☐ km ☐ m

6 Write each length as metres.

a 5 km 200 m = ☐ m
b 1 km 341 m = ☐ m
c 4 km 750 m = ☐ m
d 8 km 5 m = ☐ m
e 10 km 30 m = ☐ m
f 7 km 38 m = ☐ m

7 Estimate how many of your paces would equal 1 kilometre. ☐

3:04 Perimeter

Perimeter
= 3 + 2 + 3 + 2 cm
= 10 cm

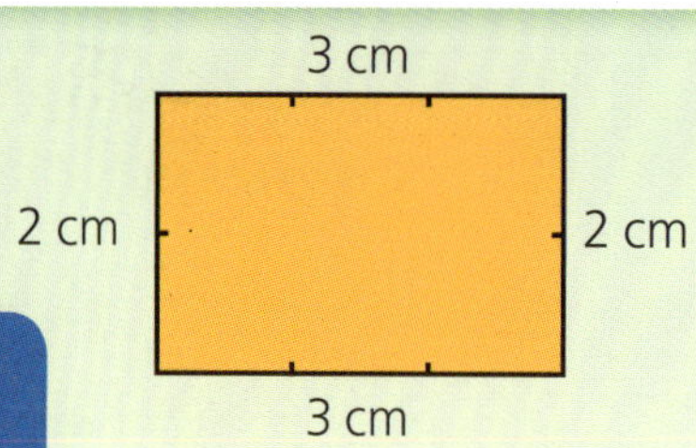

Perimeter of a rectangle = 2 × (length + width)

CONCEPT

1 Measure the length of each side and find the perimeter.

a

b

c

d

	Length	Width	Perimeter
a			
b			
c			
d			

Perimeter is the length of the boundary.

2 On grid paper, draw the following shapes and find the perimeter of each.

a A rectangle with sides 4 cm and 3 cm ☐
b A square with 4 cm sides ☐
c A rectangle with sides 5 cm and 4 cm ☐
d A square with 5 cm sides ☐
e A rectangle with sides 6 cm and 4 cm ☐
f A square with 6 cm sides ☐

3 Find the perimeter of each rectangle.

a 8 cm by 4 cm — P = ☐
b 7 cm by 5 cm — P = ☐
c 9 cm by 3 cm — P = ☐
d 7 m by 6 m — P = ☐
e 12 m by 4 m — P = ☐
f 7 m by 7 m — P = ☐

4 What is the perimeter of this page? ☐

3:05 Perimeter

CONCEPT

Perimeter = 2 × (length + width)

6 cm

2 cm

Perimeter = 16 cm

1 Find the perimeter of each rectangle.

a 7 cm, 5 cm — P = ☐

b 8 cm, 4 cm — P = ☐

c 11 cm, 4 cm — P = ☐

d 5 cm, 5 cm — P = ☐

e 9 cm, 6 cm — P = ☐

f 13 cm, 3 cm — P = ☐

2 Complete the tables.

	Length	Width	Perimeter
a	7 cm	6 cm	
b	10 cm	7 cm	
c	12 cm	10 cm	
d	15 cm	7 cm	
e	20 cm	16 cm	

	Length	Width	Perimeter
f	9 m	7 m	
g	12 m	8 m	
h	15 m	6 m	
i	20 m	8 m	
j	25 m	25 m	

3 Circle the correct perimeter for each rectangle with the following measurements.

a Length 10 cm, Width 8 cm
Perimeter: 18 cm 80 cm 36 cm

b Length 8 cm, Width 6 cm
Perimeter: 24 cm 28 cm 48 cm

c Length 9 cm, Width 5 cm
Perimeter: 14 cm 28 cm 45 cm

d Length 11 cm, Width 7 cm
Perimeter: 36 cm 77 cm 18 cm

4 Write the dimensions of a rectangle that has perimeter 18 cm.

Length = ☐ , Width = ☐

3:06 Calculating Area

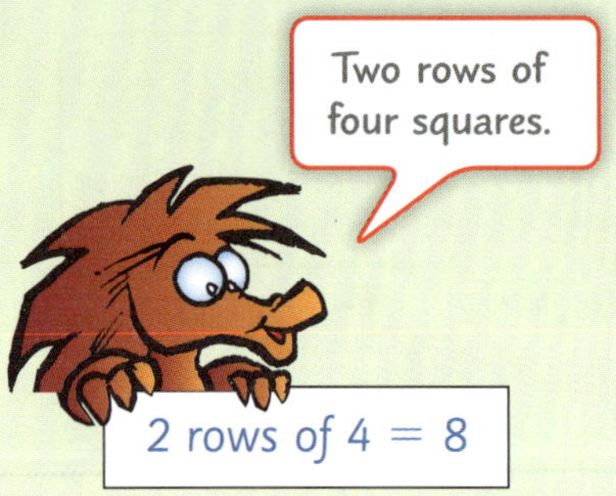

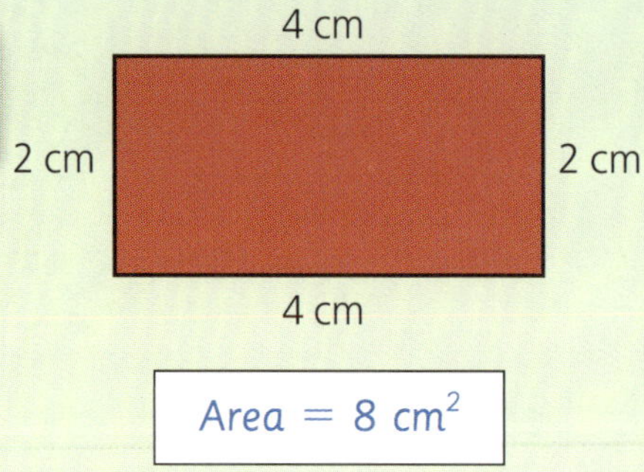

CONCEPT

1 Use the formula (Area = length × width) to find the areas. Complete the table.

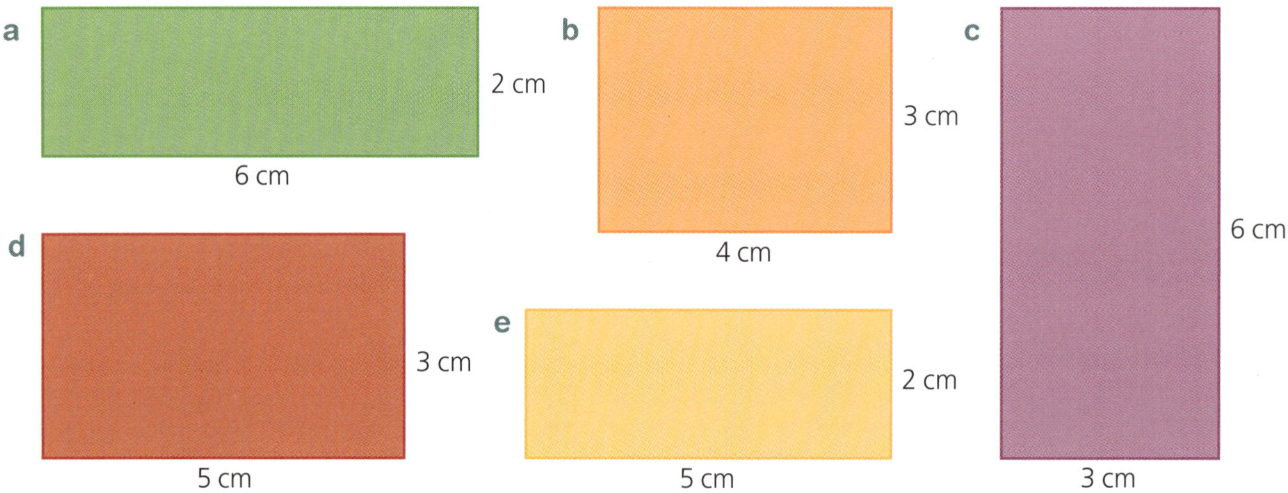

	Length	Width	Area
a	cm	cm	cm^2
b	cm	cm	cm^2
c	cm	cm	cm^2
d	cm	cm	cm^2
e	cm	cm	cm^2

A plastic square centimetre grid could be used.

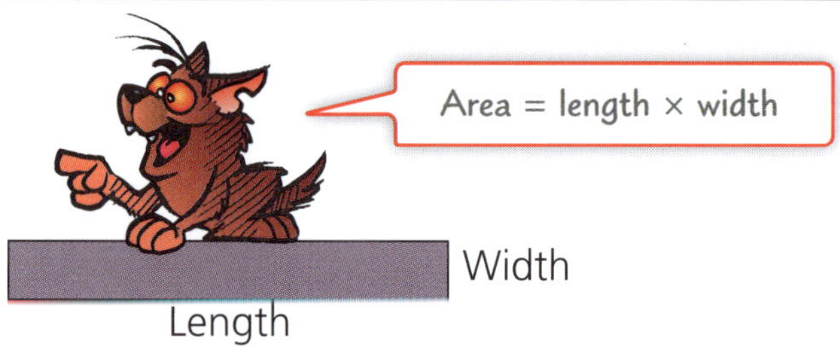

2 Circle the correct area for each rectangle with the following measurements.

a Length 6 cm, Width 4 cm
Area: 10 cm^2 24 cm^2 12 cm^2

b Length 9 cm, Width 5 cm
Area: 45 cm^2 14 cm^2 28 cm^2

c Length 7 cm, Width 5 cm
Area: 12 cm^2 70 cm^2 35 cm^2

d Length 10 cm, Width 6 cm
Area: 32 cm^2 60 cm^2 16 cm^2

3 What is the approximate area of this page? ____________
(You could use a 1 cm grid transparency.)

 ISBN 978 1 4886 2190 1

3:07 Square Metres

CONCEPT

$A = l \times w$
$= 5\,m \times 3\,m$
$= 15\,m^2$

1 Find the area of each rectangle.

a (4 m by 5 m) A = ☐ m^2

b (5 m by 2 m) A = ☐ m^2

c (3 m by 2 m) A = ☐ m^2

d (7 m by 2 m) A = ☐ m^2

e (2 m by 2 m) A = ☐ m^2

2 Complete the tables.

	Length	Width	Area
a	9 m	4 m	m^2
b	7 m	6 m	m^2
c	10 m	8 m	m^2
d	8 m	7 m	m^2
e	9 m	6 m	m^2

	Length	Width	Area
f	7 cm	6 cm	cm^2
g	6 cm	5 cm	cm^2
h	9 cm	7 cm	cm^2
i	8 cm	5 cm	cm^2
j	10 cm	8 cm	cm^2

3 Circle the correct area for each rectangle with the following measurements.

a Length 9 m, Width 7 m
Area: $16\,m^2$ $63\,m^2$ $97\,m^2$

b Length 8 m, Width 6 m
Area: $48\,m^2$ $14\,m^2$ $80\,m^2$

4 Calculate the area of each shape.

a rectangle with 6 m and 4 m sides ☐

b rectangle with 9 m and 8 m sides ☐

c rectangle with 12 m and 6 m sides ☐

d square with 6 m sides ☐

5 Use newspaper to make an area of 1 square metre.
Calculate the area of the floor in a room. ☐

 • *AUSTRALIAN SIGNPOST MATHS 5* • ISBN 978 1 4886 2190 1

3:08 Hectares

1 hectare $= 10\,000\,m^2$

The short way to write hectare is **ha**.

CONCEPT

1 Calculate how many square metres (**m²**) there are in:

a 2 ha ☐ b 5 ha ☐ c 7 ha ☐ d 4 ha ☐
e 6 ha ☐ f 9 ha ☐ g 3 ha ☐ h 8 ha ☐

2 Calculate how many hectares (**ha**) there are in:

a $10\,000\,m^2$ ☐ b $50\,000\,m^2$ ☐ c $30\,000\,m^2$ ☐ d $60\,000\,m^2$ ☐
e $40\,000\,m^2$ ☐ f $20\,000\,m^2$ ☐ g $90\,000\,m^2$ ☐ h $70\,000\,m^2$ ☐

3 Would you use square metres (**m²**) or hectares (**ha**) to find the area of:

a an airport? ☐ b a sandpit? ☐ c a door? ☐
d a large farm? ☐ e a National Park? ☐ f the kitchen floor? ☐
g a city block? ☐ h a large beach? ☐ i a squash court? ☐

4 Groups of students measured different areas of their playground. Use a calculator to find if each area is less than, equal to or more than a hectare. Complete the table.

		Length	Width	Area (m^2)	Compared to 1 ha
a	Group 1	100 m	100 m		
b	Group 2	350 m	20 m		
c	Group 3	200 m	50 m		
d	Group 4	450 m	30 m		
e	Group 5	400 m	25 m		

A hectare can be any shape.

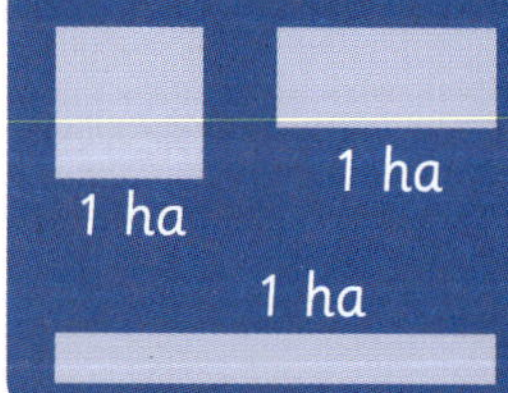

ACTIVITY

- Use suitable measuring instruments and markers to measure a 1 ha rectangle within the school grounds.
- Use a calculator to check that the area of your rectangle is $10\,000\,m^2$.

 • *AUSTRALIAN SIGNPOST MATHS 5* • ISBN 978 1 4886 2190 1

3:09 Area

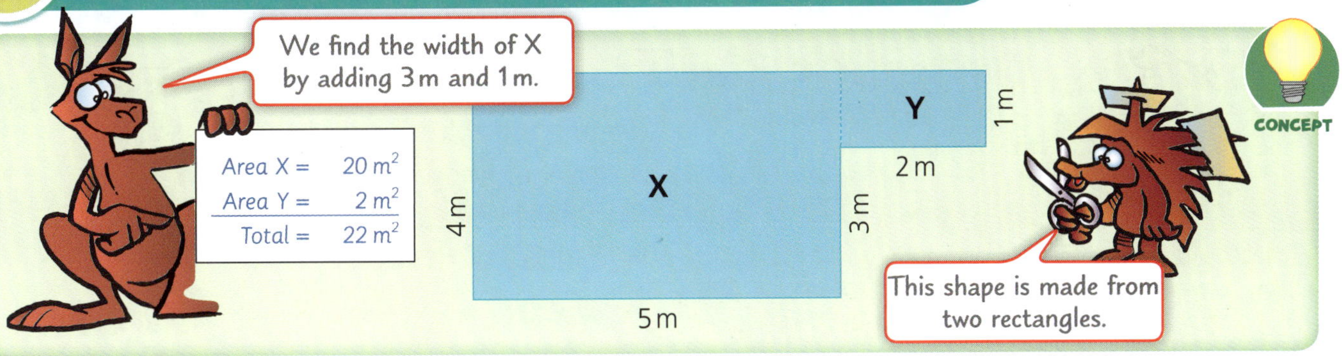

1 Find the area of each shape.

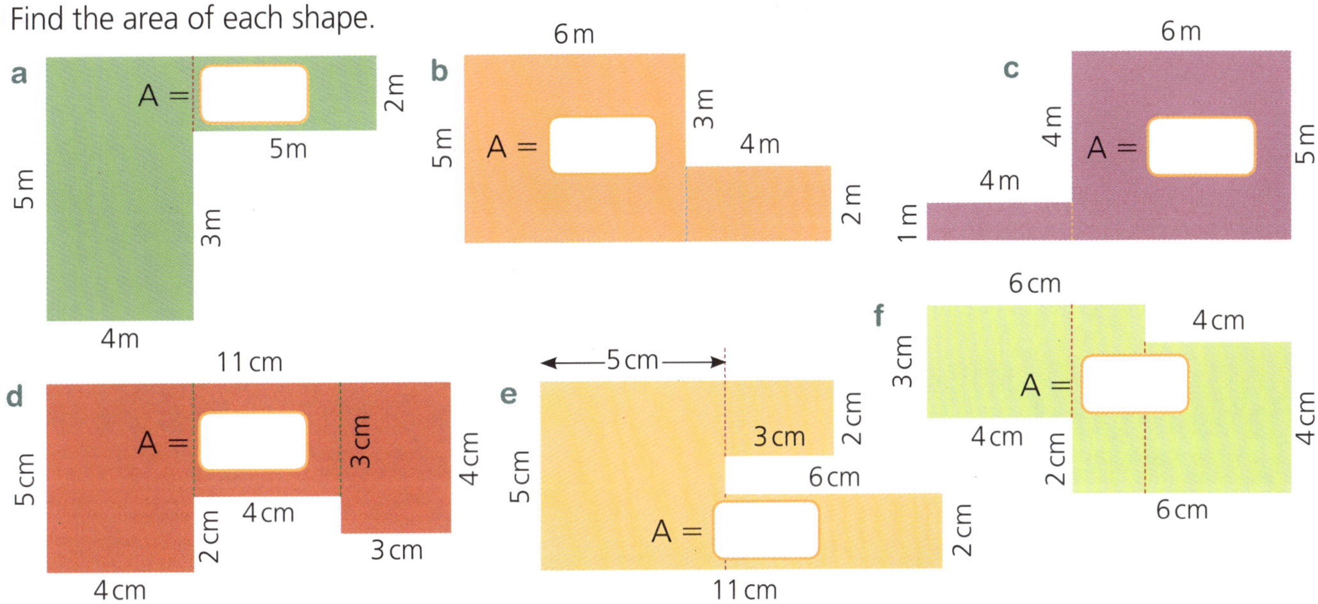

2 Write the dimensions (in whole cm) of rectangles that have an area of:

a $12\,cm^2$ ______, ______, ______

b $20\,cm^2$ ______, ______, ______

3 Write the most appropriate unit of measure (cm^2, m^2, ha) for:

a a classroom floor = 150 ______ b a farm = 250 ______ c a city park = 25 ______

d a postage stamp = 6 ______ e a garden = 600 ______ f a whiteboard = 8 ______

Sometimes we need to change areas from square metres to hectares.

- We have to divide by 10 000. (There are $10\,000\,m^2$ in 1 ha.)
- For the area on the right, we enter it on a calculator as: $300 \times 200 \div 10\,000 = 6$.

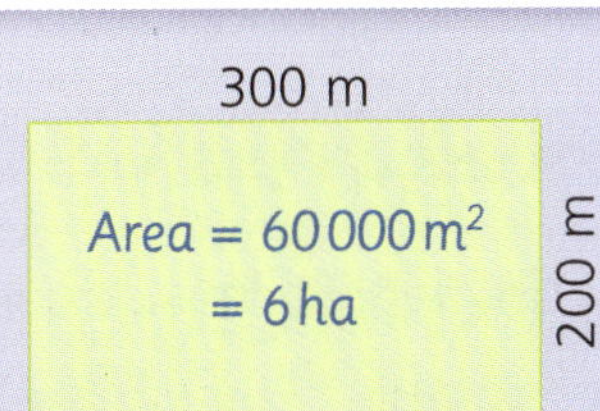

ICT

4 Find the area (in hectares) of these rectangles.

a Length 400 m, Width 200 m ______

b Length 800 m, Width 700 m ______

3:10 Cubic Centimetres

Volume = $20\,cm^3$

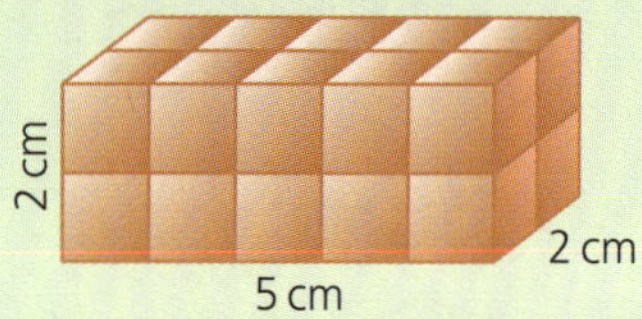

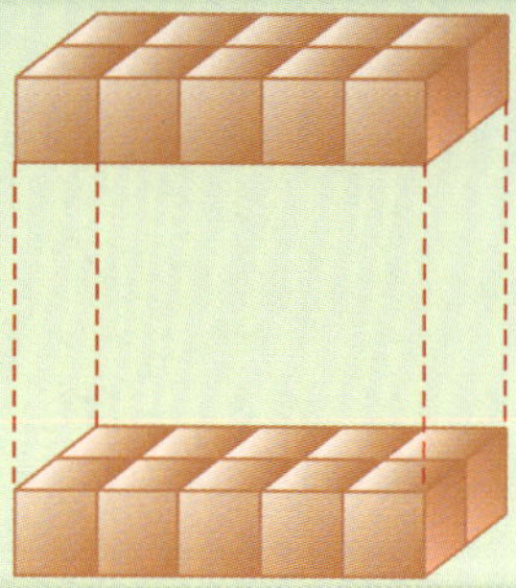

CONCEPT

To find the volume:
- work out how many cubes in one layer
- multiply by the number of layers.

1 Find the volume of each prism.

a

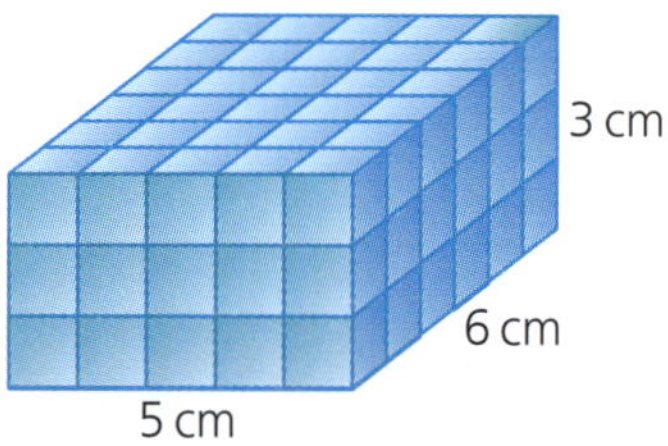

V = ______

b

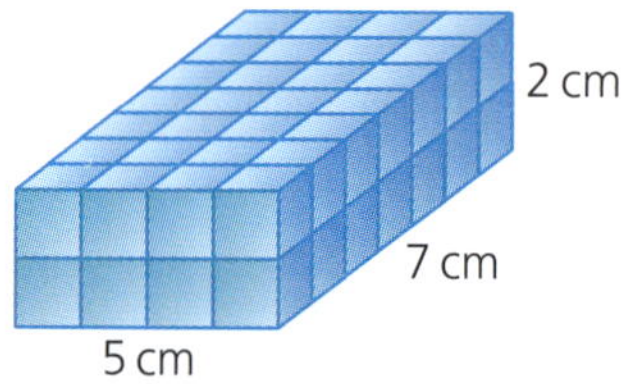

V = ______

c

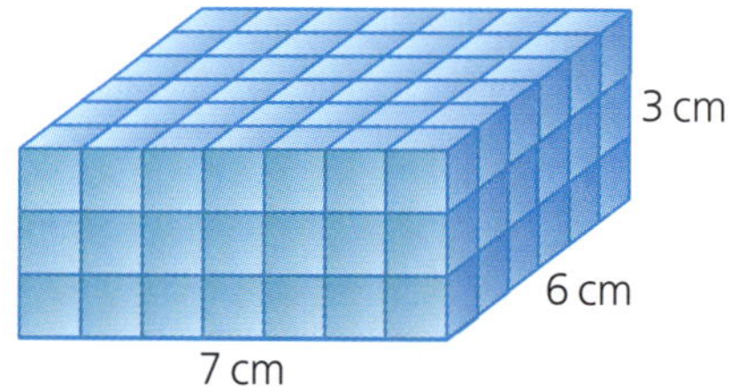

V = ______

d

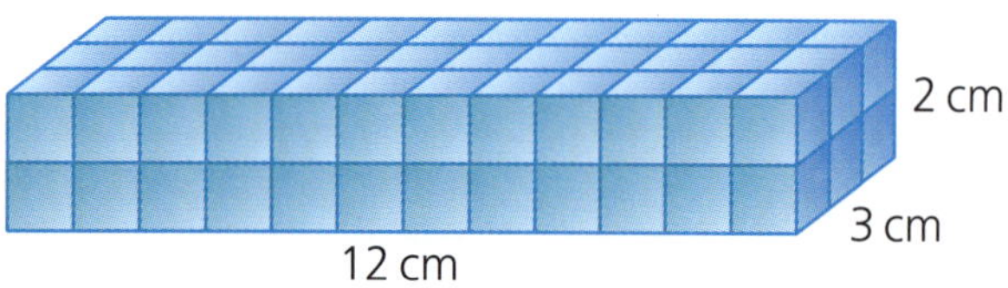

V = ______

e

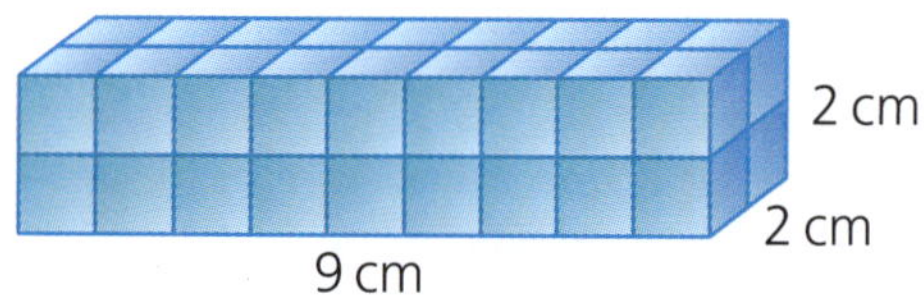

V = ______

2 Draw in the cubes and find the volume of each prism.

a

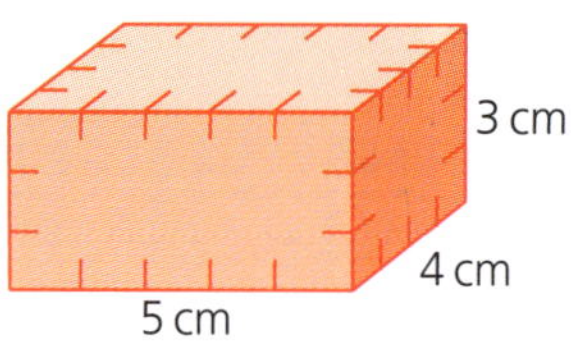

V = ______

b

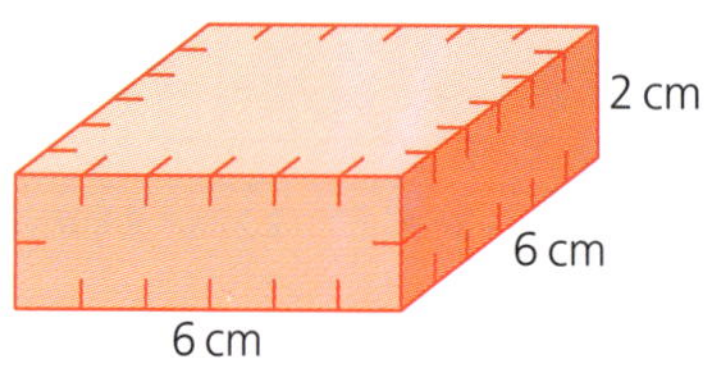

V = ______

c

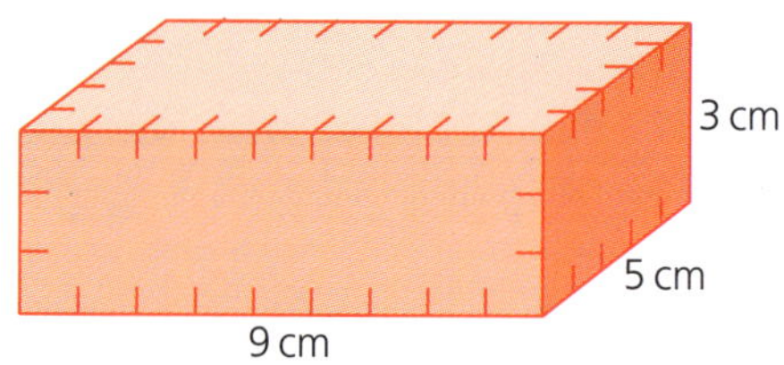

V = ______

3 Draw diagrams of these prisms and find the volume of each one.

a rectangular prism with sides 9 cm, 5 cm and 6 cm ______

b rectangular prism with sides 7 cm, 4 cm and 3 cm ______

c cube with 5 cm sides ______

d cube with 6 cm sides ______

e rectangular prism, length 8 cm, width 6 cm and height 4 cm ______

 • *AUSTRALIAN SIGNPOST MATHS 5* • ISBN 978 1 4886 2190 1

3:11 Cubic Centimetres

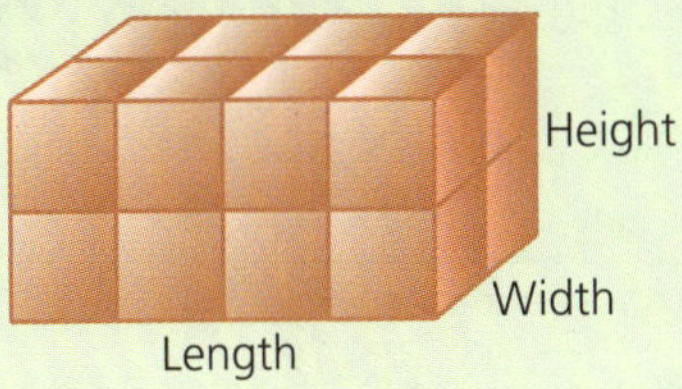

CONCEPT

Length	Width	Height	Volume
4 cm	2 cm	2 cm	16 cm^3

1 Use cubes to model these prisms and then complete the table.

a
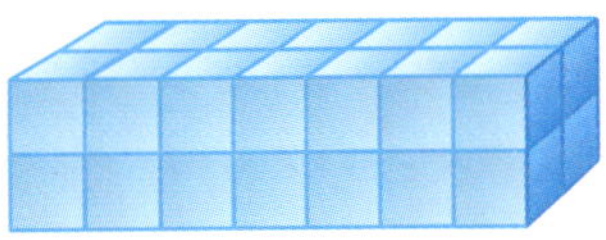

b
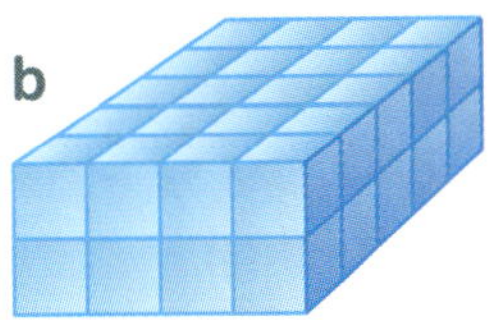

c
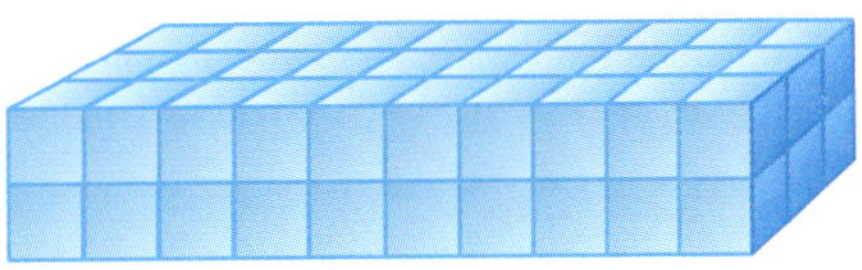

d
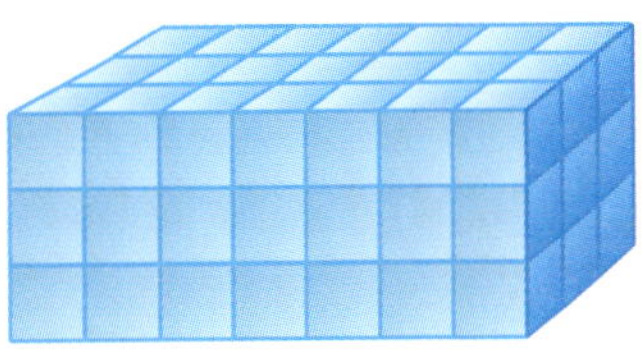

e
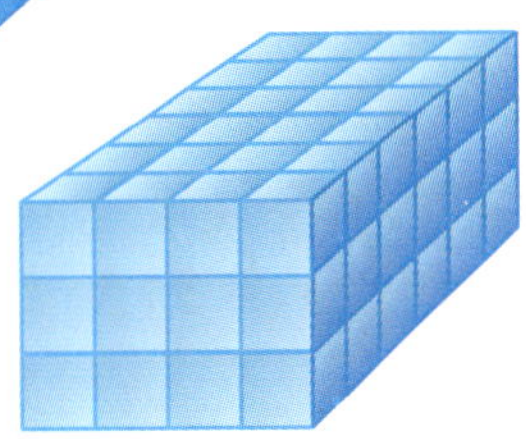

f
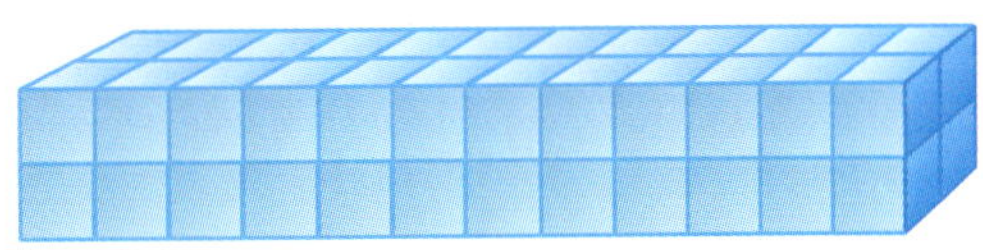

	Length	Width	Height	Volume
a	cm	cm	cm	cm^3
b	cm	cm	cm	cm^3
c	cm	cm	cm	cm^3
d	cm	cm	cm	cm^3
e	cm	cm	cm	cm^3
f	cm	cm	cm	cm^3

2 On 1 cm grid paper, draw the net of an open rectangular prism as shown. Cut and fold the net to make an open prism.

Stack cubes inside the prism to measure the volume.

V = ______ cm^3

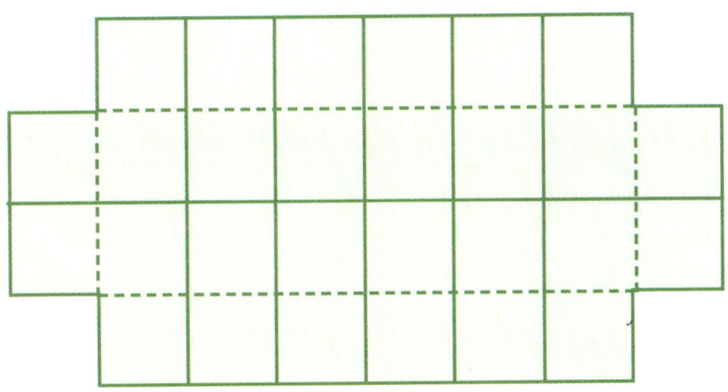

3 Write the dimensions of a rectangular prism that has a volume of 40 cm^3.

length = ______, width = ______, height = ______

4 Here are three views of a 3D object. Use cubes to build the shape and find its volume.

Volume = ______ cm^3

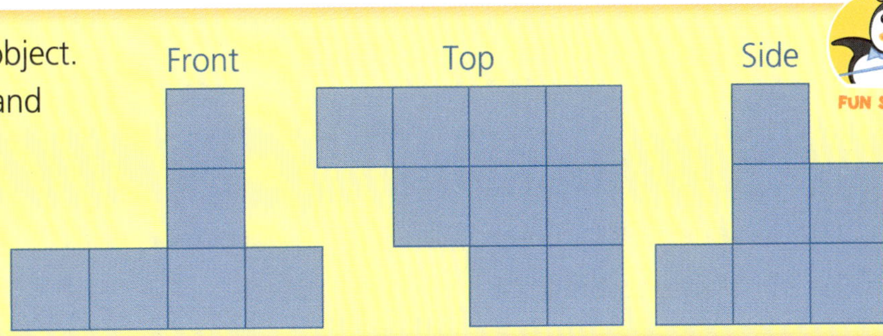

FUN SPOT

3:12 24-Hour Time

1 These watches show 24-hour time. Use **am** or **pm** to rewrite each time.

a 

b 

c 

d 

e 

f 

g 

h 

i 

j 

2 Write each time in 24-hour form.

a 
pm

b 
am

c 
pm

d 
pm

e 
am

3 Use 24-hour time to write the time one hour after:

a 3:00 pm
b 7:00 am
c 6:00 pm
d 9:00 pm
e 1:00 am
f 8:00 pm
g 11:00 am
h 11:00 pm
i 4:00 pm

- Use the internet to research where 24-hour time is used.
- Explain why you think 24-hour time is used.

ICT

3:13 Using 12- and 24-Hour Time

CONCEPT

00:00 (or 0000)	04:00 (or 0400)		15:00 (or 1500)	24:00 (or 2400)
Midnight (or 12 am)	4 am	Noon (or 12 pm)	3 pm	Midnight (or 12 am)

5 pm = 17:00 or 1700

1030 = 10:30 or 10:30 am

1 Write each time in two other ways.

a 5 pm = 17:00 or ______

b 0800 = 08:00 or ______

c 0300 = ______ or ______

d 2 am = ______ or ______

e 7 pm = ______ or ______

f 0100 = ______ or ______

g 1100 = ______ or ______

h 0400 = ______ or ______

i 10 pm = ______ or ______

j 6 pm = ______ or ______

k 0900 = ______ or ______

l 12 am = ______ or ______

2 What is the time:

a 2 hours after 11 am? ______

b 2 hours after 13:00? ______

c 3 hours before 7 am? ______

d 3 hours before 12:00? ______

e 4 hours after 9 am? ______

f 4 hours after 15:00? ______

g 3 hours before 6 pm? ______

h 2 hours before 08:00? ______

i 5 hours before 11 pm? ______

j 4 hours before 10:00? ______

FUN SPOT

Start at 00:00 and colour a pathway to finish at 24:00.

	07:00	09:00	04:00	09:00	10:00	11:00	12:00	14:00	20:00	21:00	23:00	
	06:00	02:00	03:00	11:00	09:00	10:00	13:00	15:00	19:00	22:00	20:00	
Start	**00:00**	01:00	04:00	07:00	08:00	11:00	14:00	17:00	18:00	23:00	**24:00**	**Finish**
	02:00	08:00	05:00	06:00	07:00	13:00	15:00	16:00	19:00	20:00	21:00	

 • *AUSTRALIAN SIGNPOST MATHS 5* • ISBN 978 1 4886 2190 1

3:14 Cubic Metres

1 m³

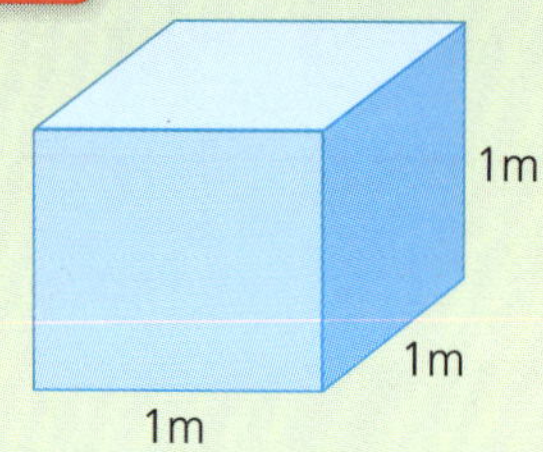

CONCEPT

Constructing a Cubic Metre

ACTIVITY

- Cut out 6 one-metre squares from cardboard.
- Tape the pieces together to form a net of a cube like the one shown.
- Fold and tape the net together to form a cube.
- Use your cube to help you to estimate the volumes of various objects.
- Write four objects for each category in the table.

Less than 1 m³				
About 1 m³				
More than 1 m³				

1 Use the abbreviated form to write:

a 3 cubic metres ☐ b 7 cubic metres ☐ c 2 cubic metres ☐

d 10 cubic metres ☐ e 15 cubic metres ☐ f 9 cubic metres ☐

g 4 cubic metres ☐ h 50 cubic metres ☐ i 12 cubic metres ☐

2 Would you use **cm³** or **m³** to measure the volume of:

a a classroom? ☐ b a shoe box? ☐ c a department store? ☐

d a wool shed? ☐ e an office? ☐ f a brick? ☐

3 Choose the most suitable unit of measure (**cm³** or **m³**) for each measurement below.

a carport = 75 ☐ b laundry = 40 ☐ c DVD case = 168 ☐

d packing crate = 2 ☐ e swimming pool = 100 ☐ f mobile phone = 150 ☐

g shop = 2 650 ☐ h classroom = 240 ☐ i desk drawer = 15 000 ☐

 ISBN 978 1 4886 2190 1

3:15 Volume of Rectangular Prisms

CONCEPT

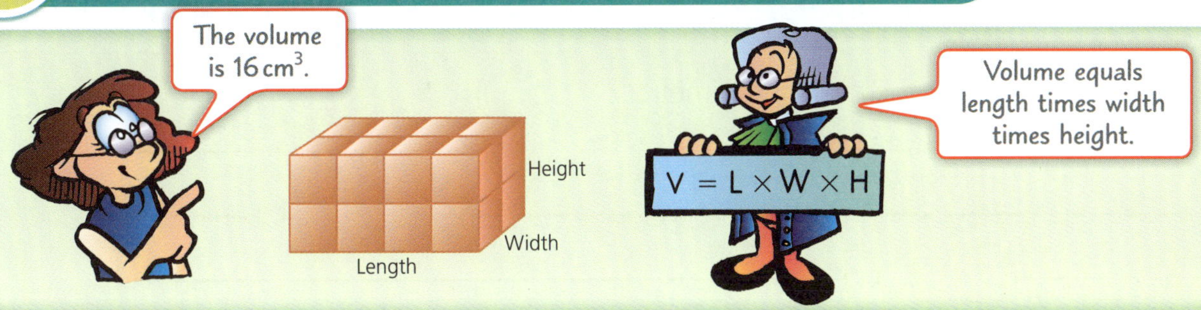

1 Calculate the volume of each rectangular prism.

a
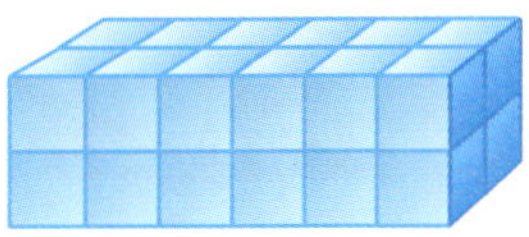
b
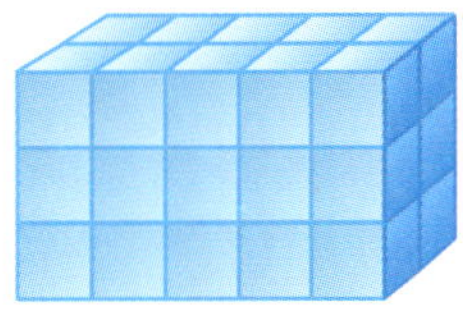
c
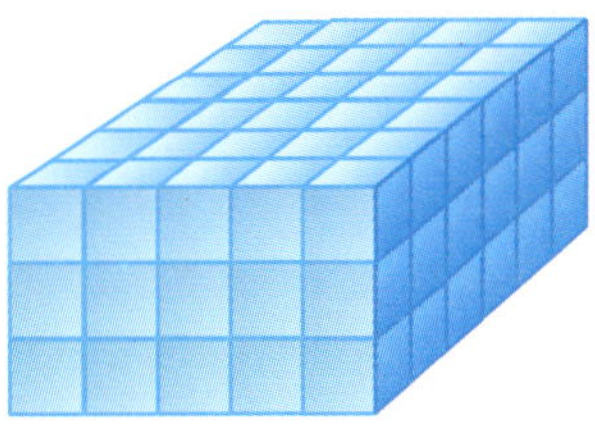
d
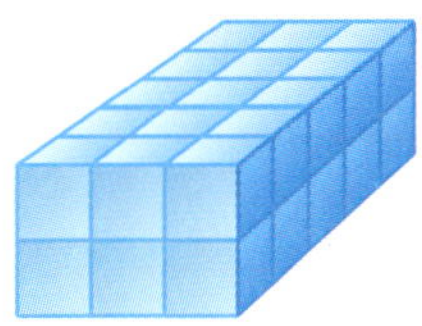
e

f
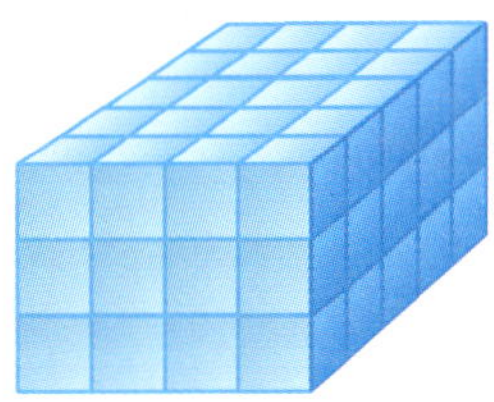
g
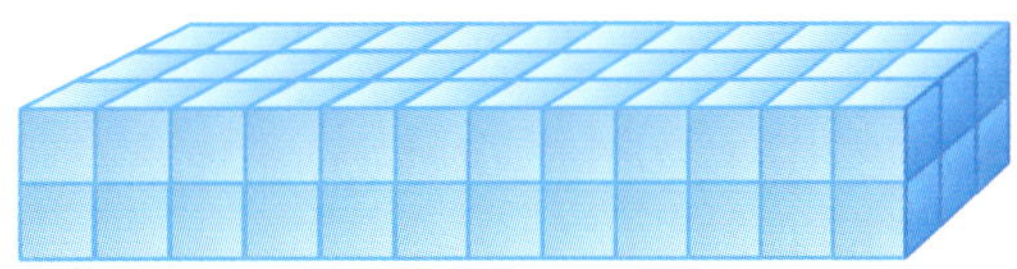

	Length	Width	Height	Volume
a	cm	cm	cm	cm^3
b	cm	cm	cm	cm^3
c	cm	cm	cm	cm^3
d	cm	cm	cm	cm^3
e	cm	cm	cm	cm^3
f	cm	cm	cm	cm^3
g	cm	cm	cm	cm^3

2 Calculate the volume of each rectangular prism.

a
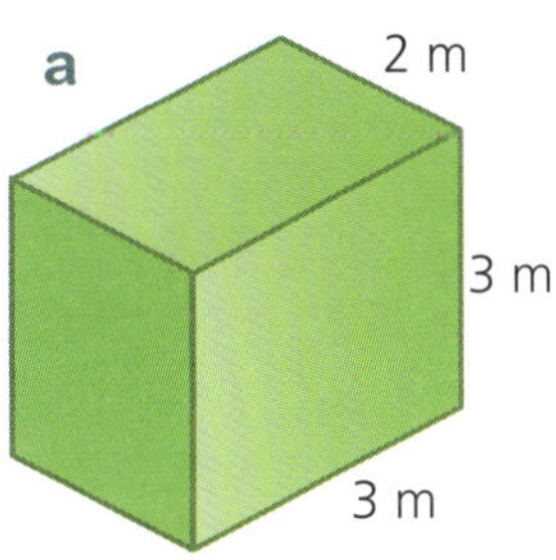

V = ☐ m^3

b
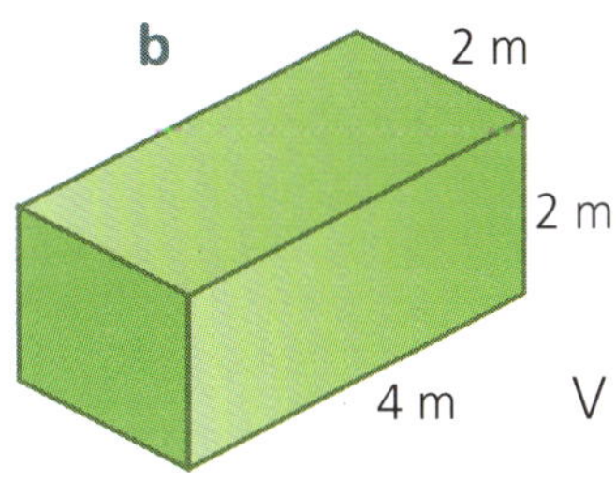

V = ☐ m^3

c
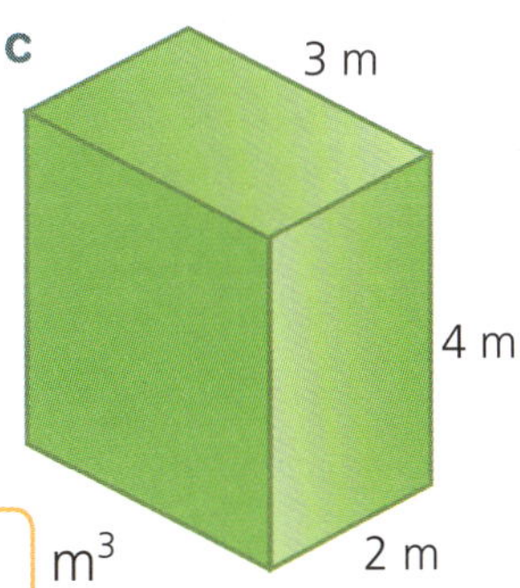

V = ☐ m^3

d
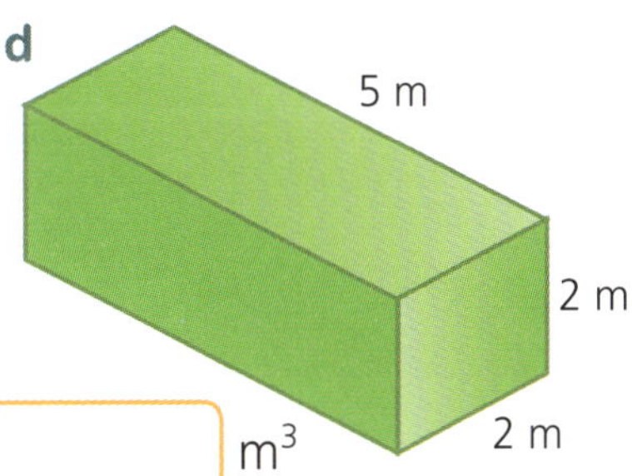

V = ☐ m^3

3:16 Millimetres

1 Measure each interval to the nearest millimetre.

2 Arrange the lengths in ascending order.

3 Choose the best unit (**km**, **m**, **cm**, **mm**) to measure the:

a length of a brick
b length of a road
c width of a window
d length of this book
e height of a boy
f width of a toothpick
g length of a football field
h width of a street

4 Use the decimal form to write each in centimetres.

a 6 cm 4 mm
b 8 cm 3 mm
c 3 cm 6 mm
d 4 cm 1 mm
e 9 cm 2 mm
f 5 cm 9 mm

5 Use the decimal form to write each in centimetres.

a 39 mm
b 65 mm
c 43 mm
d 82 mm
e 76 mm
f 91 mm
g 162 mm
h 158 mm

6 What is the perimeter of each shape? Answer correct to the nearest millimetre.

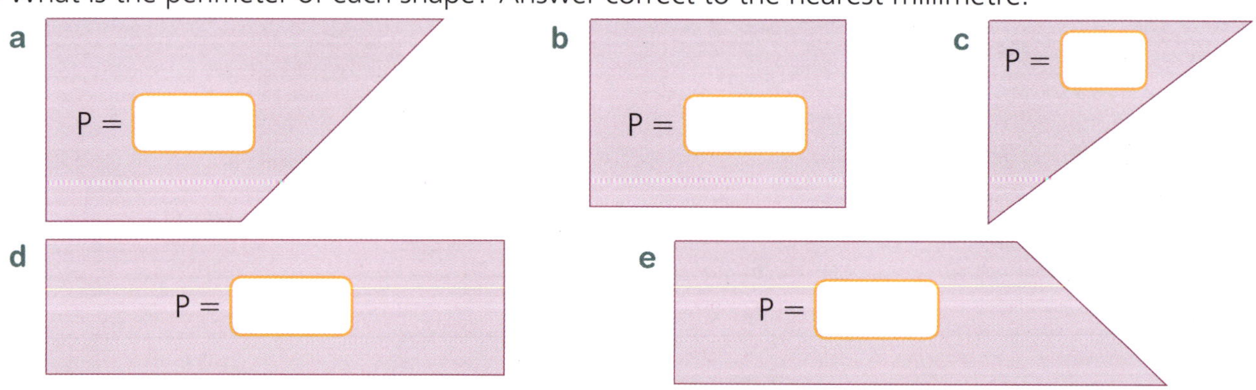

7 Calculate the perimeter for each shape.

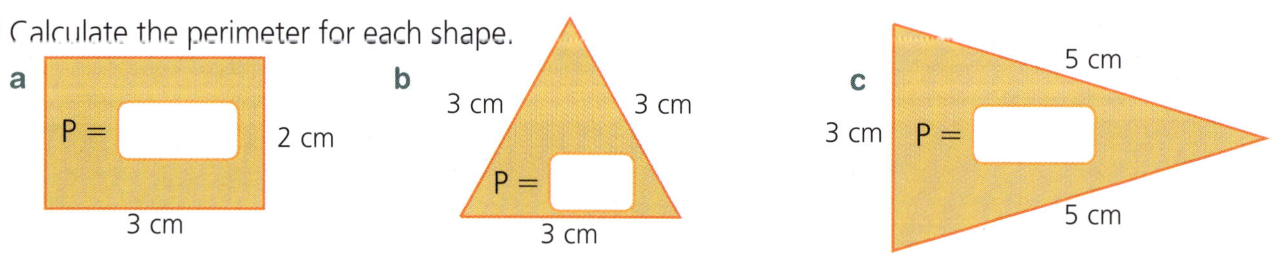

d Discuss shortcut methods to find the perimeters of these shapes.

 • *AUSTRALIAN SIGNPOST MATHS 5* • ISBN 978 1 4886 2190 1

3:17 Perimeter

1 Measure the length of each interval to the nearest millimetre (in centimetres).

a ______ cm b ______ cm

c ______ cm d ______ cm

e ______ cm f ______ cm

2 Measure the perimeter of each shape in centimetres.

a P = ______ cm

b P = ______ cm

c P = ______ cm

d P = ______ cm

3 On this 5 mm grid paper, draw the following shapes.

a square with 3·5 cm sides

b rectangle with sides 2·5 cm and 4·5 cm

c rectangle with sides 3·5 cm and 5 cm

ACTIVITY

- Measure the perimeters of three different books to the nearest millimetre.
- Arrange the perimeters in size from smallest to largest.
- Measure the perimeters of other objects such as tables or windows.
- Discuss the best unit of measure for this purpose.

Perimeter of a rectangle = 2 × length + 2 × width.

3:18 24-Hour Time

1 Write each time in 24-hour time.

a 
pm

b 
am

c 
pm

d 
am

e 
pm

f 
pm

g 
am

h 
am

2 Use **am** or **pm** to write:

a 19:51 b 02:37 c 06:13 d 16:24

e 05:49 f 13:06 g 18:27 h 08:59

i 10:33 j 17:42 k 09:16 l 03:01

3 Use 24-hour time to write:

a 3:54 pm b 9:32 am c 12:38 pm d 10:09 pm

e 6:18 am f 5:47 pm g 8:14 am h 1:26 am

i 4:23 pm j 2:31 pm k 11:56 pm l 7:28 am

4 Find the difference in time between:

a 3:21 am and 5:06 am

b 12:34 pm and 5:19 pm

c 6:48 am and 11:23 am

d 10:56 am and 2:07 pm

ACTIVITY

- Use a stopwatch to find the time it takes a partner to run 30 metres.
- Use a calculator to determine the speed in metres per second, e.g. 30 m in 5 seconds = 6 m/s.
- Time a variety of activities to complete the table.
- Round your answer, e.g. 1·6666 = 1·7.

Time	Distance	Average speed
5 s	30 m	6 m/s
	100 m	
10 s		
30 s		
	10 m	
40 s		

  • *AUSTRALIAN SIGNPOST MATHS 5* • ISBN 978 1 4886 2190 1

3:19 Problems Involving Time

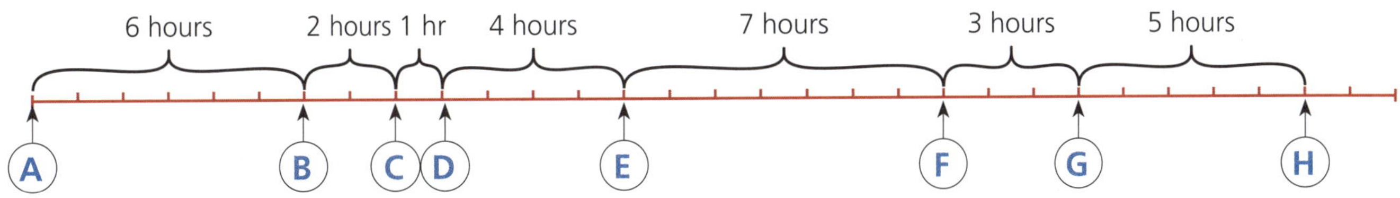

1. I leave **A** at 7 am, what time do I arrive at:
   a **B**? ______ b **D**? ______ c **E**? ______

2. I leave **A** at 13:30, when do I arrive at:
   a **B**? ______ b **D**? ______ c **C**? ______

3. I leave **D** at 4:30 am, what time do I arrive at:
   a **E**? ______ b **F**? ______ c **H**? ______

4. I leave **D** at 14:30, when do I arrive at:
   a **E**? ______ b **B**? ______ c **A**? ______

5. I leave **H** at 12:00 am (midnight), what time do I arrive at:
   a **F**? ______ b **E**? ______ c **D**? ______ d **B**? ______

6. I leave **H** at 00:00, when do I arrive at:
   a **G**? ______ b **F**? ______ c **C**? ______ d **B**? ______

7. Answer these problems.
   a Our trip took $3\frac{1}{2}$ hours. When did we arrive if we left at 4:30 pm? ______
   b We left home at 13:30. The trip took 5 hours. When did we reach our destination? ______

8. Tell a story about the time line in Question 1.

9. Start at 12 am (midnight) and colour a pathway to the finish.

	3.00 am	4.00 am	7.00 am	8.00 am	9.00 am	6.00 pm	2.00 pm	8.00 pm	9.00 pm	10.00 pm	9.00 pm	
	4.00 am	7.00 am	6.00 am	4.00 am	10.00 am	7.00 pm	3.00 pm	7.00 pm	2.00 pm	11.00 pm	**12.00 am**	**Finish**
	1.00 am	3.00 am	5.00 am	7.00 am	11.00 am	12.00 pm	8.00 pm	6.00 pm	3.00 pm	4.00 pm	7.00 pm	
Start	**12.00 am**	1.00 am	4.00 am	9.00 am	6.00 am	1.00 pm	7.00 pm	5.00 pm	2.00 pm	5.00 pm	8.00 pm	
	3.00 am	2.00 am	3.00 am	10.00 am	5.00 am	2.00 pm	3.00 pm	4.00 pm	1.00 pm	6.00 pm	9.00 pm	

3:20 Tonnes

CONCEPT

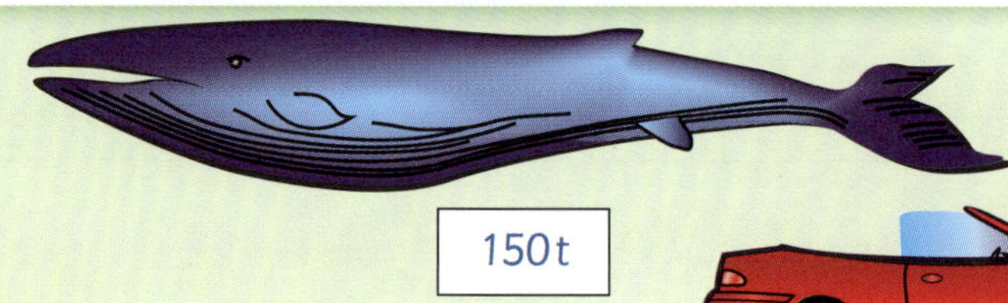

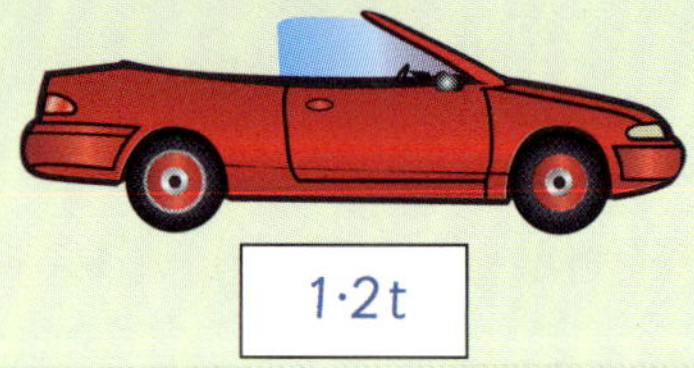

We use tonnes to measure large masses.
The abbreviation for tonne is **t**.

1 Would you use kilograms (**kg**) or tonnes (**t**) to measure the mass of:

a a skateboard? ☐ **b** a tractor? ☐ **c** a man? ☐

d a bag of apples? ☐ **e** a wooden bed? ☐ **f** a truck? ☐

2 Write the following in kilograms (**kg**).

a 3 t ☐ **b** 6 t ☐ **c** 9 t ☐ **d** 5 t ☐

e 7 t ☐ **f** 10 t ☐ **g** 4 t ☐ **h** 40 t ☐

i 15 t ☐ **j** 25 t ☐ **k** 60 t ☐ **l** 70 t ☐

3 Write the following in tonnes (**t**).

a 1 000 kg ☐ **b** 5 000 kg ☐ **c** 2 000 kg ☐ **d** 9 000 kg ☐

e 4 000 kg ☐ **f** 6 000 kg ☐ **g** 10 000 kg ☐ **h** 8 000 kg ☐

i 20 000 kg ☐ **j** 17 000 kg ☐ **k** 30 000 kg ☐ **l** 45 000 kg ☐

4 Complete these tables.

	kilograms	tonnes
a	3 500	
b		4·5
c		7·5

	kilograms	tonnes
d	8 500	
e	6 500	
f		9·5

Weighty Problems

5 **a** A loaded truck was found to have a mass of 19 t. When unloaded, the same truck had a mass of 3·7 t. What was the mass of the load? ☐

b Twenty-five trucks each carried 17 t of sand. What was the total mass of the sand carried? ☐

 ISBN 978 1 4886 2190 1

3:21 Grams and Kilograms

1 Find how many grams (**g**) there are in:

a 7 kg ☐ **b** 4 kg ☐ **c** 8 kg ☐

d 1 kg 625 g ☐ **e** 3 kg 750 g ☐ **f** 2 kg 125 g ☐

g 3·5 kg ☐ **h** 9·5 kg ☐ **i** 7·5 kg ☐

2 Rewrite the following as kilograms (**kg**) and grams (**g**).

a 2475 g ☐ **b** 1295 g ☐ **c** 6456 g ☐

d 3865 g ☐ **e** 5125 g ☐ **f** 4796 g ☐

g 9350 g ☐ **h** 7642 g ☐ **i** 8923 g ☐

3 Tick the most suitable unit of mass to measure the following.

	Mass	g	kg	t
a	the mass of a snail			
b	the mass of a semi-trailer			
c	the mass of a girl			
d	the mass of a man			
e	the mass of an elephant			
f	the mass of a lunch box			
g	the mass of a train			
h	the mass of a pencil case			

4 Write **true** or **false** for each statement.

a 3 t > 3100 kg ☐ **b** 12 kg < 15000 g ☐ **c** 600 g > 600 kg ☐

d 2·5 kg > 2600 g ☐ **e** 6400 kg < 6·5 t ☐ **f** 700 kg < 7 t ☐

ACTIVITY

- Compare different measuring devices, e.g. standard balance scales, kitchen scales, bathroom scales, spring scales.
- Discuss the purposes for which each is used.
- Practise using each kind of measuring device.

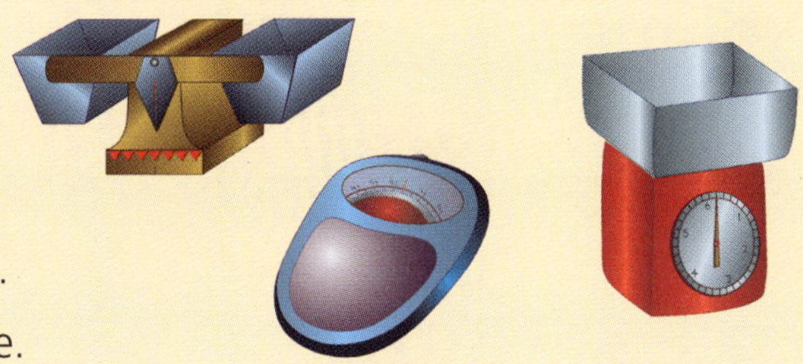

3:22 Using Measurement Scales

CONCEPT

When we use a measuring instrument, we read the scale to the nearest unit mark.

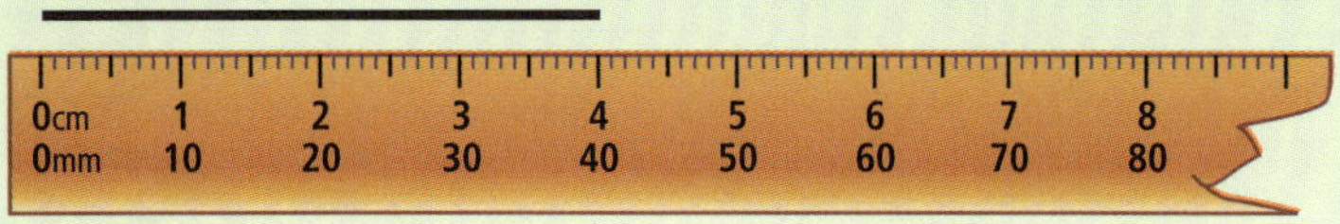

The measured length would be recorded as 4 cm.

1 Record the measure of the bar using each of the rulers.

a

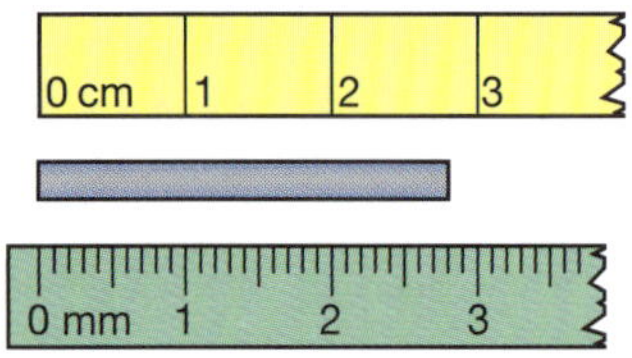

☐ cm

☐ mm

b

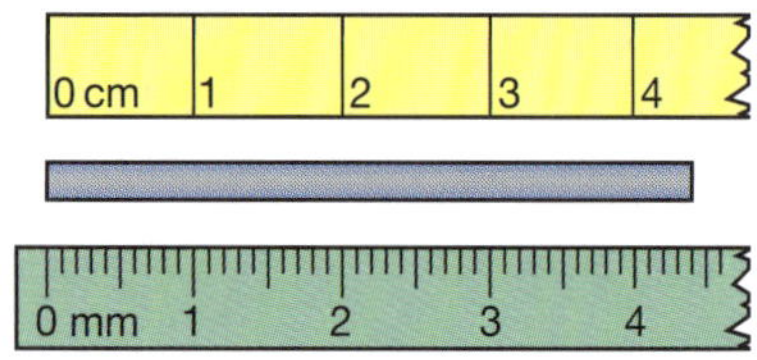

☐ cm

☐ mm

2 Record the measure shown on each scale.

a

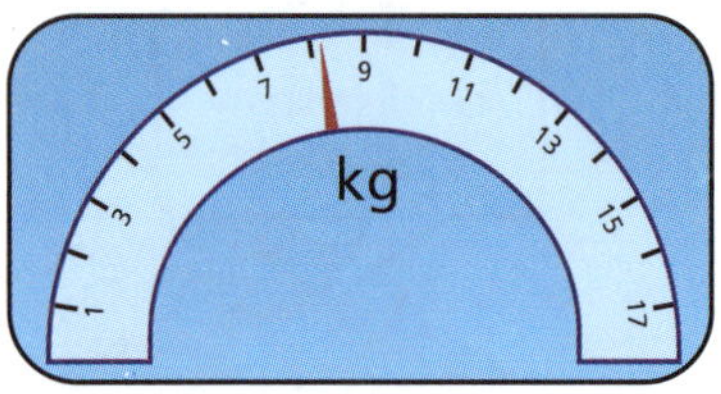

The measure is ☐ kg.

b

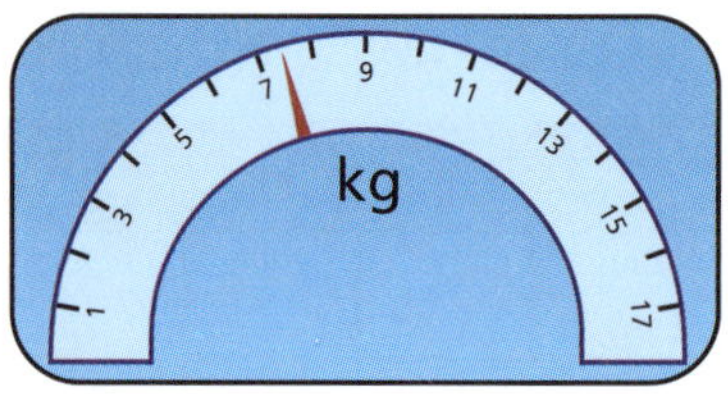

The measure is ☐ kg.

c

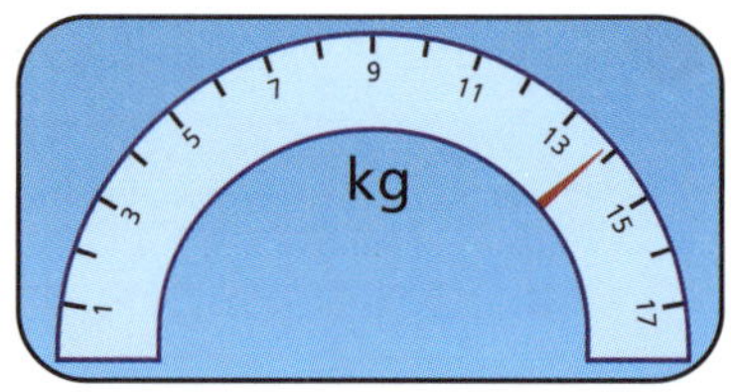

The measure is ☐ kg.

d

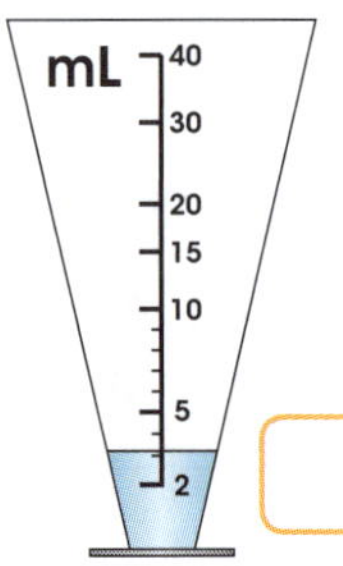

☐ mL

e

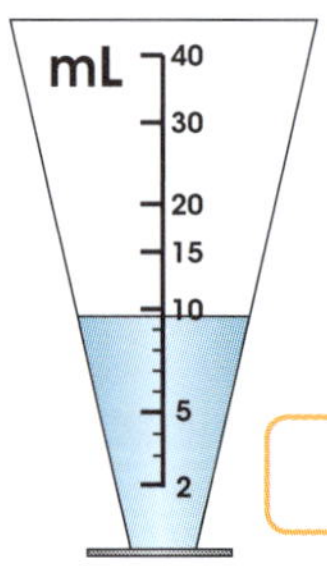

☐ mL

f

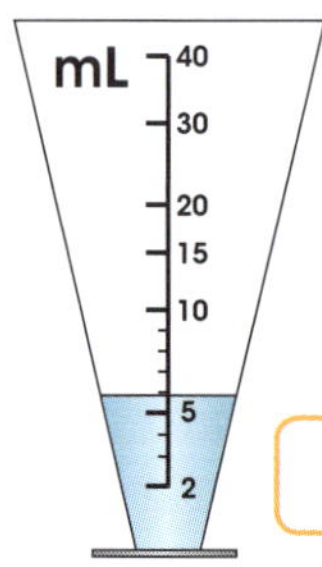

☐ mL

g

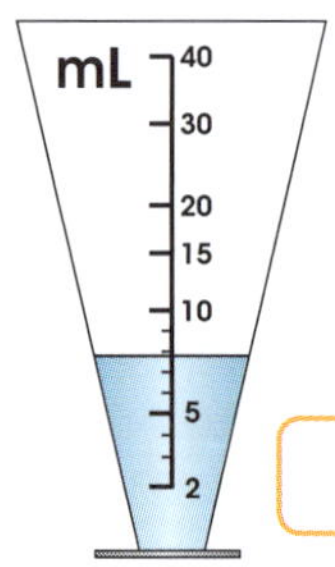

☐ mL

h

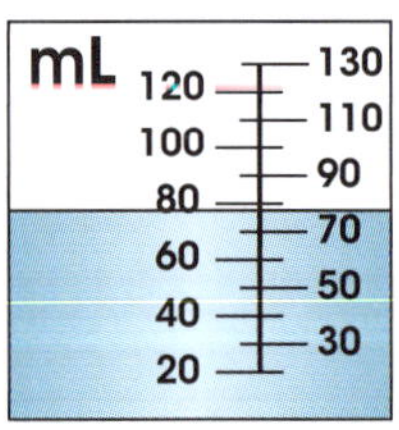

☐ mL

i

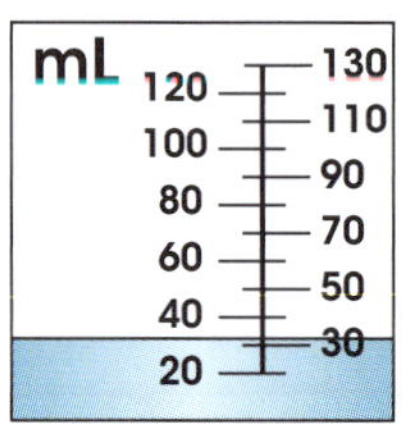

☐ mL

j

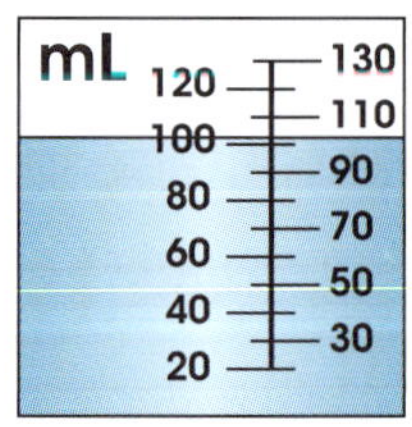

☐ mL

k

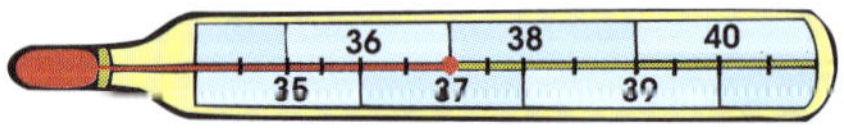

☐ °C

l

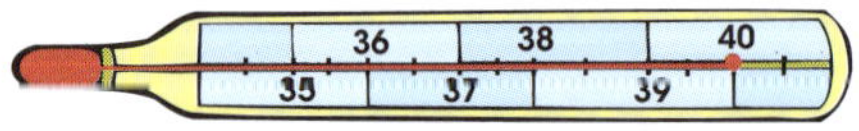

☐ °C

m

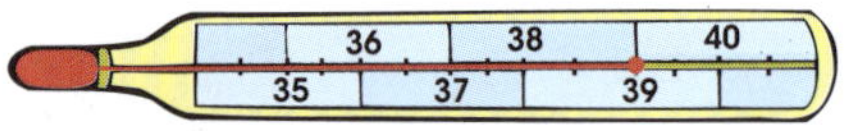

☐ °C

n

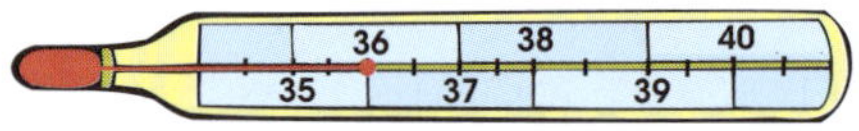

☐ °C

3:23 Converting Measurements

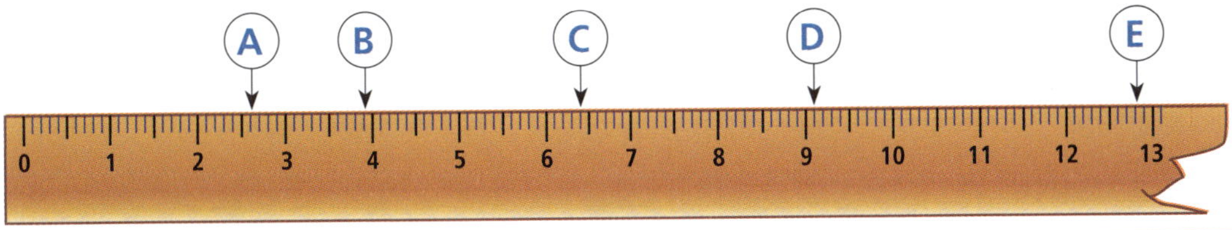

1 Record the lengths shown on the ruler.

A ____ mm ____ cm   B ____ mm ____ cm   C ____ mm ____ cm   D ____ mm ____ cm   E ____ mm ____ cm

2 Use the decimal form to write these as centimetres.

a 49 mm ____   b 64 mm ____   c 83 mm ____   d 51 mm ____
e 92 mm ____   f 75 mm ____   g 108 mm ____   h 123 mm ____

3 Use the decimal form to write these as metres.

a 251 cm ____   b 829 cm ____   c 375 cm ____   d 642 cm ____
e 563 cm ____   f 925 cm ____   g 1 021 cm ____   h 1 165 cm ____

4 Write these as centimetres.

a 3·16 m ____   b 8·31 m ____   c 4·65 m ____   d 9·54 m ____
e 5·27 m ____   f 2·95 m ____   g 7·45 m ____   h 10·75 m ____

5 Choose the most suitable unit (**km**, **m**, **cm**, **mm**) to measure:

a the length of a plane ____
b the width of a road ____
c the length of a river ____
d the length of a classroom ____
e the thickness of a match ____
f the distance to Perth ____
g the length of your thumb ____
h the width of Victoria ____

ACTIVITY

- Find a variety of objects in the classroom to measure.
- Use the different units of measure and complete the table.

Object Measurement	mm	cm	m
pencil sharpener length	26 mm	2·6 cm	—

3:24 Stopwatches

- Examine and operate various types of stopwatches.
- Discuss the following:
- When would it be important to measure the time to the nearest **hundredth** of a second?
- How accurately can you measure time?

ACTIVITY

With a group of 4 or more students timing, find how long it takes to hop ten metres.

Compare your recorded times.
Are they exactly the same? Why not?

1 Circle the faster time in each pair.

a 00:04·75 / 00:04·70	**b** 00:31·46 / 00:30·40	**c** 01:39·74 / 01:40·06	**d** 03:59·60 / 03:59·50
e 06:29·56 / 06:29·83	**f** 13:19·14 / 13:26·99	**g** 25:36·43 / 25:27·86	**h** 37:19·63 / 37:19·62

A stopwatch allows accurate measurement of time intervals.

2 Write the faster times from Question 1 parts **a**, **b**, **c** and **d** in full, e.g. 3 minutes 19·56 seconds.

a

b

c

d

3 Write the difference in time between:

a 01:27·41 and 01:28·06

b 00:48·23 and 00:45·19

c 56:03·86 and 53:17·53

d 06:59·32 and 07:00·53

e 09:01·74 and 09:01·67

f 37:14·99 and 37:12·96

g 59:46·21 and 59:41·37

h 00:25·16 and 00:23·17

 • *AUSTRALIAN SIGNPOST MATHS 5* • ISBN 978 1 4886 2190 1

3:25 Exploring Perimeter, Area and Volume

1 Explore the connection between the perimeter and the area of a rectangle.

a John drew four rectangles that had a perimeter of 24 mm. Complete the table below for each rectangle drawn.

Rectangle	Perimeter	Length	Height	Area
	24 mm			
	24 mm			
	24 mm			
	24 mm			

b Draw rectangles that have a perimeter of 30 mm. Complete the table below for each rectangle.

Perimeter	Length	Height	Area
30 mm			
30 mm			
30 mm			
30 mm			

2 Explore what happens to the area if we double the dimensions of a rectangle.

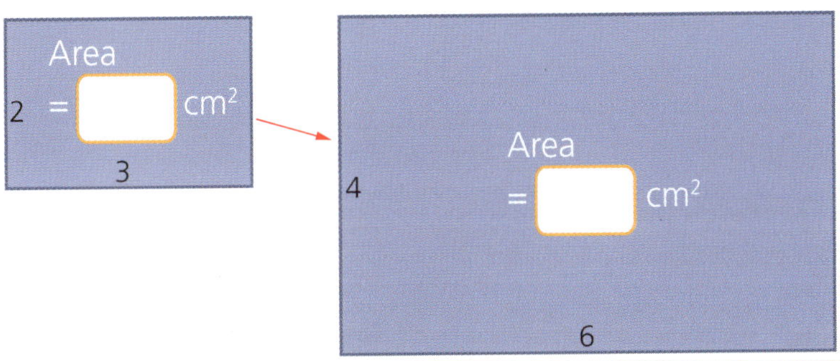

a When we doubled the dimensions, we multiplied the area by ______.

b Try doubling the dimensions of other rectangles. The area is multiplied by ______.

3 Explore what happens to the volume if we double the dimensions of a prism.

a

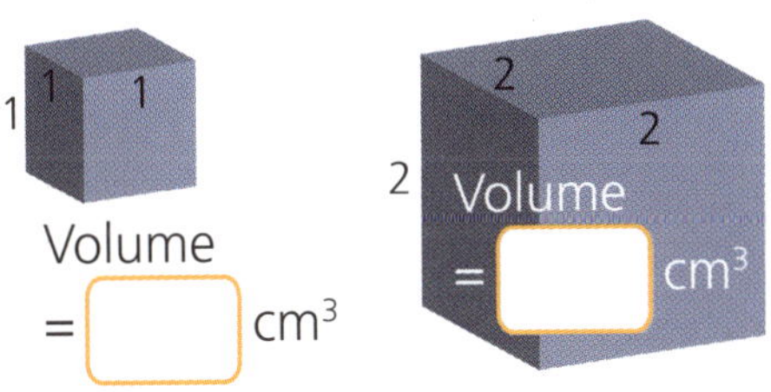

b

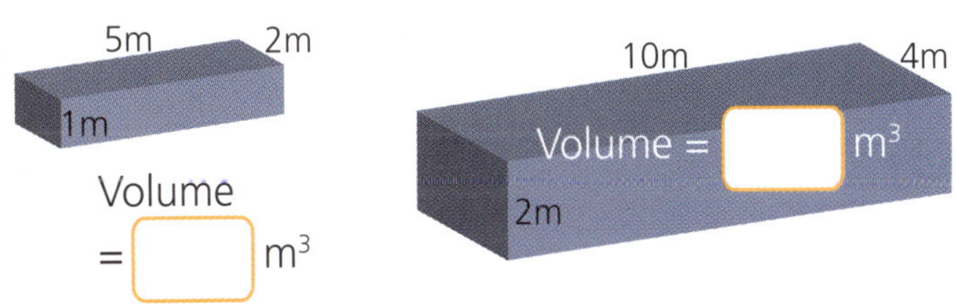

c When we double the dimensions, we multiply the volume by ______.

4 Use 72 place-value ones blocks to build different rectangular prisms.

See if you can build five. Write the length, width and height of each one you build.

Complete the formula.

Volume = ______

Making a rectangular prism			
Volume	**Length**	**Width**	**Height**
72 blocks	72	1	1
72 blocks			
72 blocks			
72 blocks			
72 blocks			
72 blocks			

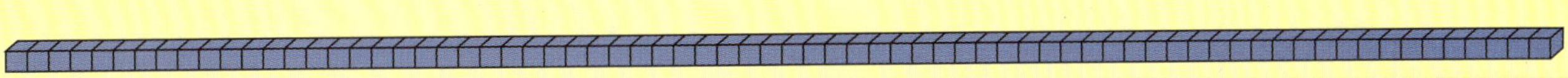

 • *AUSTRALIAN SIGNPOST MATHS 5* • ISBN 978 1 4886 2190 1

4:01 Review of 3D Space

1 Write **true** or **false** for these.

a A football has the shape of a cube.

b In any pyramid, every face is a triangle except perhaps the base.

c In any prism, every side is a rectangle except perhaps the two ends.

d The cross-section of any sphere is a circle.

2 What 3D object can be made from each net?

a

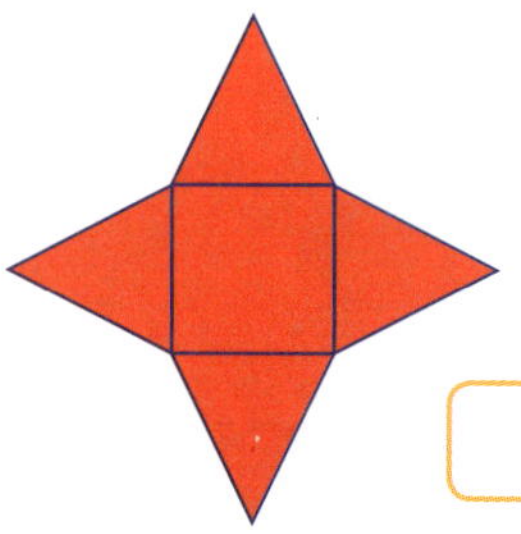

b

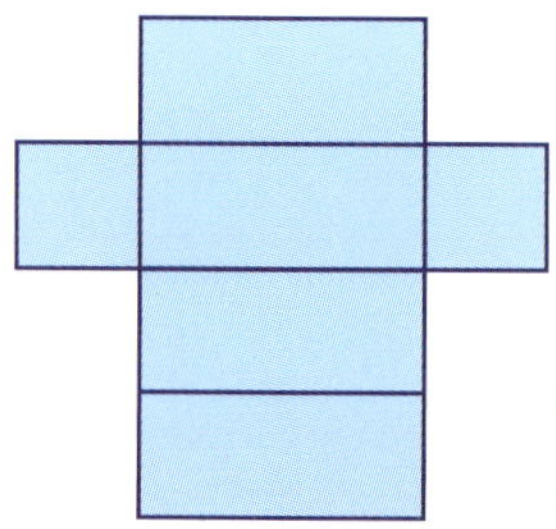

3 Use the method shown below to draw two prisms and two pyramids on your own paper.

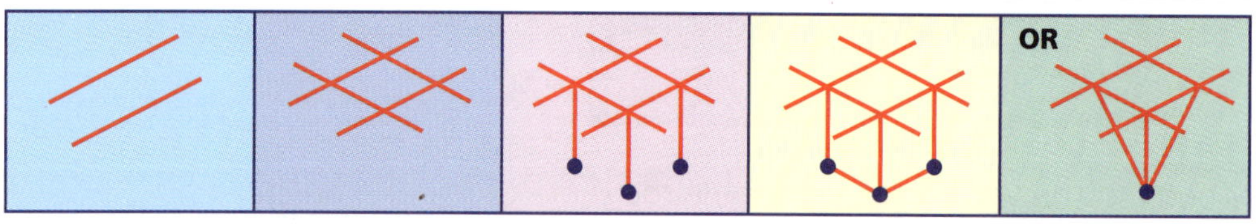

The two sides of a ruler can be used to draw parallel lines.

4 Sarah carefully cut along the edges of a box so that it could be flattened out. She made a net of the box.

a How many faces has a box?

b How many corners has a box?

c How many edges has a box?

Find a cardboard box. Cut along the edges and flatten it out to make its net.

5 For each object, draw what you would see if you looked from the front and from the top.

a

Object	Front	Top
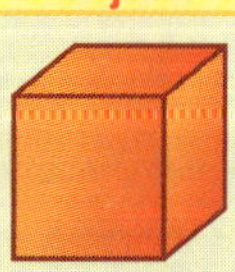		
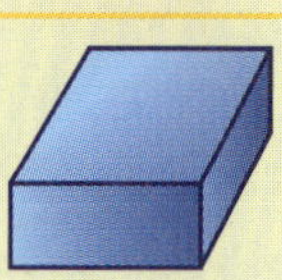		

b

Object	Front	Top

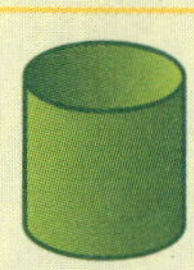		

© PEARSON AUSTRALIA 2018 • *AUSTRALIAN SIGNPOST MATHS 5* • ISBN 978 1 4886 2190 1

4:02 Prisms and Pyramids

CONCEPT

Prisms and pyramids are named according to the shape of their bases.

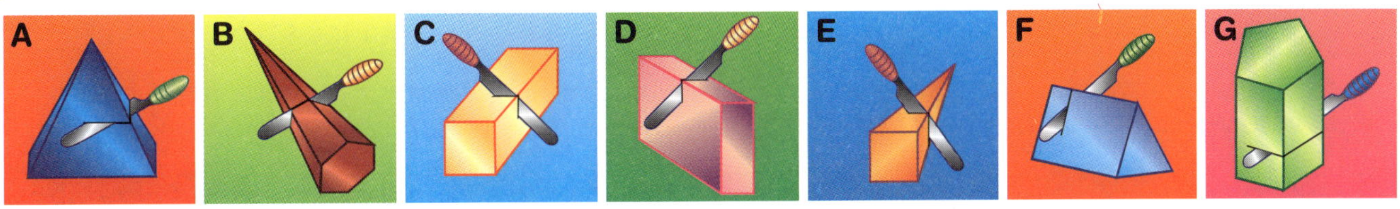

1 Which object is:

a a triangular prism? ☐ b a hexagonal pyramid? ☐ c a square prism? ☐

d a rectangular pyramid? ☐ e a rectangular prism? ☐ f a square pyramid? ☐

2 Name the shape of the cross-section of:

a A ☐ b B ☐ c C ☐ d G ☐

3 Complete these tables.

Object	Number of Corners	Number of Edges	Number of Faces
A			
B			
C			

Object	Number of Corners	Number of Edges	Number of Faces
D			
E			
F			

4 What object can be made from each set of faces?

a ☐

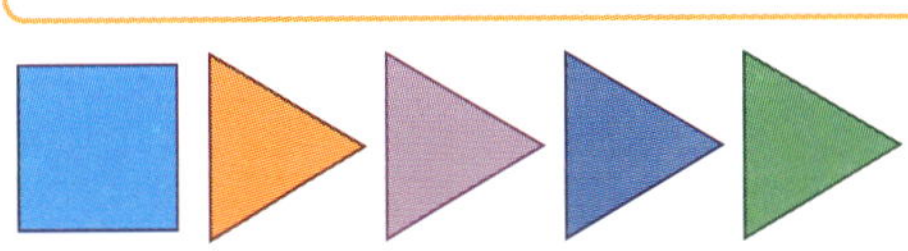

b ☐

● Use shapes like those in Question 4 to construct models of the two 3D objects.

4:03 Translations, Reflections and Rotations

CONCEPT

We can repeat the shape if we:

- slide the shape, or
- flip the shape, or
- turn the shape

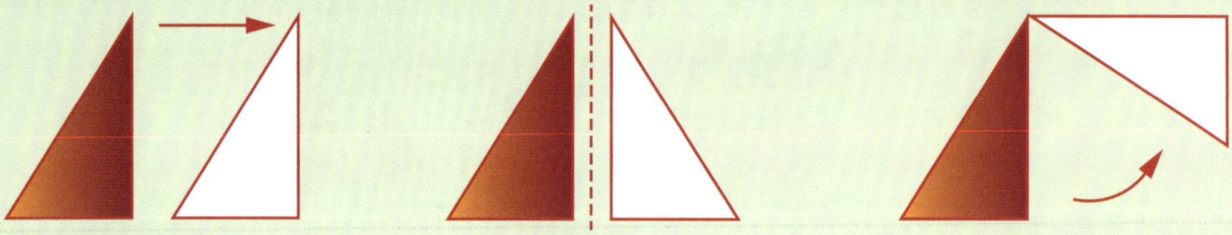

Translation (Slide the shape.)

Reflection (Flip the shape over.)

Rotation (Turn the shape around.)

1 Write **yes** or **no** to show how transformations affect the way the triangle above is moved.

Transformation	Figure Changes Position	Figure Changes Shape	Figure Changes Size	Figure Changes Orientation
translation				
reflection				
rotation				

***Orientation** is the direction something is facing or pointing.*

2

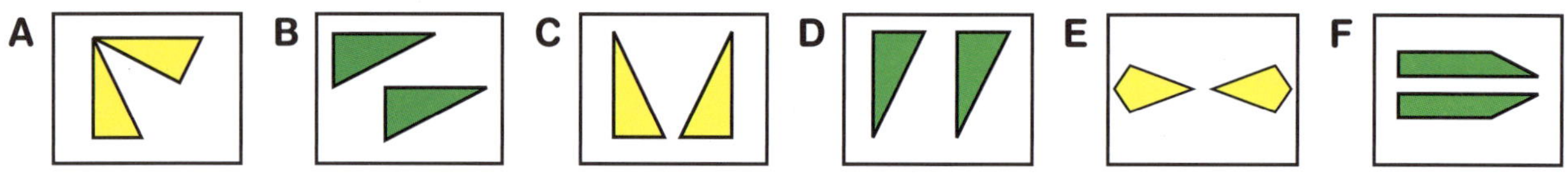

a Which pictures above show a translation?

b Which pictures above show a reflection?

For a rotation, the direction of movement is anticlockwise, unless told otherwise.

3

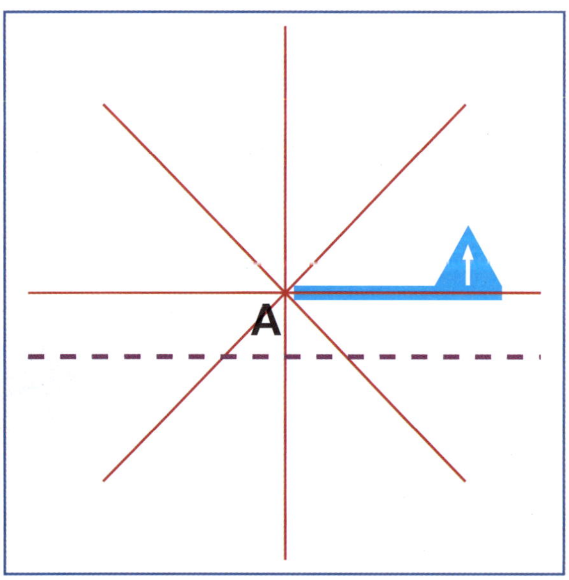

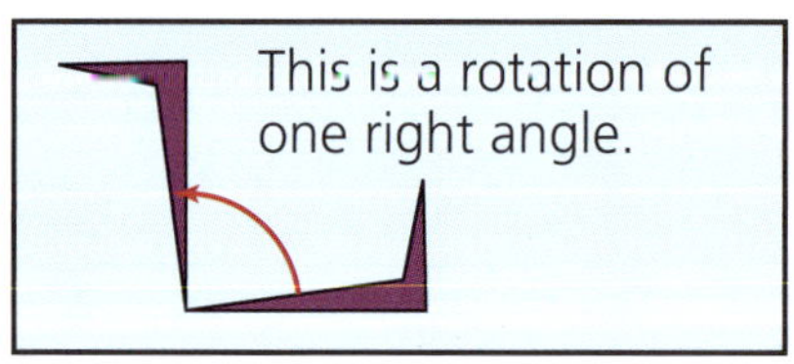

Draw the blue flag in its new position after:

a a reflection (flip) about the dotted line

b a translation (slide) of 2 cm up

c a rotation (turn) of one right angle around **A** (anticlockwise)

d a rotation of two right angles around **A**.

  • ISBN 978 1 4886 2190 1

Translations, Reflections and Rotations

A 
B 
C 
D 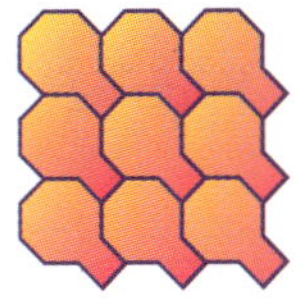
E 
F 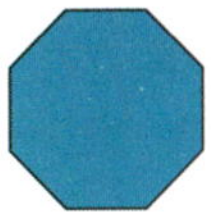

1 Which picture above is an example of:

a a reflection (flip)? ☐ b a translation (slide)? ☐ c a rotation (turn)? ☐

d a tessellation? ☐ e a pentagon? ☐ f an octagon? ☐

2 How many axes of symmetry has picture:

a A? ☐ b B? ☐ c C? ☐ d E? ☐ e F? ☐

3 a

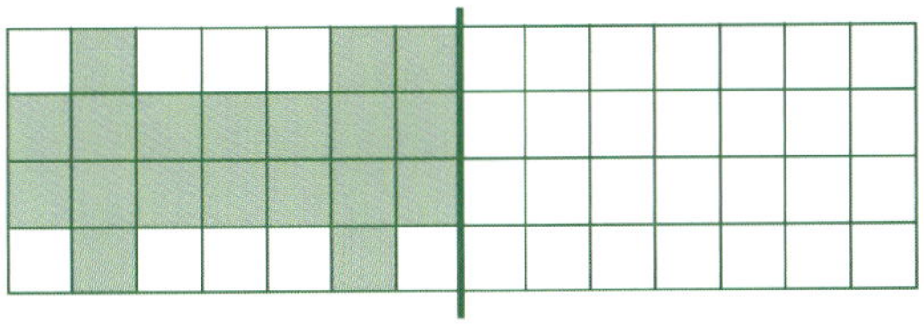

Shade squares to reflect this shape about the green line.

b Rotate this shape clockwise through a right angle.

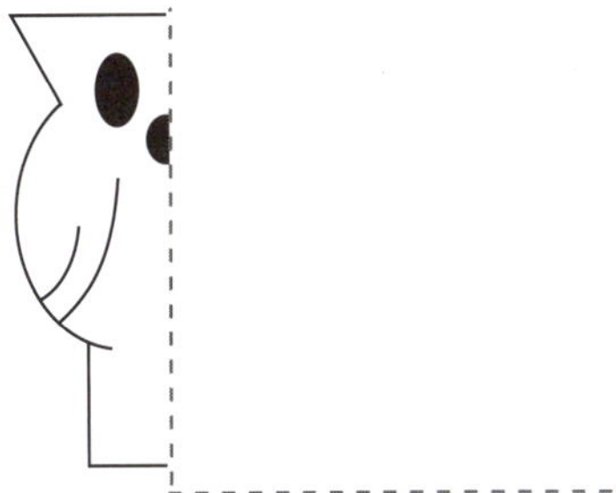

c Reflect this shape about the line.

d Translate this shape 4 cm right.

4 Complete these symmetrical designs.

a

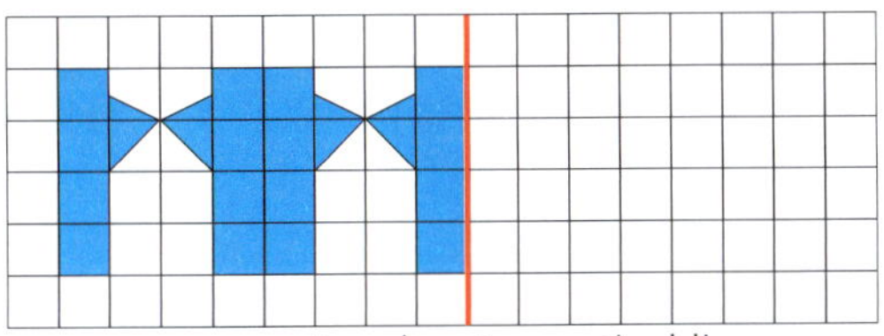

symmetry about a vertical line

b

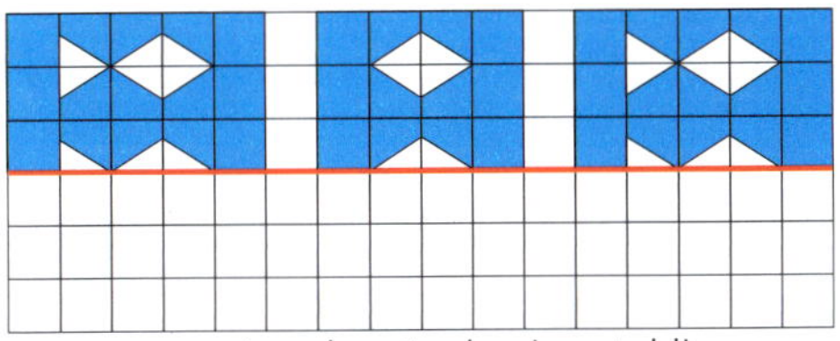

symmetry about a horizontal line

c Make your own symmetrical design.

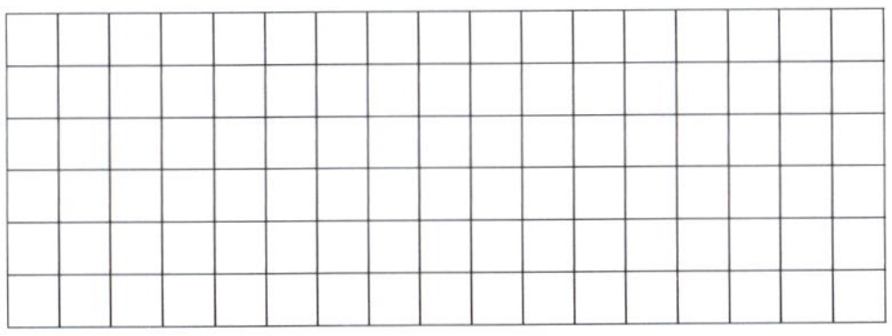

d Design a school logo using transformations.

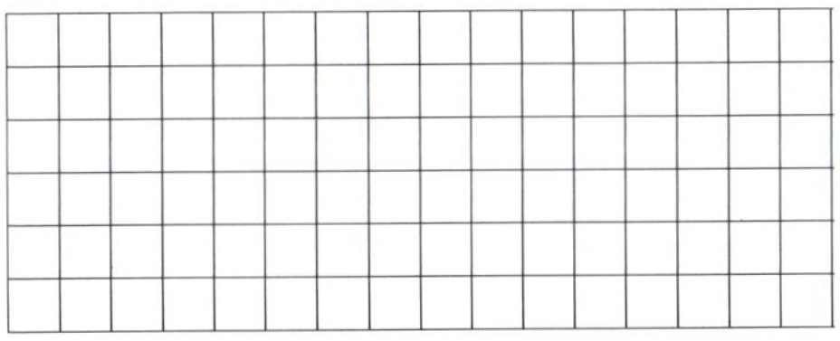

Use a computer program to make designs using translations, reflections and rotations.

ICT

4:05 Nets

1. Match each object to a net. Circle the net that does **not** match a three-dimensional object.

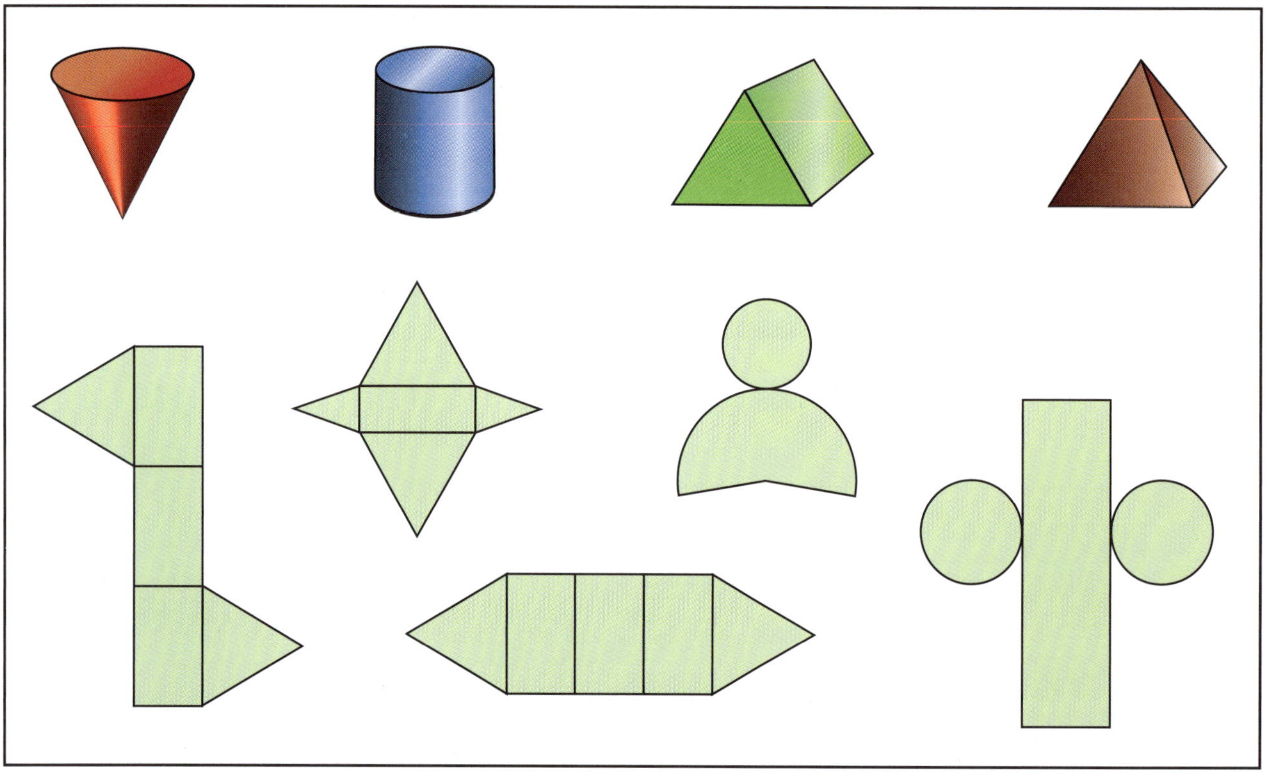

2. a This is the net of a ______.

   b Copy this net onto cardboard using twice these measurements.

   c Make the object.

INVESTIGATION

- Cut up a box to make its net. Compare your net with those of other students.

3. Circle the nets below that would be the net of a box.

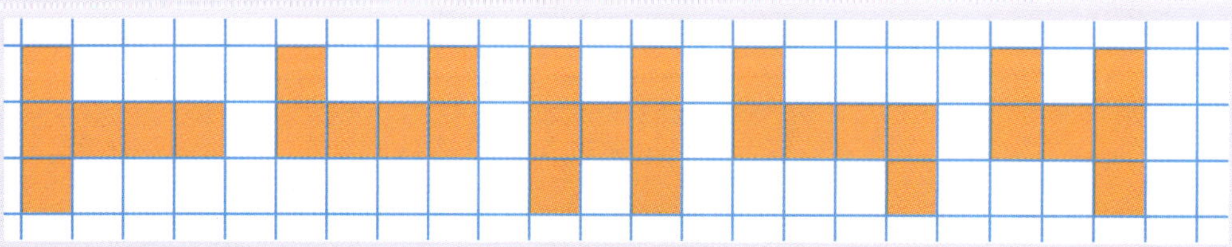

4:06 Describing Position

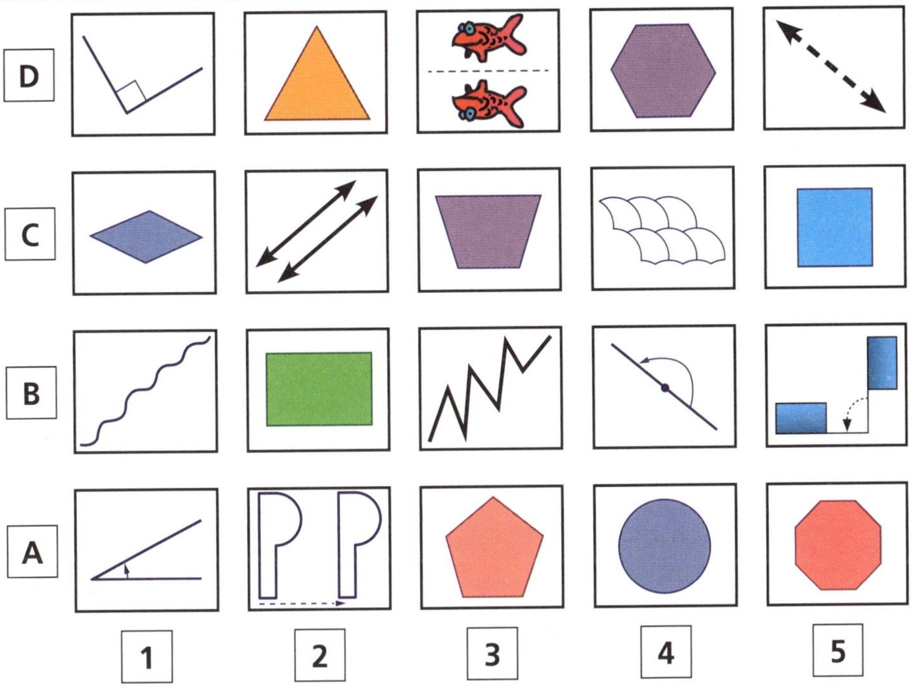

Quadrilaterals have 4 straight sides.

1 Write the coordinates of the picture that shows:

a parallel lines ____ **b** a straight angle ____ **c** the right angle ____

d the smallest angle ____ **e** a circle ____ **f** a rectangle ____

g a rhombus ____ **h** a trapezium ____ **i** a pentagon ____

j a hexagon ____ **k** an octagon ____ **l** a tessellation ____

m a flip (reflection) ____ **n** a slide (translation) ____ **o** a turn (rotation) ____

p each of the quadrilaterals ____ **q** a zigzag line ____

2 The models below are made of geostrips. (You could make them.)

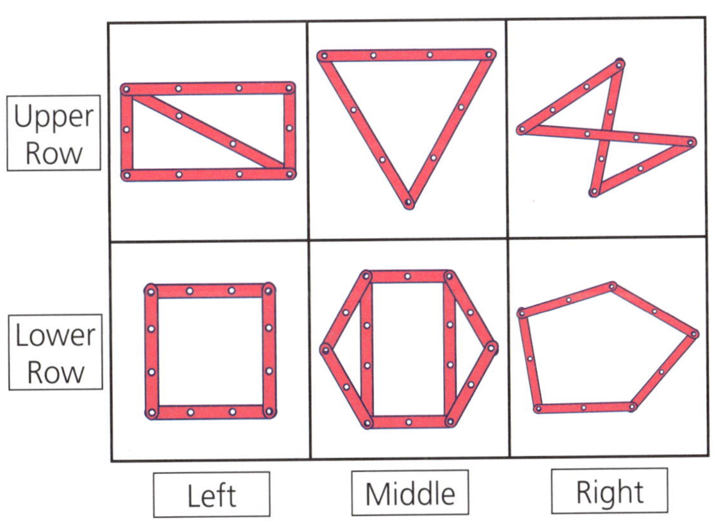

What shape is the model with position:

a middle upper? ____

b right lower? ____

c left lower? ____

What is the position of:

d the two rigid shapes? ____

e the pentagon? ____

Is the pentagon rigid or non-rigid? ____

 • *AUSTRALIAN SIGNPOST MATHS 5* • ISBN 978 1 4886 2190 1

Measuring Angles Using a Protractor

CONCEPT

- Angles are measured in degrees and shown by the symbol °. This measure gives the amount of turning.
- A protractor has:
  1. a base line
  2. a centre point where the 90° line meets the base line
  3. two scales, one on the inside and one on the outside. Each scale goes from 0° to 180°.

1 Write down the size of each angle and say whether it is **acute** or **obtuse**.

a 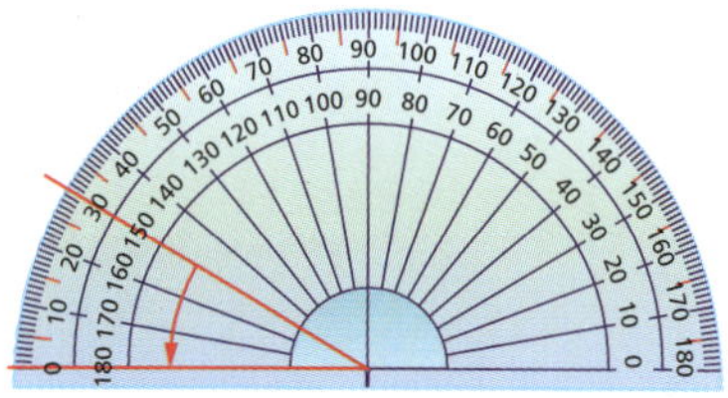

b 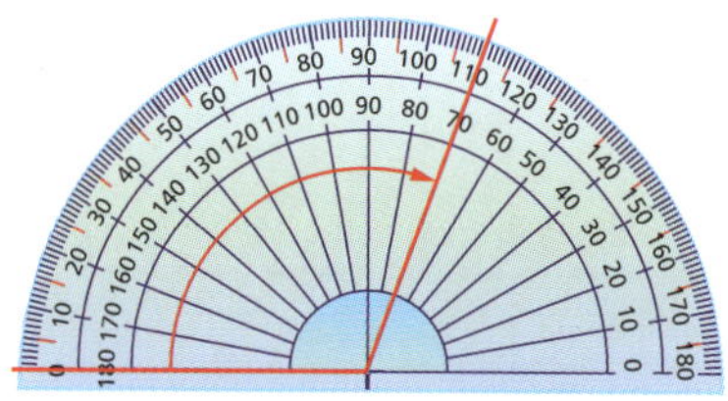

c 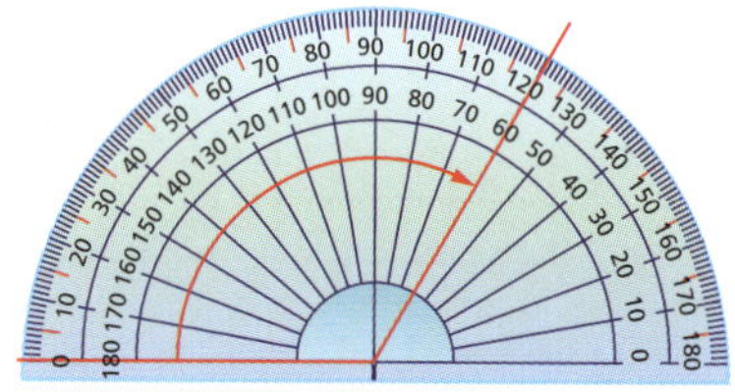

d 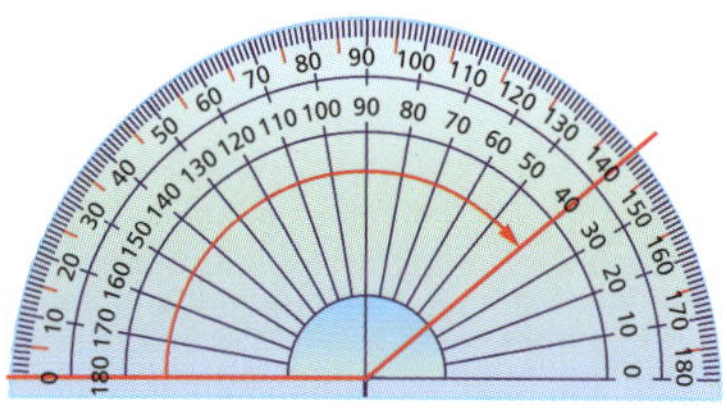

e 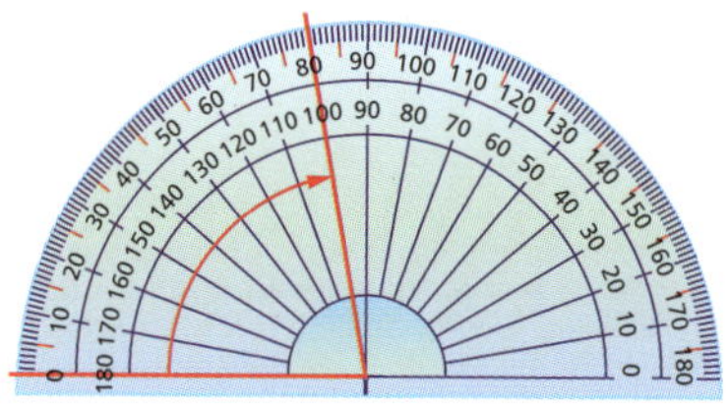

f 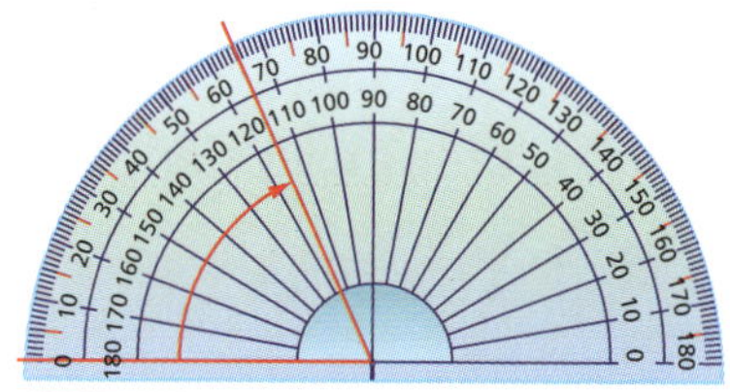

g 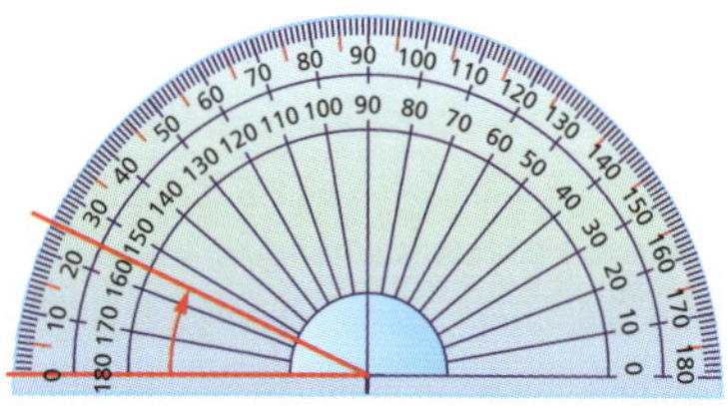

h 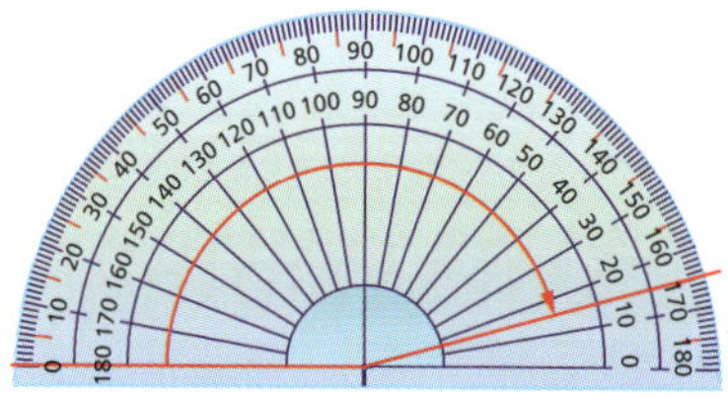

i 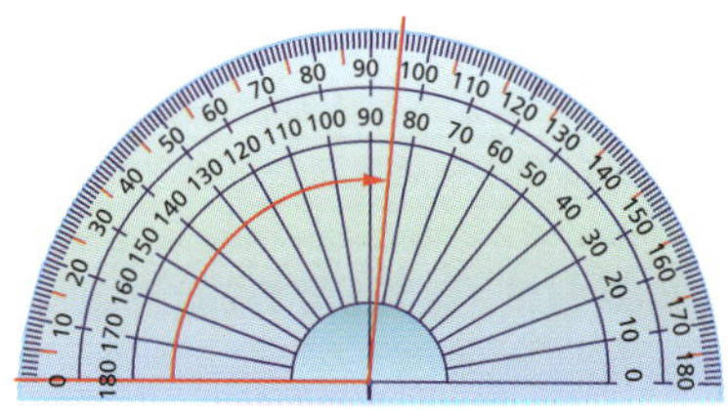

2 How many degrees has a:

a right angle?

b straight angle?

c revolution?

3 Describe how you would use a protractor to measure this angle.

4:08 Angle Types in Degrees

1 Write the size of each angle. (The size is the amount of turn.)

Protractors can measure the amount of turning, from either side.

a 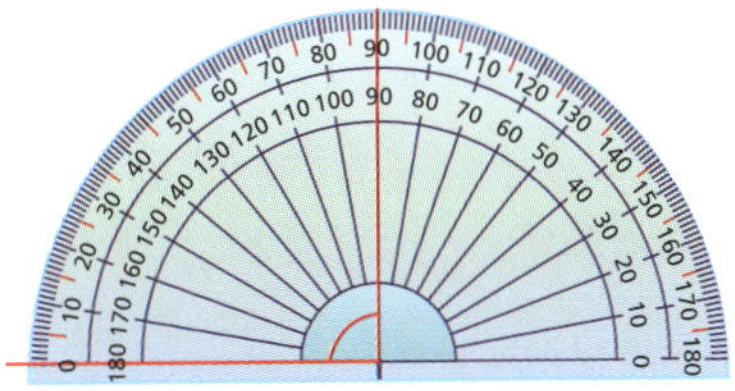

b 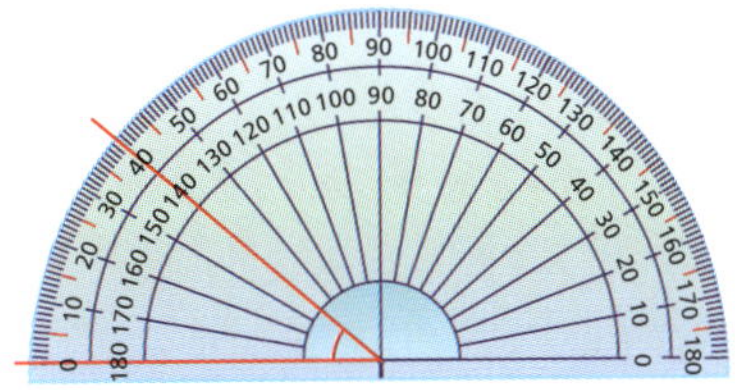

c 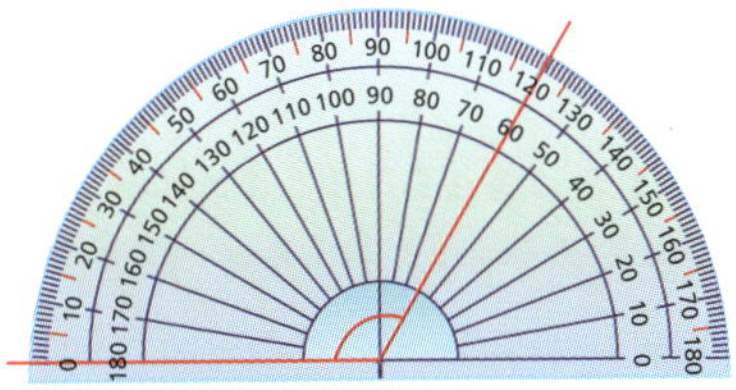

d 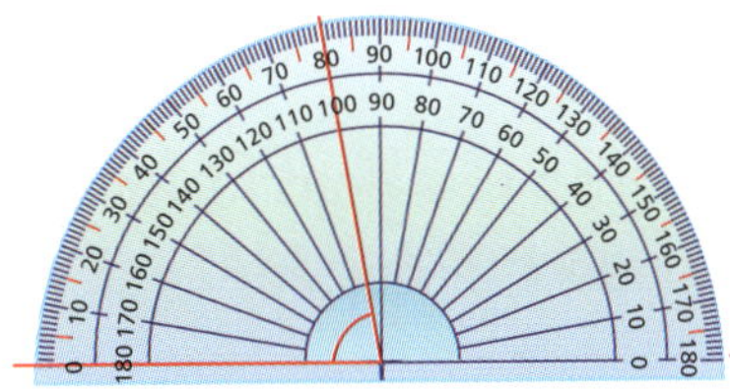

e 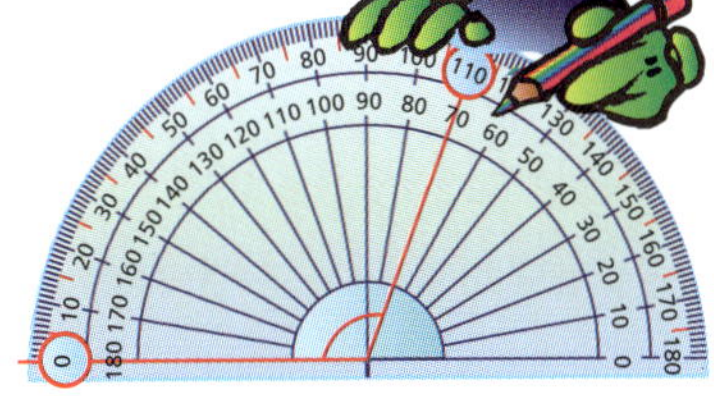

f 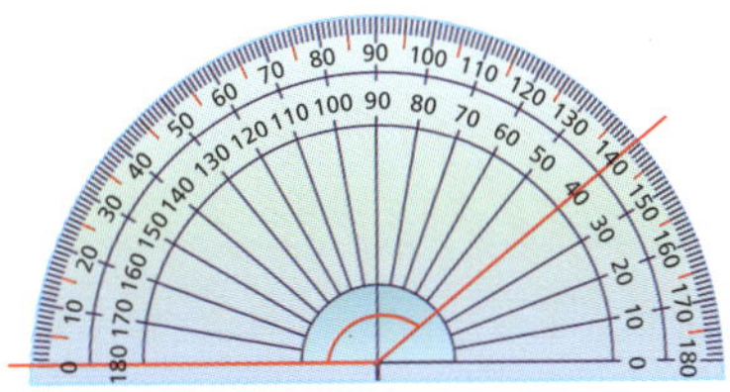

g 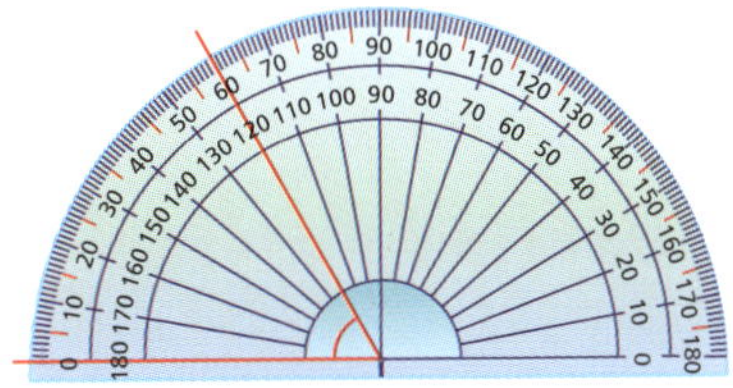

h 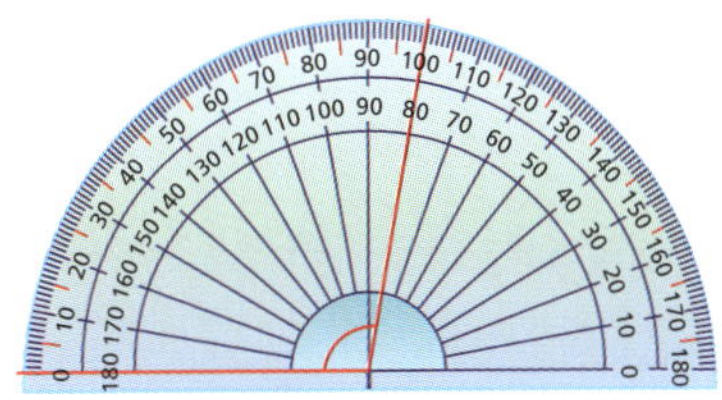

2 Complete the table below by writing the type of angle. (If you need help, see page 129.)

Picture	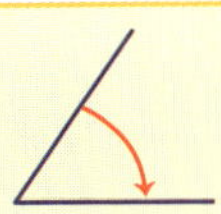	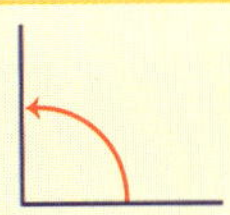	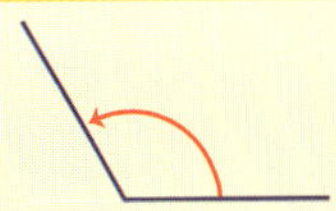	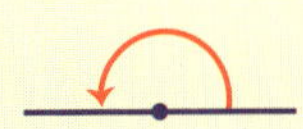	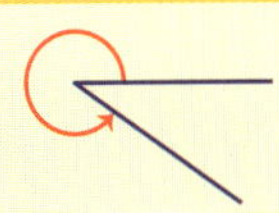	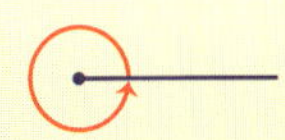
Type	acute					revolution (full turn)
Size	between 0° and 90°	90°	between 90° and 180°	180°	between 180° and 360°	360°

3 Give the type of angle if its size is:

a 35°	b 180°	c 360°	d 100°
e 75°	f 200°	g 90°	h 150°
i 5°	j 350°	k 340°	l 160°
m 15°	n 60°	o 194°	p 110°

4 Draw an example of each angle type included in Question 2.

 • *AUSTRALIAN SIGNPOST MATHS 5* • ISBN 978 1 4886 2190 1

4:09 Using a Protractor

ACTIVITY

- Cut out a photocopy of this protractor. Also cut out the inside part, so you can see the arms of the angle you want to measure.
- You may have to turn the protractor to place one arm on the baseline.

1 Estimate the size of the angle, then use a protractor to measure its size correct to the nearest 10°. Draw the arms longer if necessary.

a

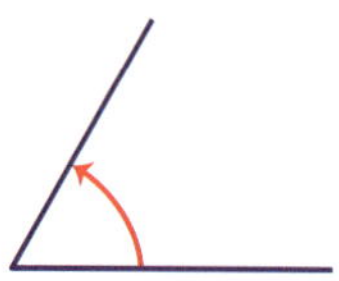

b

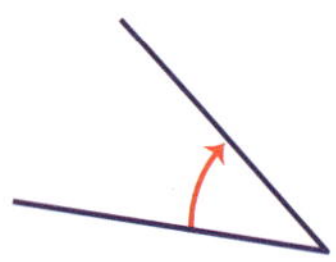

c

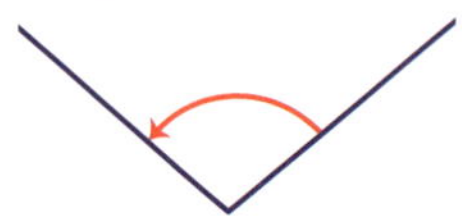

d

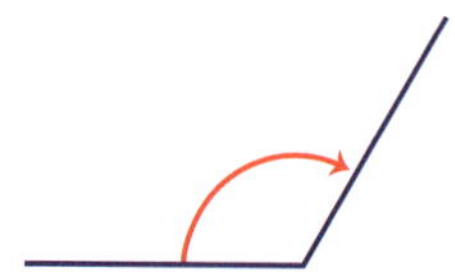

e

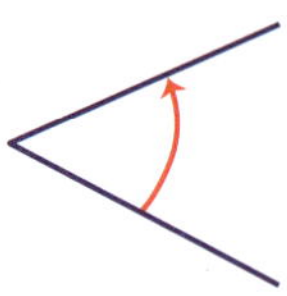

f

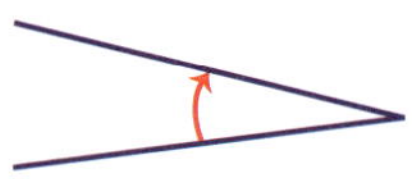

g

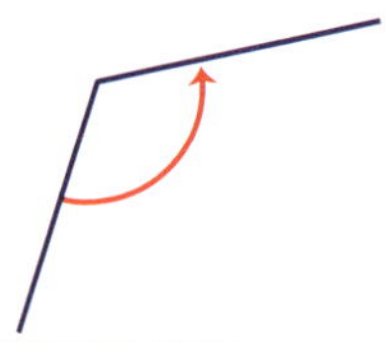

h

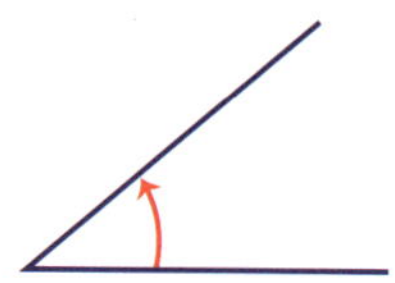

i

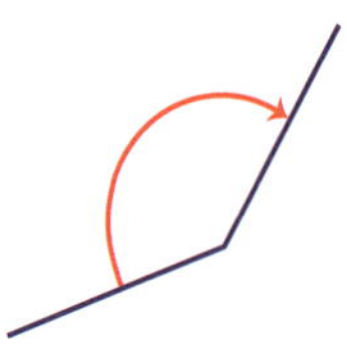

j

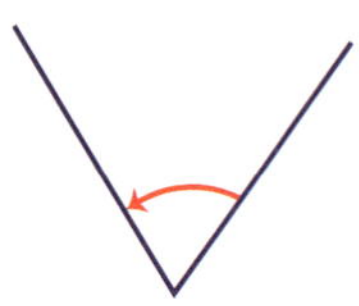

k

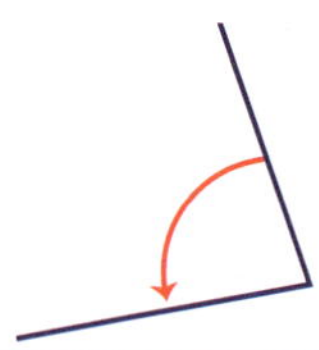

l

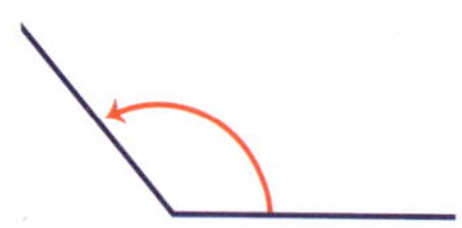

Drawing Angles

ACTIVITY

- Draw one arm of the angle and place this arm of the angle along the baseline, with its end at the centre.
- Follow the scale from zero to the measure you want. Mark the spot, remove the protractor and join the point to the end of the arm.
- Use this method to draw the following angles.

40°	60°	20°	90°
100°	10°	120°	140°
170°	70°	180°	50°

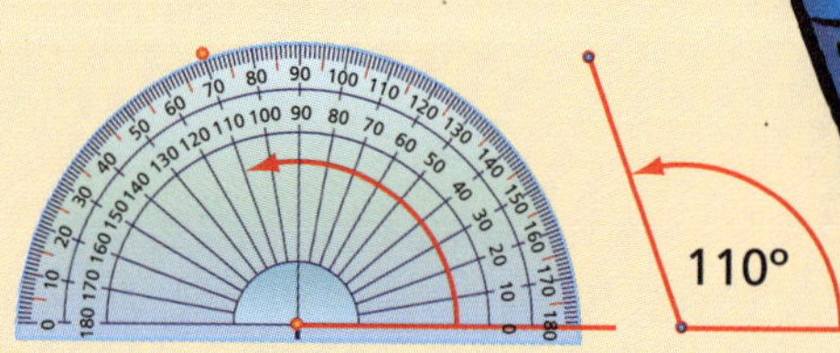

 • *AUSTRALIAN SIGNPOST MATHS 5* • ISBN 978 1 4886 2190 1

4:10 Classifying Angles

Picture	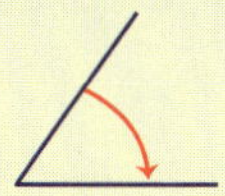	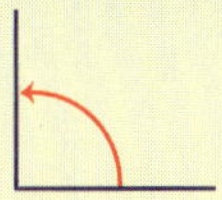	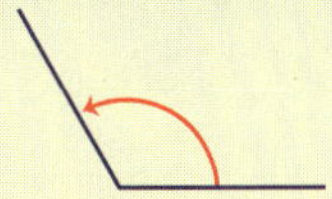	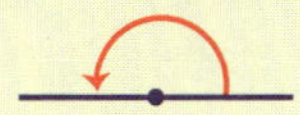	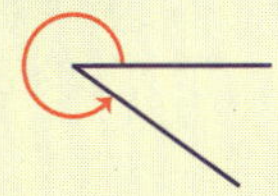	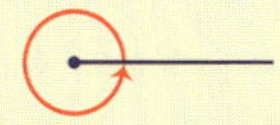
Type	acute	right	obtuse	straight	reflex	revolution (full turn)
Size	between 0° and 90°	90°	between 90° and 180°	180°	between 180° and 360°	360°

1 Name each type of angle.

a 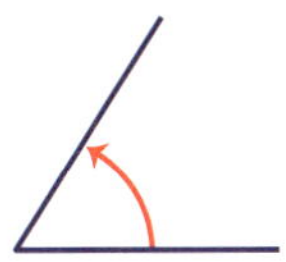

b 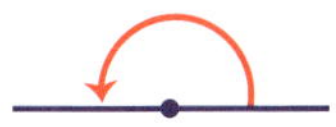

c 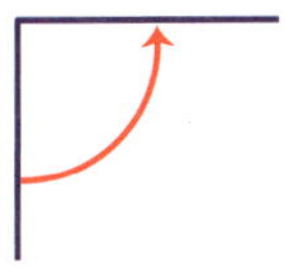

d 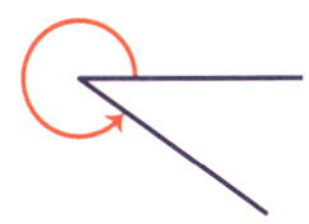

e 

f 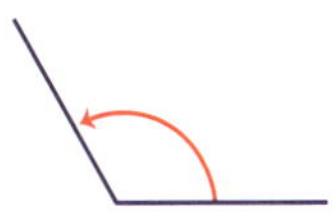

g 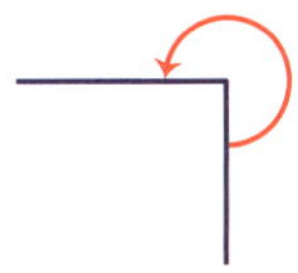

h 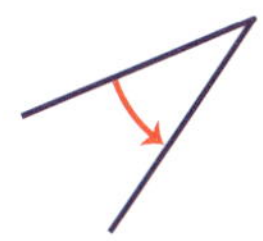

2 The diagram on the right is called a **flow chart**. Flow charts help to organise our thinking. Does it work for all angles up to one revolution?

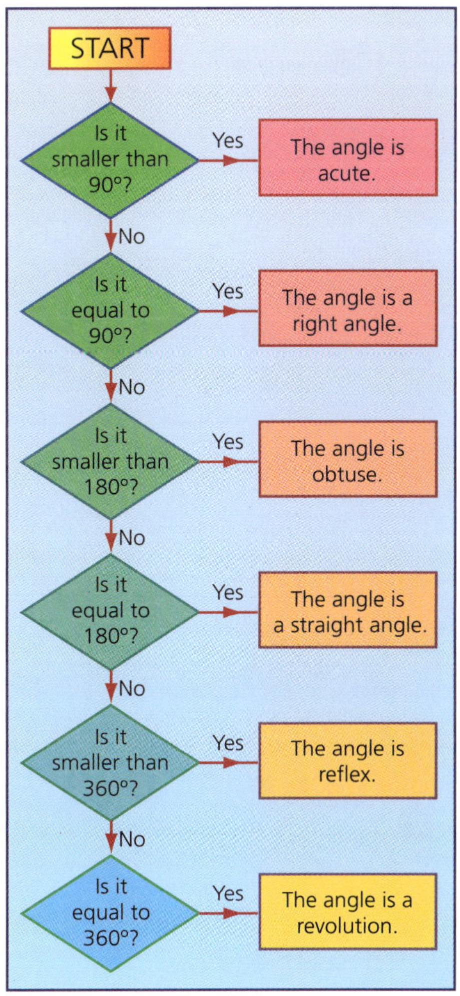

3 Give the angle type if its size is:

a 2° 　 **b** 360°

c 180° 　 **d** 120°

e 90° 　 **f** 307°

g 139° 　 **h** 92°

4 Give the name of a shape that has:

a all angles right angles

b all angles obtuse angles

c all angles acute angles

5 Draw an example of:

a a triangle that has an obtuse angle

b a triangle that has a right angle.

 • *AUSTRALIAN SIGNPOST MATHS 5* • ISBN 978 1 4886 2190 1

MEASUREMENT & GEOMETRY

4:11 Compass Directions

CONCEPT

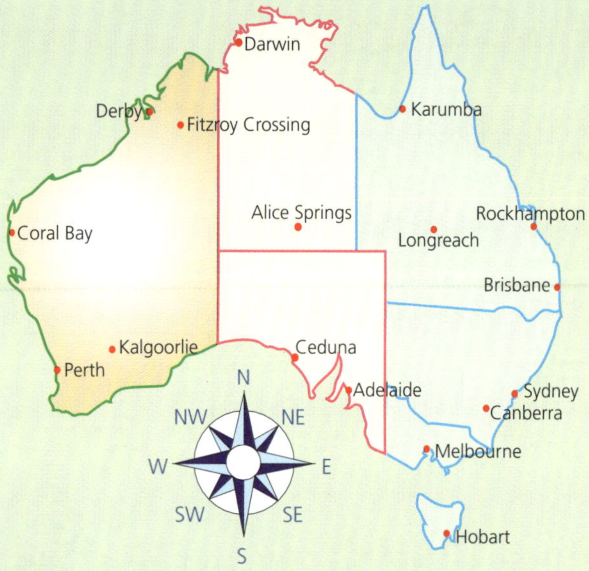

- Halfway between north and east is **north-east**.
- Halfway between north and west is **north-west**.
- Halfway between south and east is **south-east**.
- Halfway between south and west is **south-west**.

- South Australia is in the south.
- Western Australia is in the west.
- The Northern Territory is in the north.

1 Give the direction of each place from Alice Springs. (Use N, E, W, S, NE, SE, NW or SW.)

a Coral Bay ☐ **b** Ceduna ☐ **c** Rockhampton ☐ **d** Darwin ☐

e Derby ☐ **f** Canberra ☐ **g** Karumba ☐ **h** Kalgoorlie ☐

i Sydney ☐ **j** Perth ☐ **k** Longreach ☐ **l** Fitzroy Crossing ☐

2 Give the direction of Alice Springs from:

a Coral Bay ☐ **b** Ceduna ☐ **c** Rockhampton ☐ **d** Darwin ☐

e Derby ☐ **f** Canberra ☐ **g** Karumba ☐ **h** Kalgoorlie ☐

i Sydney ☐ **j** Perth ☐ **k** Longreach ☐ **l** Fitzroy Crossing ☐

3 Name the place that is:

a north of Longreach ☐ **b** east of Kalgoorlie ☐

c north-east of Adelaide ☐ **d** north-west of Kalgoorlie ☐

e north-west of Ceduna ☐ **f** south-west of Darwin ☐

g south-west of Canberra ☐ **h** south-east of Karumba ☐

4 Which is the opposite direction to:

a north? ☐ **b** east? ☐ **c** south-west? ☐

d south-east? ☐ **e** north-east? ☐ **f** north-west? ☐

4:12 Reading a Street Directory

CONCEPT

- This directory is divided into columns and rows.
- The yellow square marked on the map is **5B**, as it is in **column 5** and row **B**.

On this map 1 cm represents 250 m.

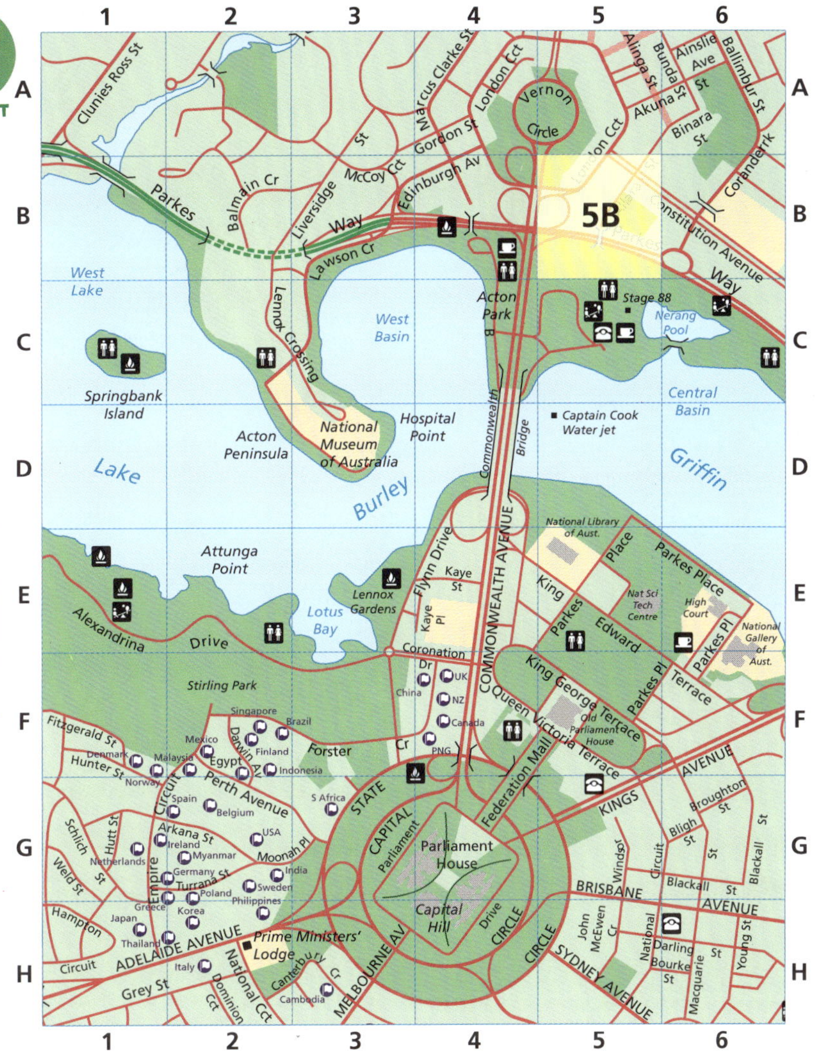

1 Use the map to name the special feature in square:

a 1C

b 5A

c 3E

2 Name the reference coordinates that could be used to find these locations.

a Vernon Circle	b National Library	c Attunga Point
d Hospital Point	e Nerang Pool	f Kaye Place

3 In which street do you finish, if you start from Alexandrina Drive (E1) and take the:

a first right, then first right, then second left, then first left?

b first left, then second right, then first right, then first right?

c first right, then second right, then second right, then first left?

ACTIVITY

Create a grid reference system for the classroom and use it to locate objects.
Describe routes from one object to another.

 • *AUSTRALIAN SIGNPOST MATHS 5* • ISBN 978 1 4886 2190 1

4:13 Rotational Symmetry

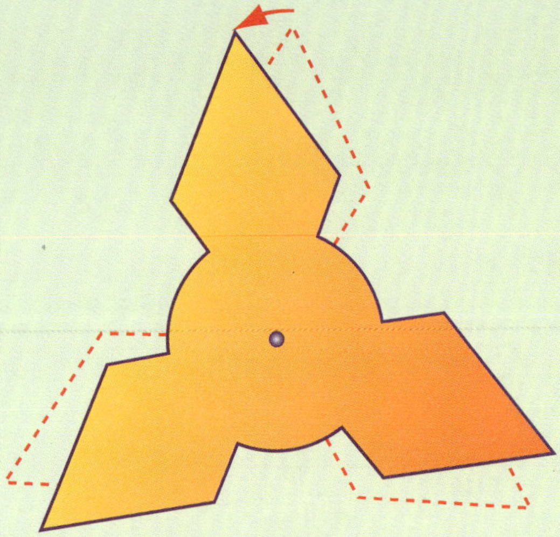

A shape is said to have **rotational symmetry** if a tracing of the shape matches it after the tracing is rotated part of a full turn around the centre.

- If it is rotated about its centre, this shape will match its original position 3 times in one revolution.
- This shape has rotational symmetry of **order 3**.

1 Shape **A**

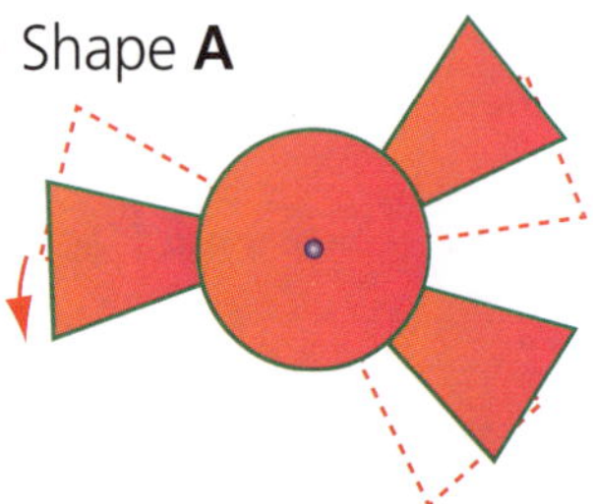

Shape **B**

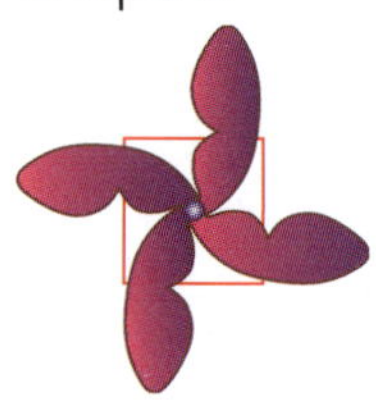

a Does shape **A** have line symmetry? ______

Does it have rotational symmetry? ______

b Does shape **B** have line symmetry? ______

Does it have rotational symmetry? ______

2 Which of the shapes below have rotational symmetry? ______

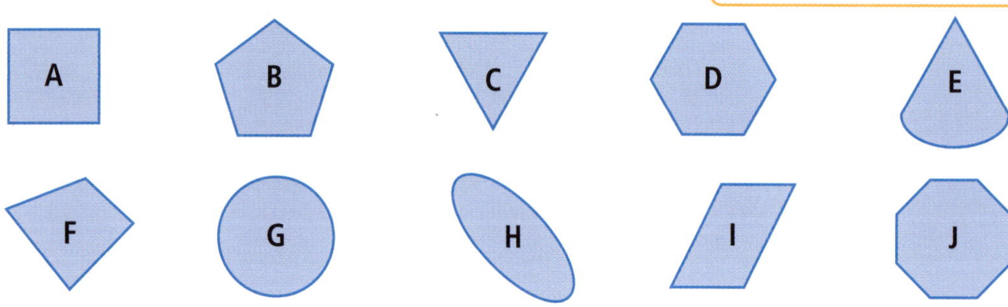

3 Which of the shapes above have line symmetry? ______

4 Do the patterns below have rotational symmetry (**Yes** or **No**)?

If so, what is the order of rotational symmetry? (It must be greater than 1.)

a 

b 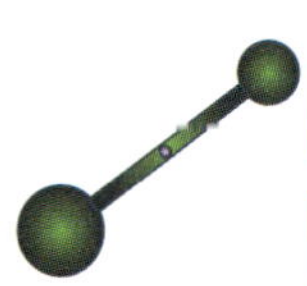

c 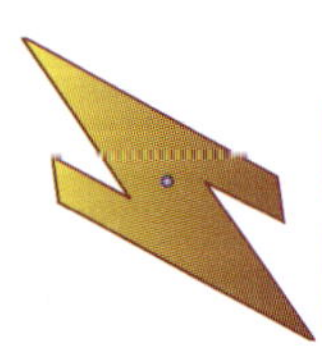

d 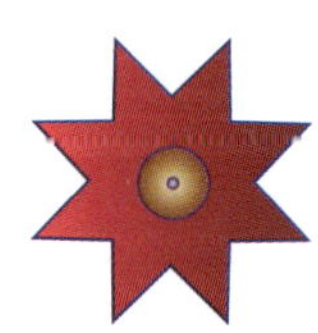

e 

f 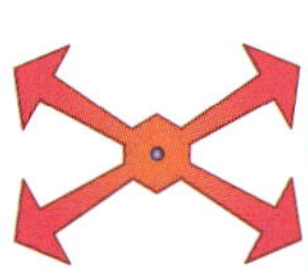

g 

h 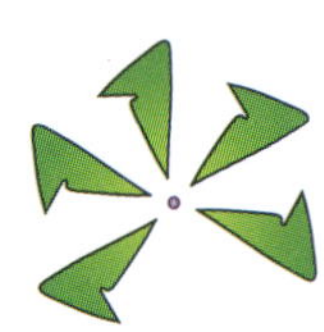

4:14 Measuring Angles of Rotation

CONCEPT

- As the hand turns from 12 to 3 it moves through an angle of 90°.
- As it turns from 2 to 3 it moves through 30°.

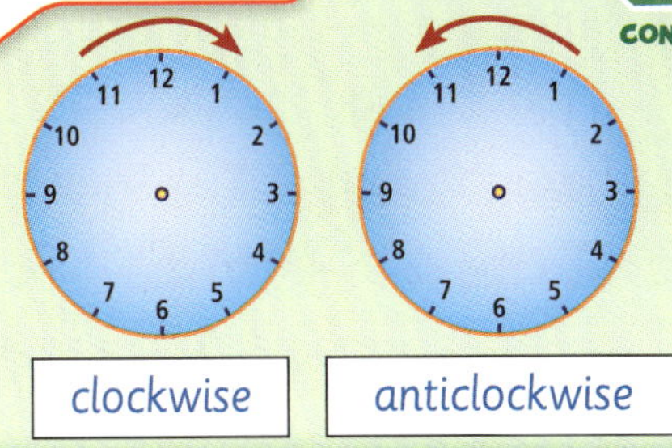

1 Use a protractor to measure the angle through which the hand has turned clockwise.

a  ☐

b  ☐

c  ☐

d 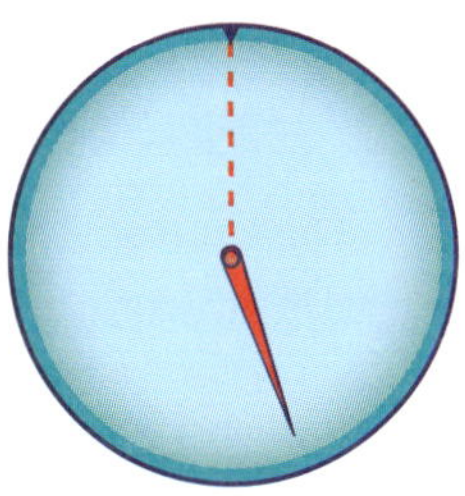 ☐

e 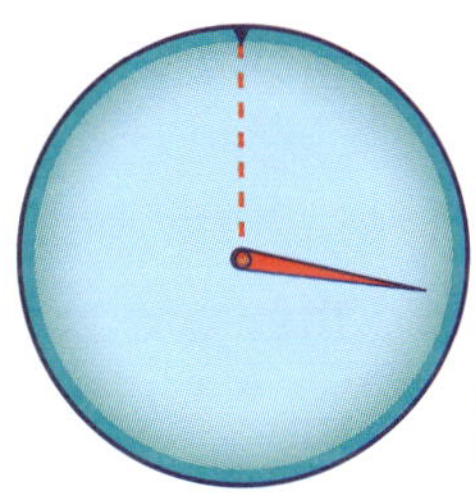 ☐

f 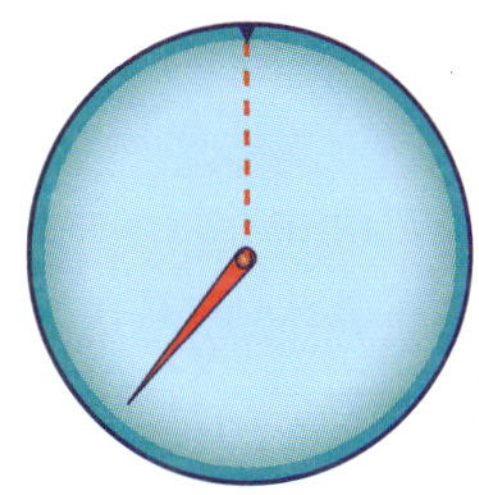 ☐

2 Estimate (E) and then measure (to the nearest 10°) the angle through which the shape has been rotated.

a 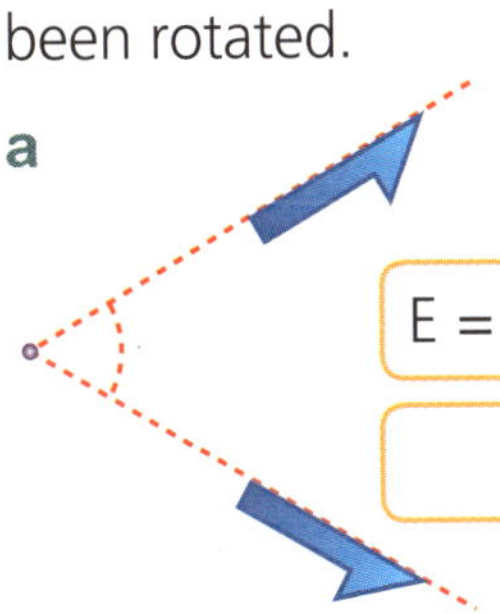
E = ☐
☐

b E = ☐
☐
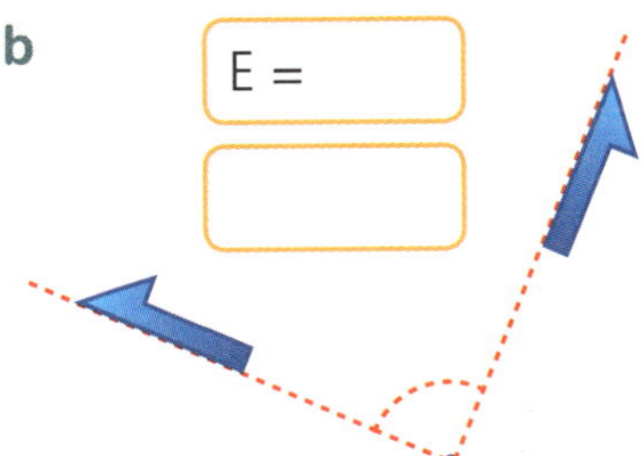

c 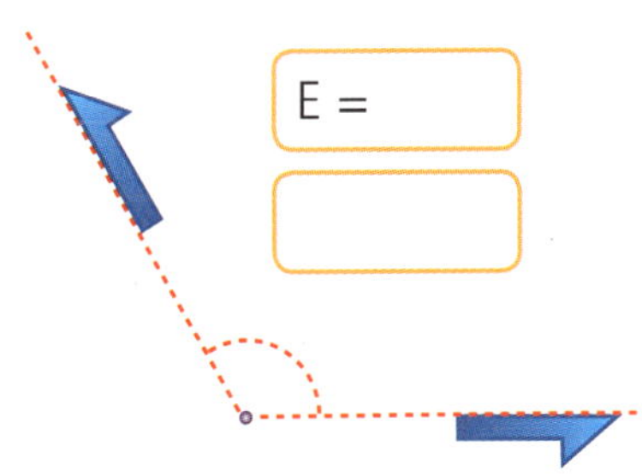
E = ☐
☐

d E = ☐ ☐
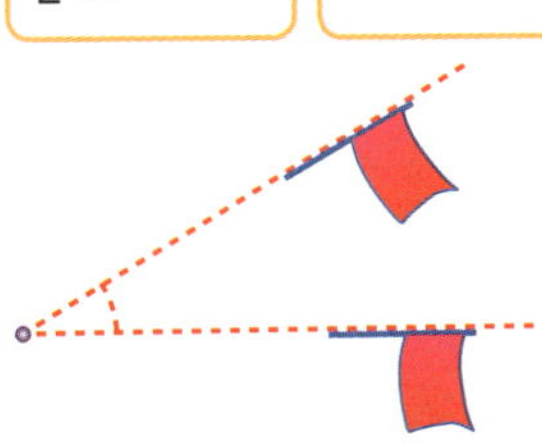

e 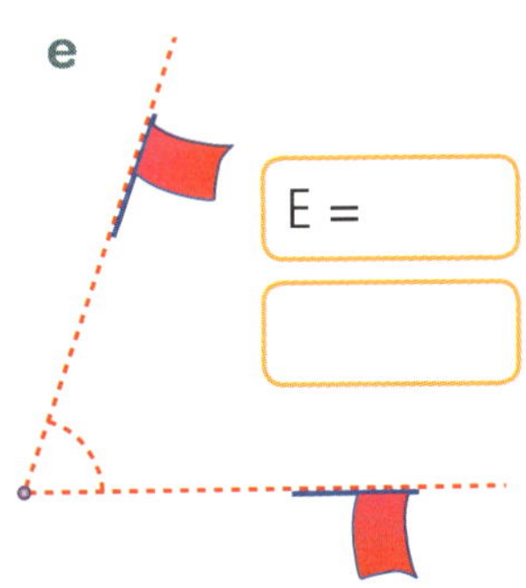
E = ☐
☐

f E = ☐ ☐
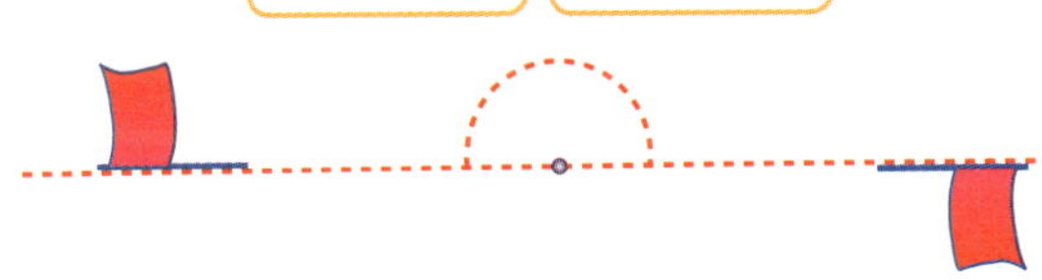

4:15 Rotational Symmetry

FUN SPOT

Turning Patterns

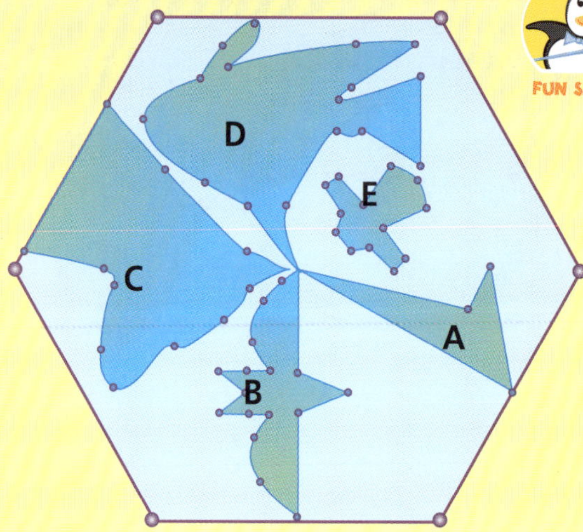

1 Put some thin stiff cardboard under this page. Push a pin through the six vertices and the centre of this hexagon so that they show on the cardboard.

2 Use the pinholes to draw a copy of the hexagon on the cardboard. Make five cardboard hexagons and cut them out.

3 On each cardboard hexagon, copy one of the shapes (**A** to **E**) using a pin and then join up the holes.

4 Cut out the shape to make a template.

5 Now follow the steps in the pictures below.

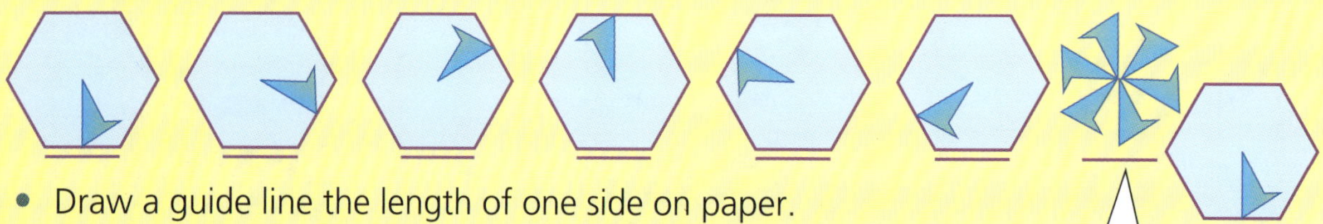

- Draw a guide line the length of one side on paper.
- Trace the inside shape each time you rotate the hexagon.

This pattern has rotational symmetry.

6 Use shape **A** to make a pattern by turning the hexagon through:

a one side at a time b two sides at a time c three sides at a time.

7 Now do the same with shapes **B** and **E**.

8 Use shapes **C** and **D** to make patterns by turning the hexagon through:

a two sides at a time b three sides at a time.

ICT

1 Use a computer to draw these regular shapes.

- All have sides of 12 mm.
- The dots represent centres.

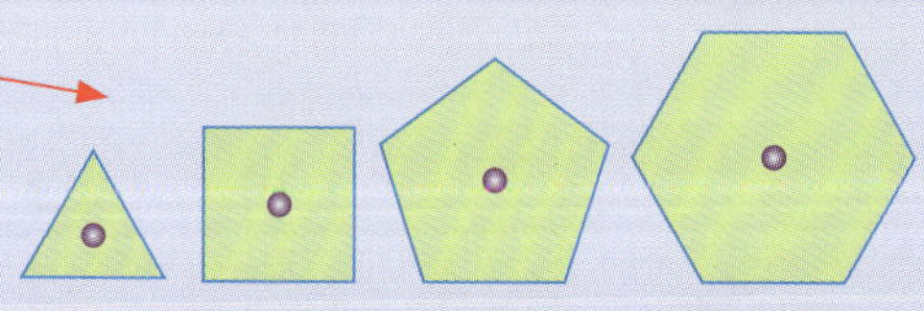

Make the shapes below by copying and pasting the shapes on the right.

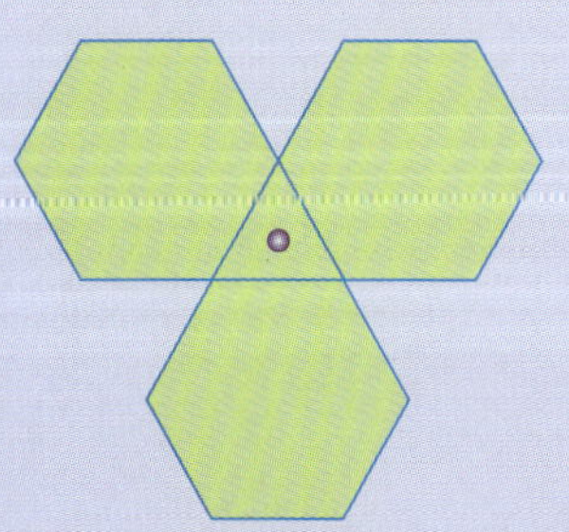

Shape **P**

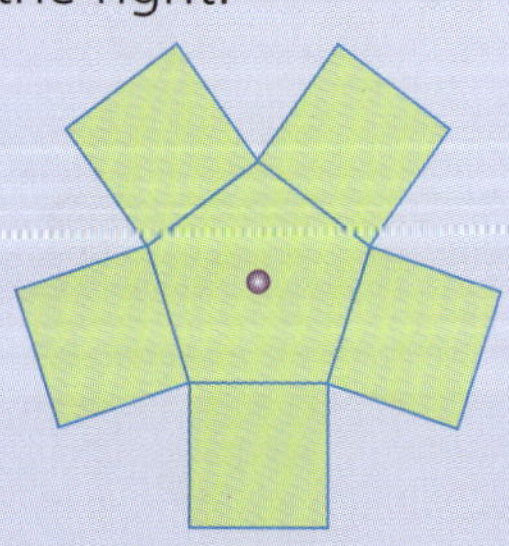

Shape **Q**

Use the rotation tool in your computer graphics software to find how many different ways a copy would fit exactly onto the original for:

a **P** ______ b **Q** ______

The number of times a copy would exactly fit the original in one revolution of turn is called the ***order of rotational symmetry****.*

4:16 Views and Nets of 3D Objects

1 Use these objects to complete the table.

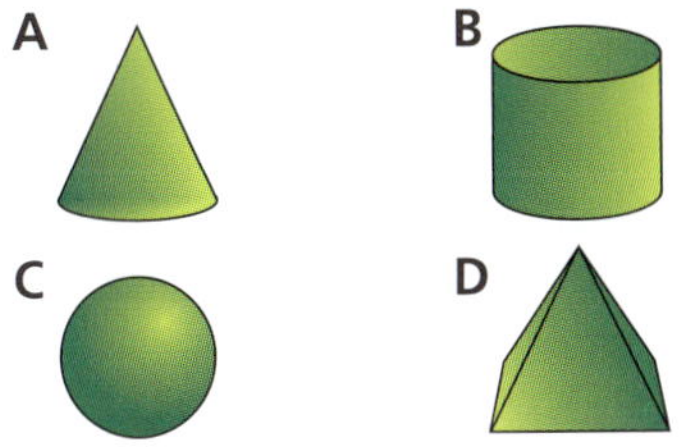

Number of:	A	B	C	D
surfaces				
edges				
vertices (corners)				
curved surfaces				
faces (flat surfaces)				

2 Which of the objects A–D could have these top views?

a b c

3 Which of the objects A–D could have these front views?

a b c

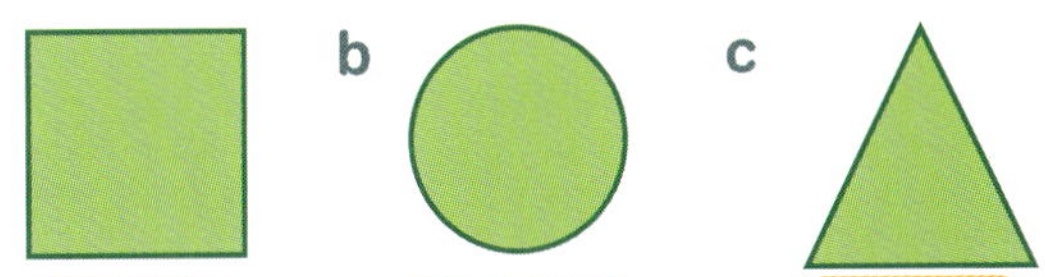

4 On the right is the net of an open cube. (It has no lid.)
Which of the nets below are of an open cube?

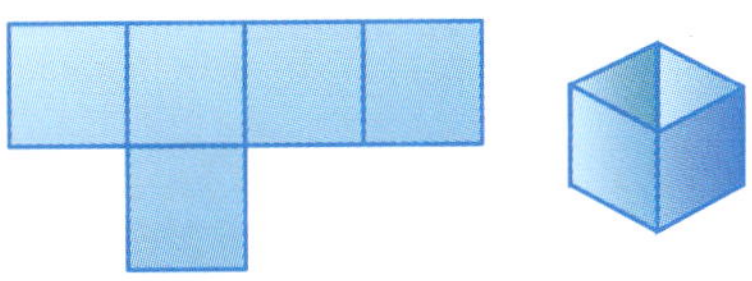

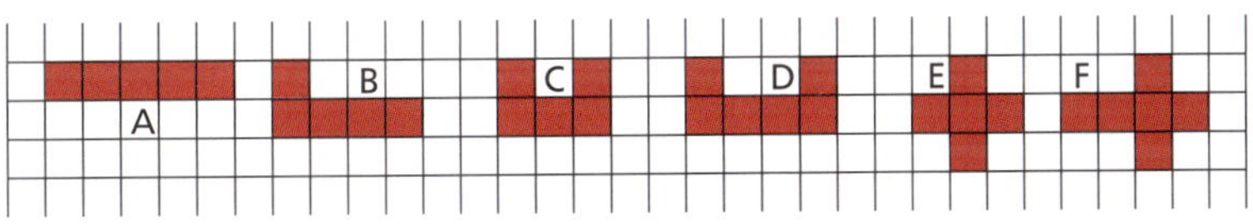

Copy one of the nets above to make an open cube of side length 3 cm.

5 Which of the nets below are of an open rectangular prism?

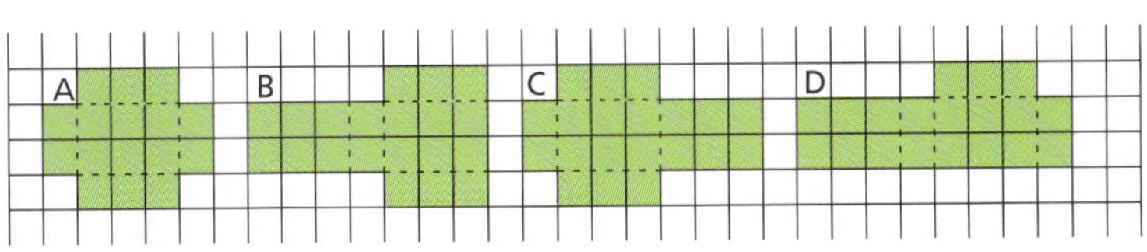

Copy one of them to make an open rectangular prism of height 2 cm and with a base 6 cm long and 4 cm wide.

6 **a** What is the name of the 3D object drawn on these isometric dots?

b For this object, find the number of faces + vertices – edges.

c On the isometric dots, draw a cube and a rectangular prism. For each object, find the number of faces + vertices – edges.

Cube Rectangular prism

Vertices are **corners**.

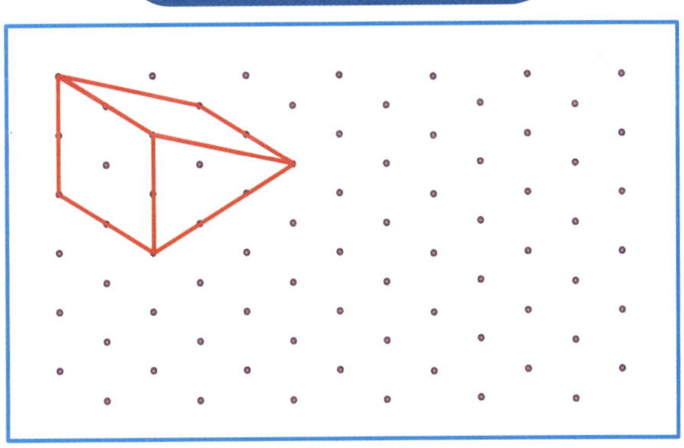

 • *AUSTRALIAN SIGNPOST MATHS 5* • ISBN 978 1 4886 2190 1

4:17 Reading a Street Directory

1. As this is map 24,
24 H7 is the square
2. marked on the map.
H7 is in **column H** and **row 7**.

Scale: 1 cm represents 200 m.

Use the internet to find aerial views of places.

1 Use this map to name the special feature in the square:

a E3 ______ **b** F1 ______ **c** H3 ______

d G4 ______ **e** H5 ______ **f** A6 ______

2 Name the reference coordinates that could be used to find:

a Parliament House ______ **b** The Domain ______ **c** NSW Art Gallery ______

d Observatory Park ______ **e** Man o' War Jetty ______ **f** Wynyard Station ______

3 **a** What two reference coordinates could be used for Pyrmont Bridge? ______

b What three reference coordinates would be best for Hyde Park? ______

4 What is the length of Pyrmont Bridge to the nearest one hundred metres? ______

5 **a** What building is north of Government House? ______

b What route would you take to travel from the Opera House to the Town Hall?

______

4:18 Using Coordinates

CONCEPT

On grids, coordinates are used to describe a position. Coordinates often refer to the **intersection** of grid lines. On a map, north is towards the top.

1. Write down the countries that have these coordinates.
   - a C14
   - b A15
   - c H10
   - d E7
   - e F9
   - f G7
   - g C11
   - h B13

2. Which of the coordinates should we use for French Guiana?

3. Give the coordinates of:
   - a Uruguay
   - b Ecuador
   - c Suriname
   - d Paraguay

4. Give three sets of coordinates for Colombia.

5. Which different sets of coordinates could be used for Bolivia?

6. Give the greatest distance (to the nearest 100 km) across each country (west to east).
   - a Uruguay
   - b Bolivia
   - c Brazil
   - d Colombia

7. Which country is approximately:
   - a 1 500 km south of Colombia?
   - b 800 km west of Suriname?

8. Use a grid to draw a map of Victoria using a simple scale.

 • *AUSTRALIAN SIGNPOST MATHS 5* • ISBN 978 1 4886 2190 1

4:19 Drawing Angles

Drawing an Angle

CONCEPT

1 Draw one arm of the angle.
2 Place the protractor so that the arm of the angle is on the baseline with one end at the centre.
3 Follow the scale from zero to the measure you want. Mark the spot.
4 Join this mark to the end of the arm to make the angle.

1 Use the method shown above to draw each angle on the left of the line.

a 50° b 40° c 65°

d 120° e 105° f 147°

2 On your own paper draw an angle of:

a 30° b 90° c 135° d 45° e 165°
f 73° g 118° h 93° i 14° j 172°

3 Draw the reflex angle:

a 300° b 270°
c 190° d 355°
e 225° f 317°

A **reflex angle** is between a half turn and a full turn.

How to Draw a Reflex Angle

CONCEPT

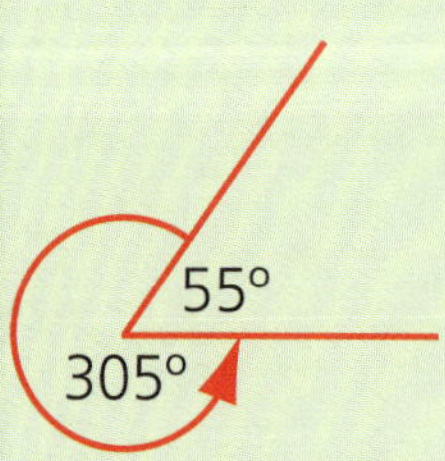

1 To draw an angle of 305°, calculate 360° – 305° (= 55°).
2 Draw the 55° angle.
3 Mark the reflex angle (305°) on the other side of 55°.

4:20 Measuring Angles Greater Than 180°

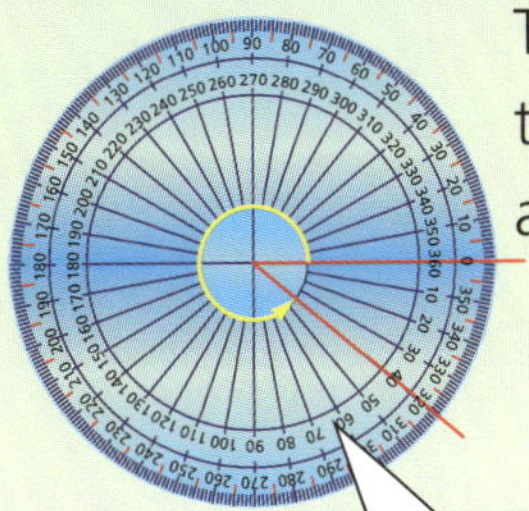

This is a 360° protractor.
The angle shown is 320°.

To measure a reflex angle, measure the smaller angle between the arms and subtract this from 360°.

This angle is:
360° – 30°= 330°.

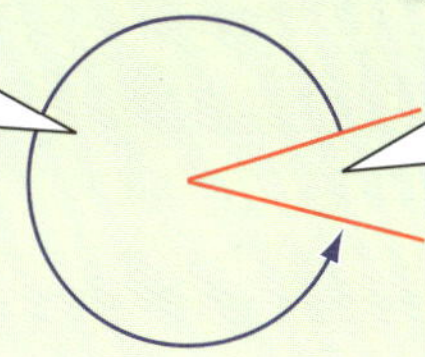

This angle measures 30°.

If the angle size is between a straight angle (180°) and a revolution (360°), it is called a reflex angle.

CONCEPT

1 Write the angle shown on each 360° protractor.

a

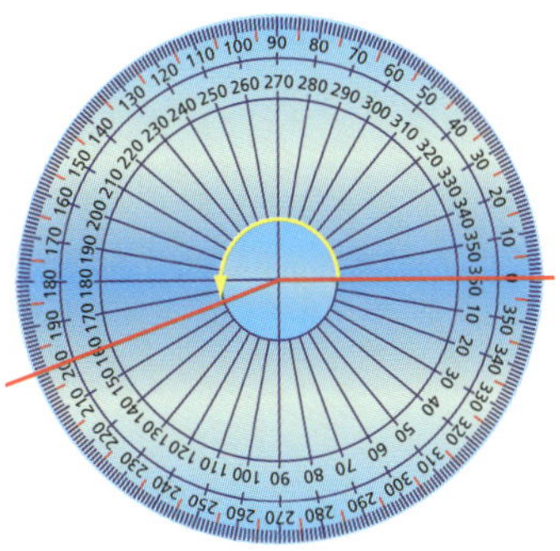

b

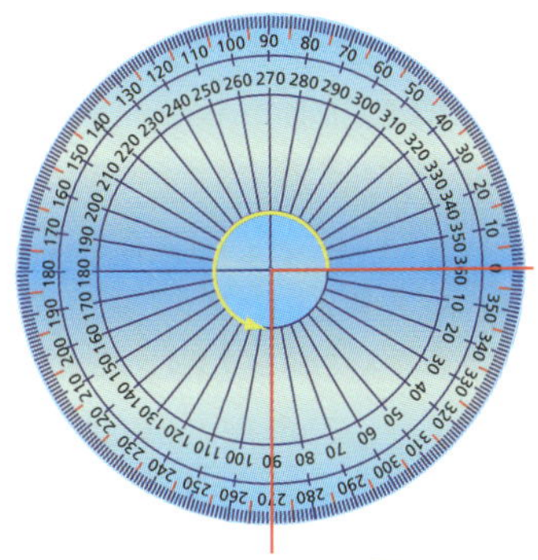

c

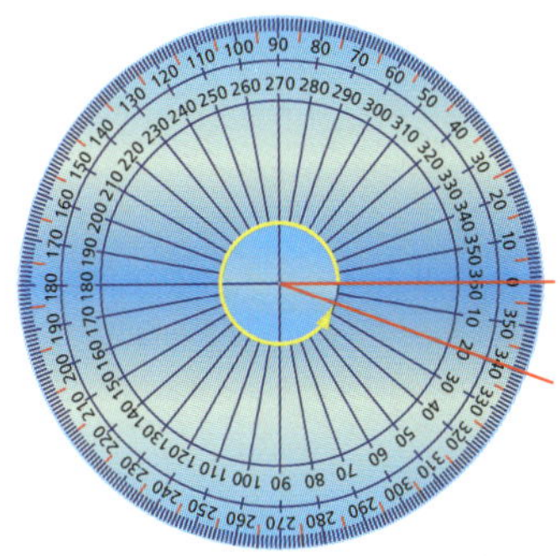

2 Angles at a point make a revolution (add up to 360°).
For each part of Question 1, find the size of the smaller angle at the point.

a   **b**   **c**

3 The size of the smaller angle is given. Find the size of each reflex angle.

a

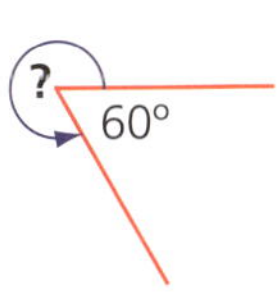

b

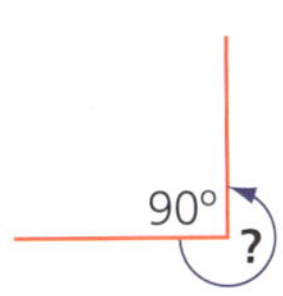

c

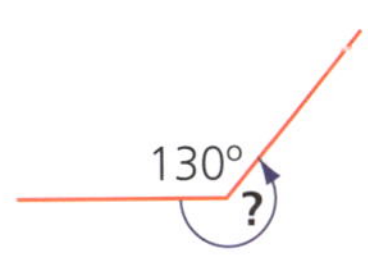

d

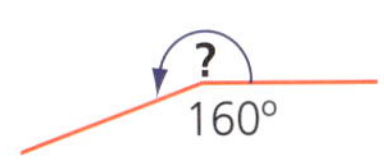

e

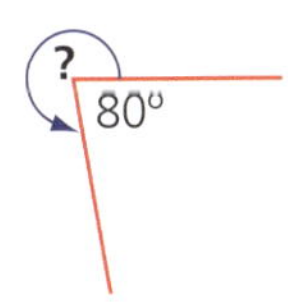

4 Measure each of these reflex angles correct to the nearest 10°.

a

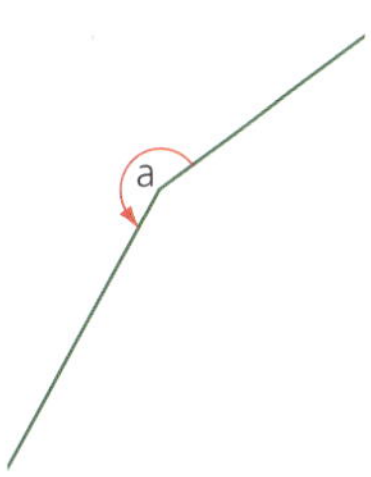

b

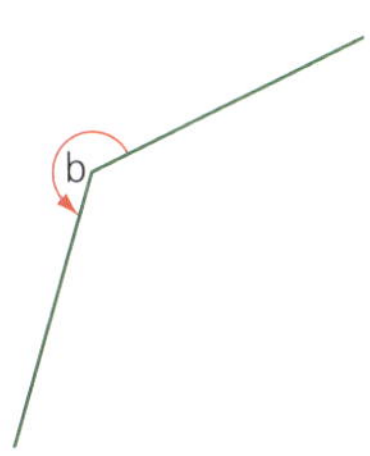

c

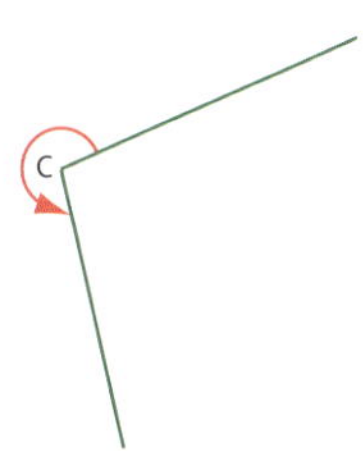

d

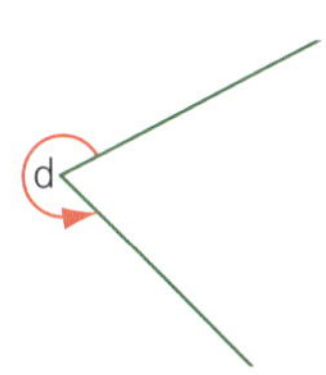

e

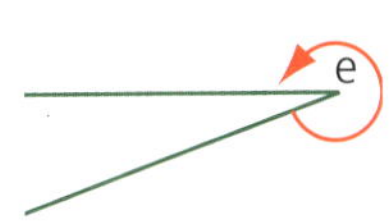

f

 • *AUSTRALIAN SIGNPOST MATHS 5* • ISBN 978 1 4886 2190 1

4:21 Enlargements and Reductions

CONCEPT

Each point on the large shape is three times as far from centre **O** as the same point on the smaller shape. We have drawn an **enlargement** of the small shape.

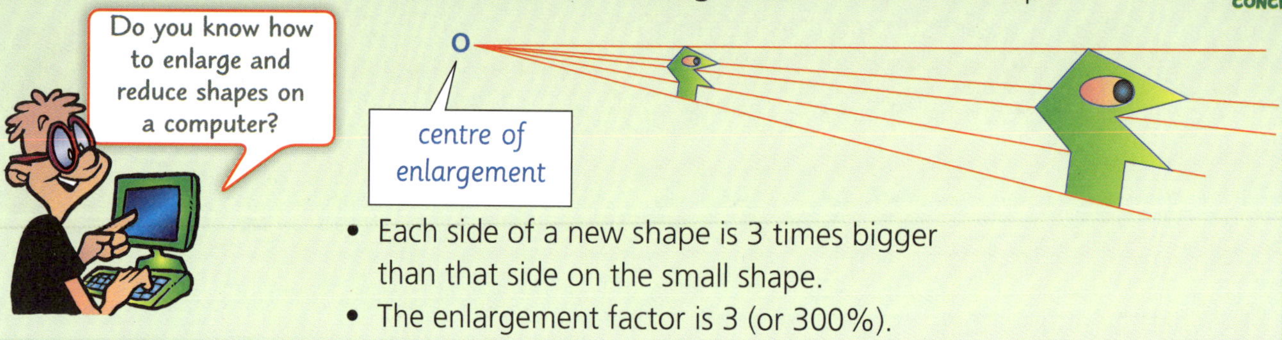

- Each side of a new shape is 3 times bigger than that side on the small shape.
- The enlargement factor is 3 (or 300%).

1 By doubling the distance of each point from the centre, draw a shape that has been enlarged by a scale factor of 2.

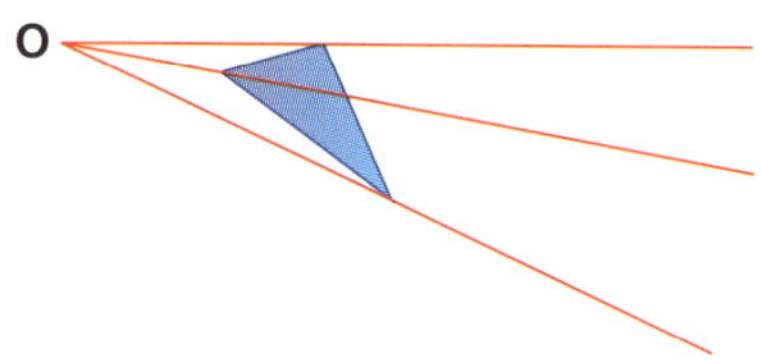

In each case the enlargement factor of the new shape is ______.

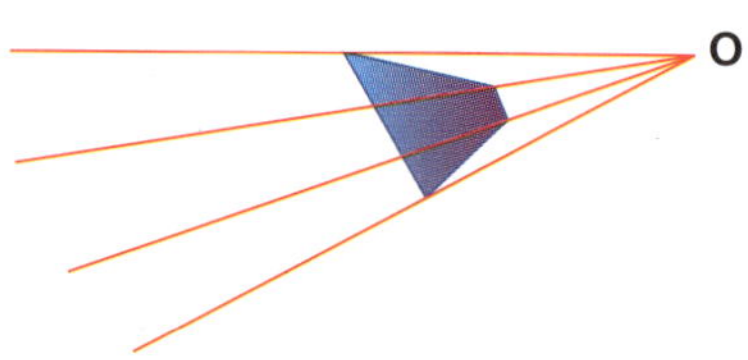

2 Here the centre of enlargement is inside the shape. Enlarge each shape by a scale factor of 3.

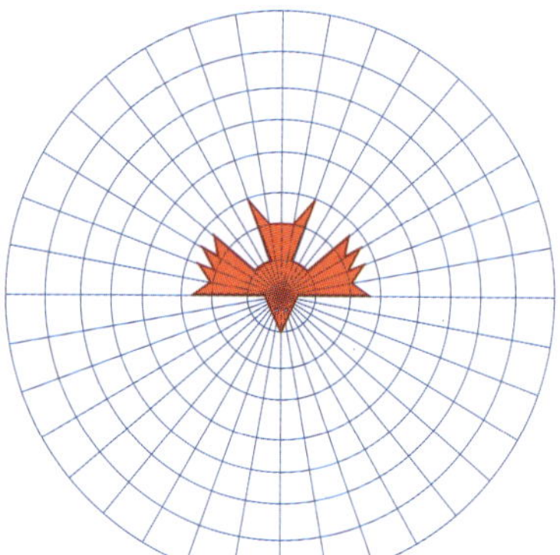

In each case the enlargement factor of the new shape is ______.

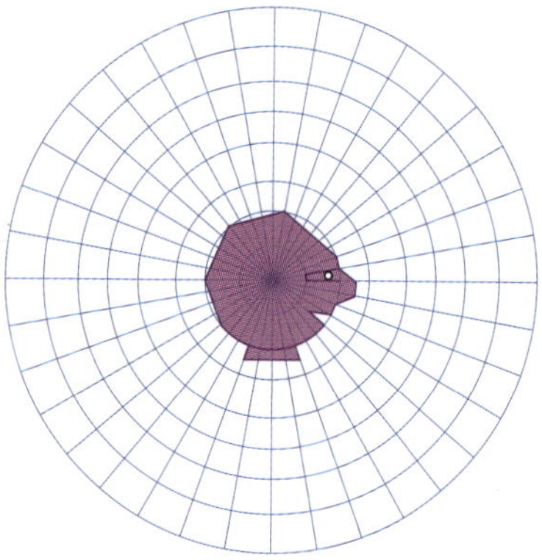

3 Each of these rectangles has the same shape.

a What is the enlargement factor of B if A is the original? ______

b What is the enlargement factor of C if A is the original? ______

c What is the enlargement factor of C if B is the original? ______

d What is the enlargement factor of A if B is the original? ______

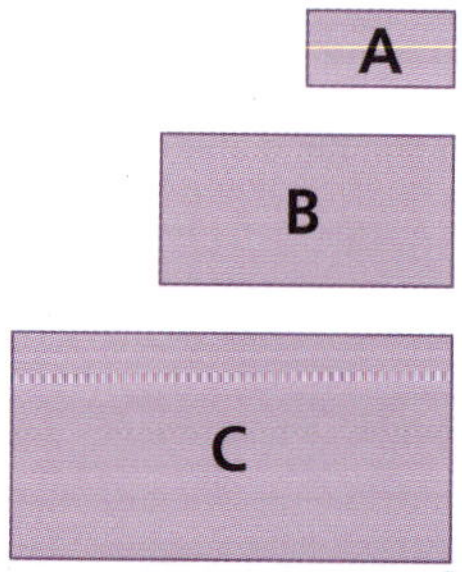

ICT

Using a graphics program on your computer, enlarge and reduce a variety of shapes.

4:22 Enlargements and Reductions

CONCEPT

If the enlargement factor is $\frac{1}{2}$, then each length is:

- $\frac{1}{2}$ of the original length or
- 50% of the original length or
- 0·5 of the original length.

$\frac{1}{2}$ = 50% = 0·5

1. Draw the alligators in the large grid by copying one square at a time.

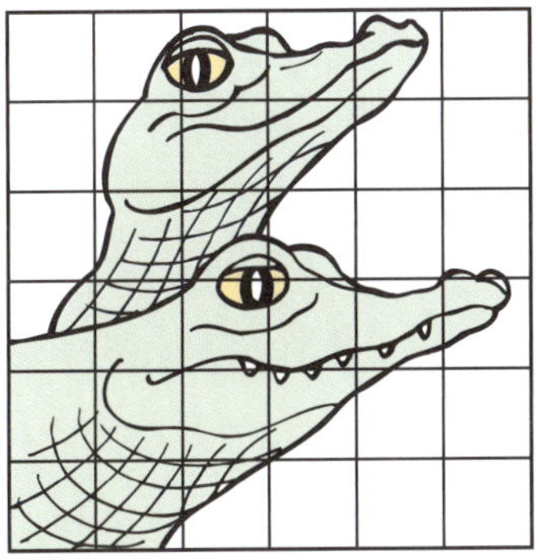

What enlargement factor has been used to make the larger drawing? ☐

2. Copy the frog onto the grid below it. What enlargement factor has been used on the copy you are making? ☐

3. Copy the camel onto the smaller grid beside it. Here the enlargement factor is $\frac{2}{3}$ or $\frac{4}{6}$. It is a reduction.

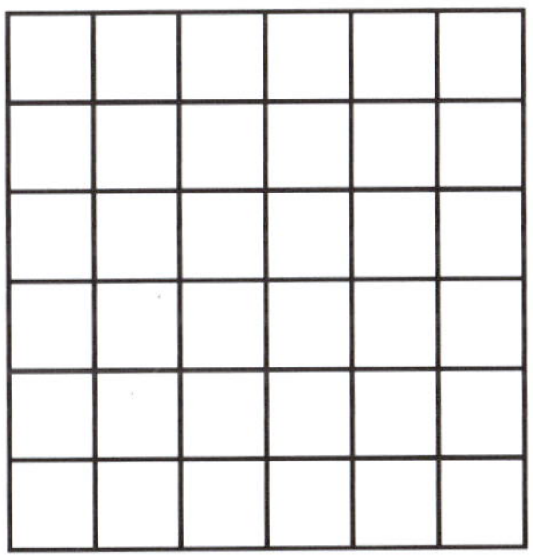

4. Draw a grid on a picture of your own and copy it.

4:23 Enlargements and Reductions

- Since the large drawing is four times as long as the ant, the scale is 4 to 1.
- This scale is written 4:1.

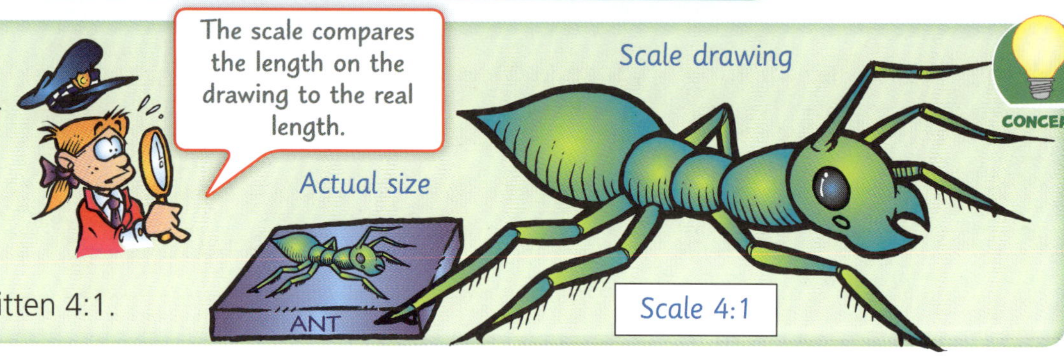

CONCEPT

1 **a** How many times longer is the leg in the drawing than the ant's leg? [ ]

b What is the length of the actual ant? [ ] What is the length of its head? [ ]

c Use your answers to find the length of the ant and the length of the head in the picture. Check by measuring the picture. Ant = [ ] Head = [ ]

2 Luke is the only boy in his family. This photograph has a scale of 1:40. Measure it in millimetres then find the real height of:

a Luke [ ]

b Luke's mother [ ]

c Luke's father [ ]

d his youngest sister [ ]

Ask: Does the answer make sense?

3 This picture of a flea is 75 times as long as the real thing.

a What is the scale used in this picture? Scale [:]

b About how long is a real flea? [ ]

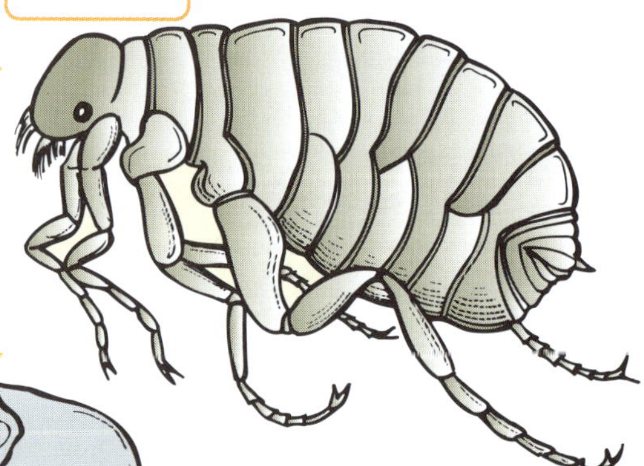

Scale = 1:60

Scale = 1:35

Scale = 1:100

Scale = 1:750

4 Each picture above has been drawn to the scale shown. By measuring each one to the nearest millimetre, find:

a the height of the kangaroo [ ]

b the length of the fox [ ]

c the height of the elephant [ ]

d the length of the blue whale [ ]

5:01 Reading Graphs

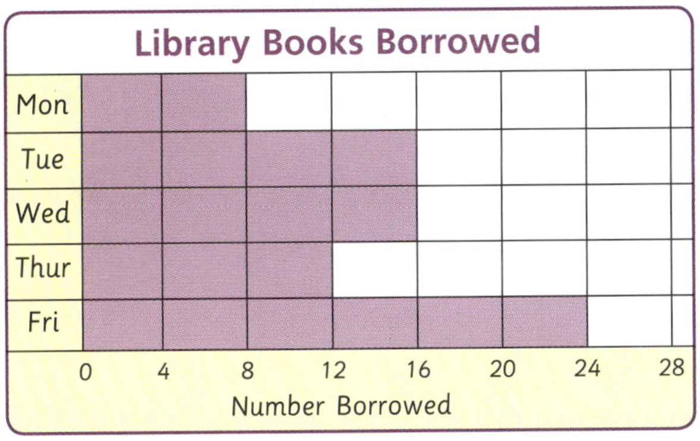

1 **a** How many books were borrowed on Thursday?

b On what day was the least number of books borrowed?

c On what days was the same number borrowed?

d One rectangle represents how many borrowed books? Scale: 1 unit =

e What was the total number of books borrowed during this week?

f Why would more books be borrowed on Friday?

2 **a** Which month had:

i most sales?

ii least sales?

b How many sales occurred in:

i summer (Dec, Jan, Feb)?

ii autumn?

iii winter? **iv** spring?

v the first half of the year? **vi** the second half of the year?

c What was the total number of cars sold in the year?

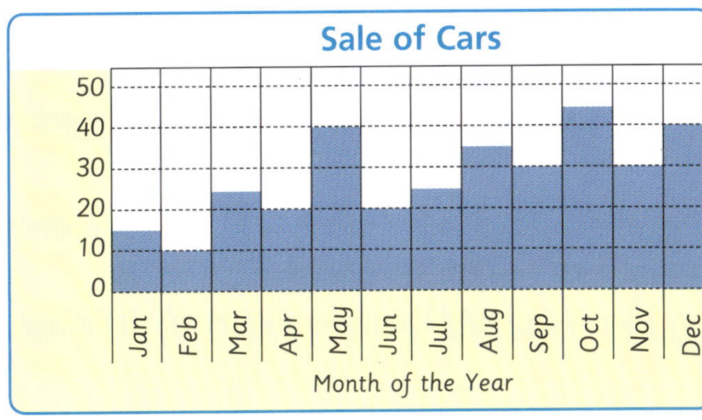

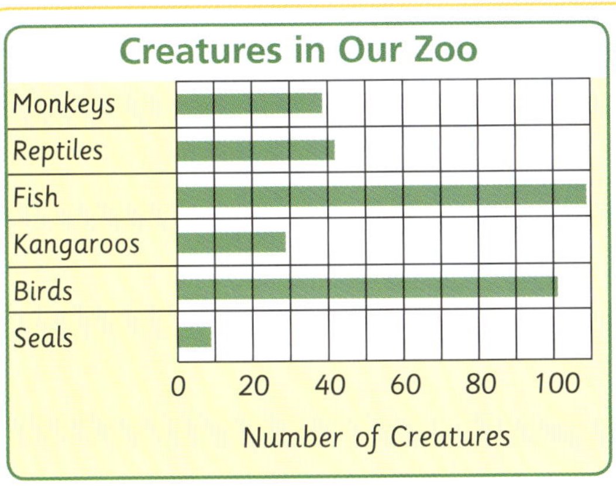

3 **a** Is it easy to tell the numbers of each creature?

b Give answers, correct to the nearest ten, for the approximate number of:

i monkeys **ii** fish

iii birds **iv** seals

c What is the approximate number of creatures in our zoo altogether?

d Why do you think there are more fish and birds in the zoo?

Noughts and Crosses

ACTIVITY

4 To win at Noughts and Crosses using a four-by-four grid you need three in a line, across, down or diagonally.

How many ways are there to get three in a line?

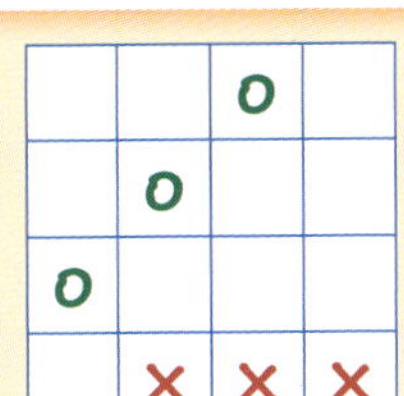

5:02 Drawing Graphs

1 Five taxi drivers recorded the distance travelled on Monday. Smith travelled 700 km, Brown travelled 600 km, Wong travelled 800 km, Ryan travelled 1 000 km and McDonald travelled 500 km. Show this information on this bar graph.

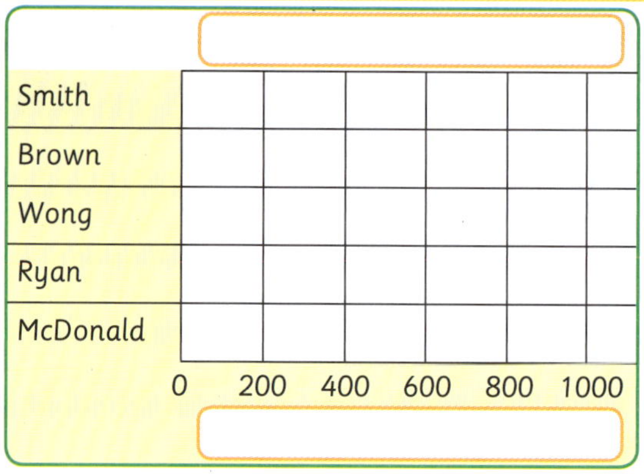

2 Year 5 teacher, Mrs Foster, had to choose three project topics for her class to do. She asked the librarian how many books covered each topic. The librarian, Mrs McSeveny, found that there were 6 books on wheat, 16 on wool, 4 on clouds, 13 on convicts, 18 on whales, 2 on ants and 8 on bees.

a Draw a column graph of this information.

b Which three topics do you think Mrs Foster would choose? Give a reason for your answer.

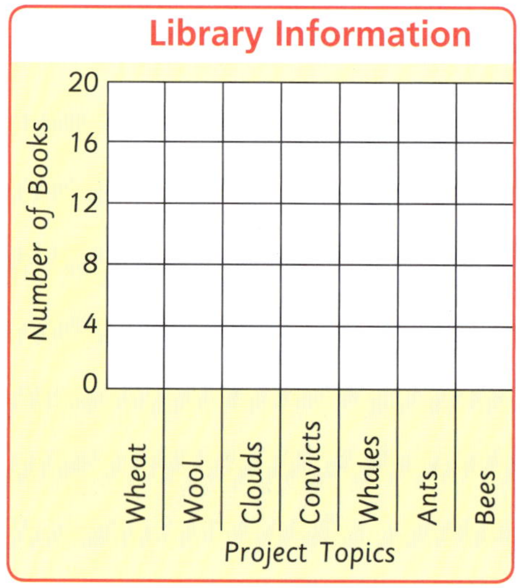

3 A tourist agency asked 2 000 people to give the main method of travel they used when going on their last holiday.

They recorded the following responses:

bus: 334; ship: 240; car: 748; plane: 121; train: 517.

Some people said that they had never had a holiday away from their home.

a Complete the graph to show the types of holiday travel. (Estimate, as best you can, the height for each column.)

b How many people had never had a holiday away from home? (You can use a calculator.)

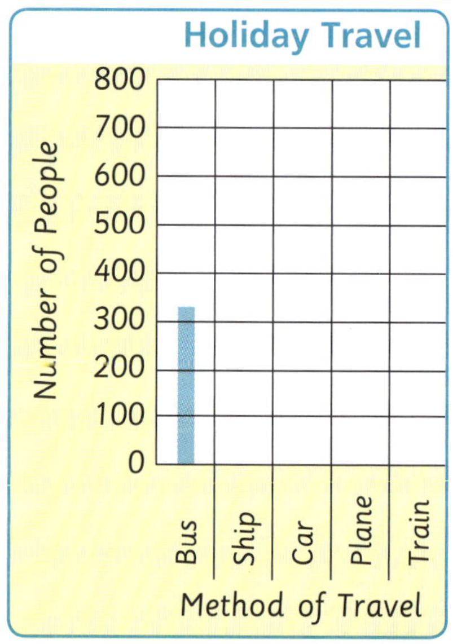

Money Blocks

ACTIVITY

4 Seven blocks were put in a line with spaces between them.

a 10c was put in the first space, 20c in the next, 30c in the next and so on. How much money was used to fill the spaces?

b How much money would be used if the blocks were arranged in a circle?

 • *AUSTRALIAN SIGNPOST MATHS 5* • ISBN 978 1 4886 2190 1

5:03 Drawing Picture Graphs

1 Elizabeth and Amber loved to watch the ships arrive and leave. They kept a record of the ships arriving each month.
Here is a summary of their records, given to the nearest ten.

Jan	Feb	Mar	Apr	May	Jun
50	60	30	20	40	60

Jul	Aug	Sep	Oct	Nov	Dec
70	80	60	50	30	10

Complete this picture graph.

Use this key:  = 20 ships

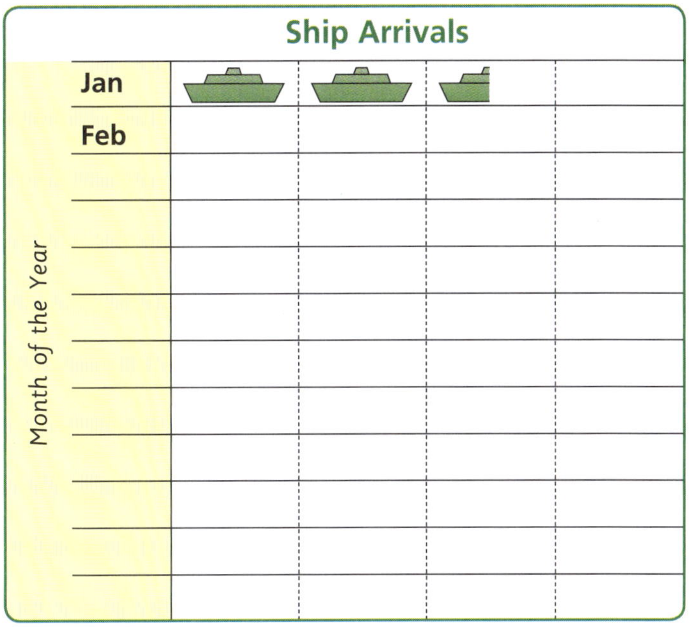

2 Use the information on the column graph to complete the picture graph on the right.

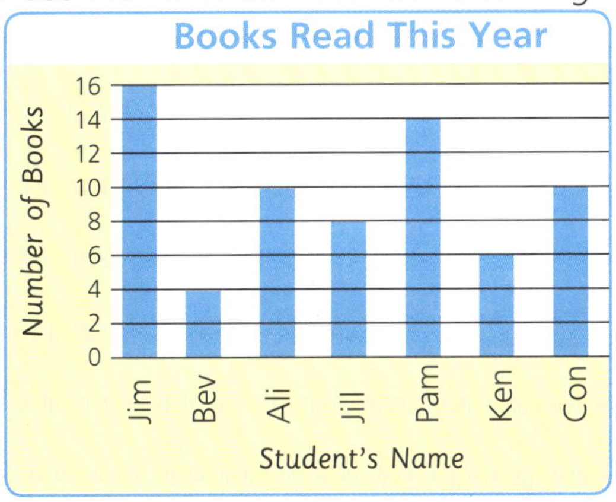

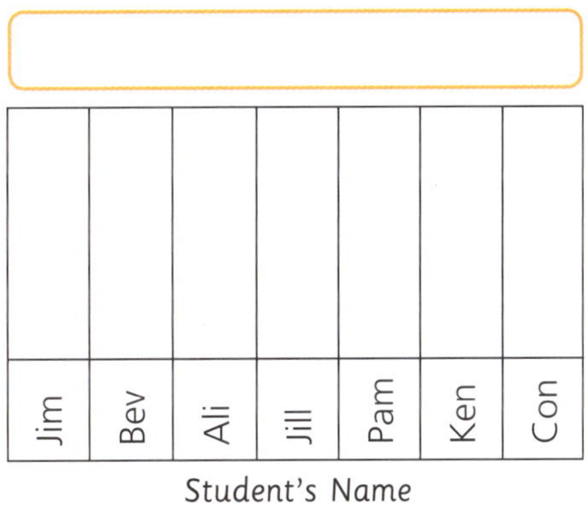

Key:  = 4 books

Which type of graph do you think is the best for this data? Why?

3 Ms Travers, the Year 5 teacher, gave stamps for good bookwork.
She kept a tally of the stamps given to her six best students.
Fill in the **Total** column in the tally below.
Use this information to complete the picture graph on the right.

Name	Tally	Total
Alana	卌 卌 卌 卌 卌 |||	
Rachel	卌 卌 卌 卌 ||	
Naomi	卌 卌 卌 卌 卌 卌 卌 卌	
Luke	卌 卌 卌 卌 卌 卌	
Heather	卌 卌 卌 卌 卌 卌 卌 |	
Sandy	卌 卌 卌 卌 卌 卌 ||	

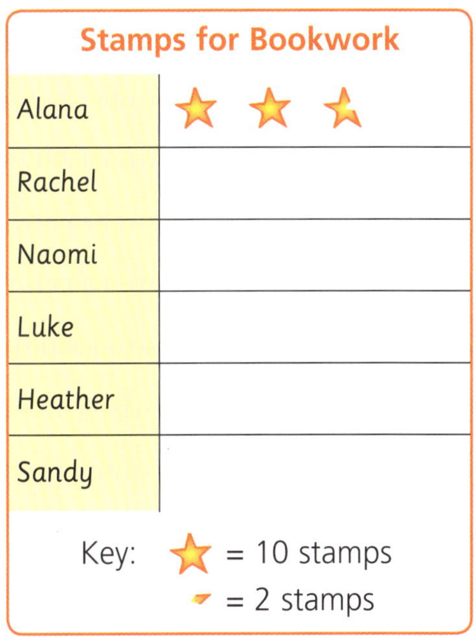

5:04 Reading Line Graphs

1 Leslie measured the mass of different amounts of sugar and graphed her findings. She joined the points to get this line graph.

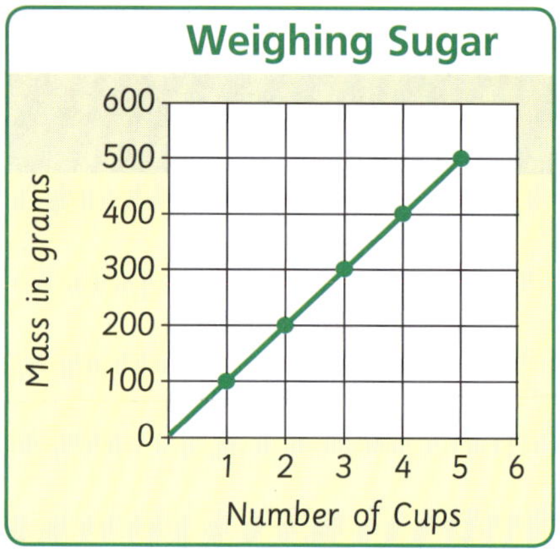

a What is the mass of:

i 2 cups? ii 4 cups?

iii $1\frac{1}{2}$ cups? iv $3\frac{1}{2}$ cups?

b How many cups of sugar have a mass of:

i 300 g? ii 500 g?

iii 100 g? iv 50 g?

c What would be the mass of 6 cups of sugar?

We can often extend a line graph to gain further information.

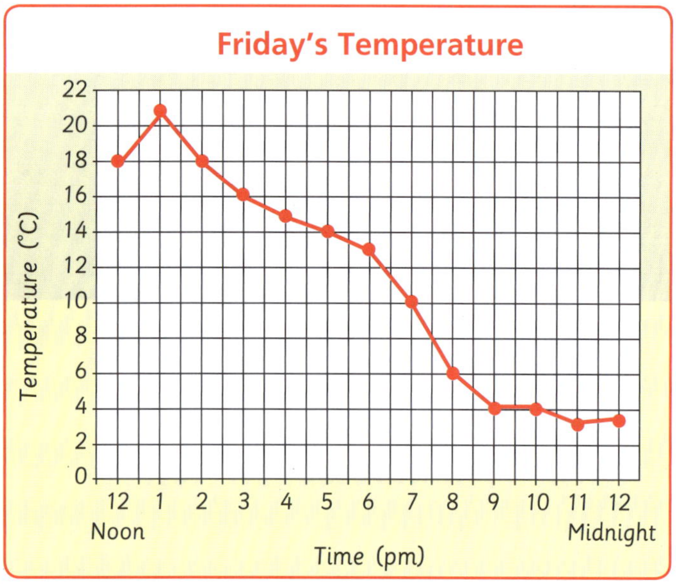

2 On Friday, Jasmine recorded the temperature every hour from noon until midnight. She then drew this graph.

a What was the temperature at:

i 2 pm? ii 11 pm?

b What was the highest temperature recorded? What was the time then?

c At what time was it:

i 10°C? ii 18°C?

d Estimate the temperature at:

i 2:30 pm ii 7:30 pm

iii 11:30 pm iv 10:30 pm

Because the temperature changes steadily, we can estimate the temperature at other times.

3 We can change 2 hours to minutes by following the arrows shown on the graph. We could change 120 minutes to hours by coming back the other way.

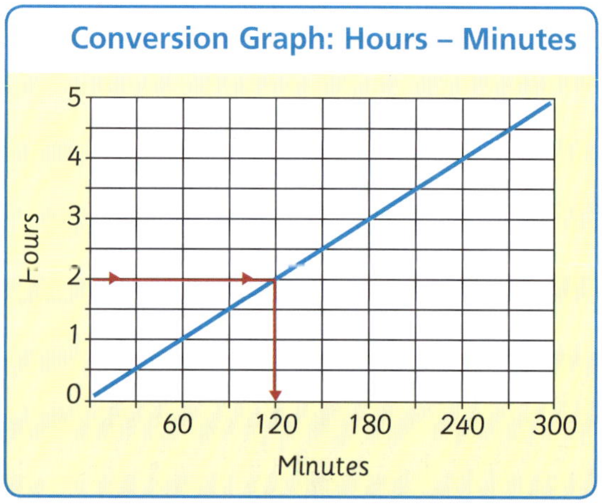

a Use the graph to convert these into minutes.

i 5 hours ii 3 hours

iii $1\frac{1}{2}$ hours iv $4\frac{1}{2}$ hours

b Use the graph to convert these to hours.

i 180 minutes ii 240 minutes iii 150 minutes iv 270 minutes

 • *AUSTRALIAN SIGNPOST MATHS 5* • ISBN 978 1 4886 2190 1

5:05 Drawing Line Graphs

1 Alan's parents measured his height on his birthday each year. Some of those measurements are shown below.

Alan's Height

0 years	2 years	4 years	6 years	8 years
50 cm	90 cm	100 cm	120 cm	130 cm

10 years	12 years	14 years	16 years	18 years
140 cm	150 cm	170 cm	175 cm	180 cm

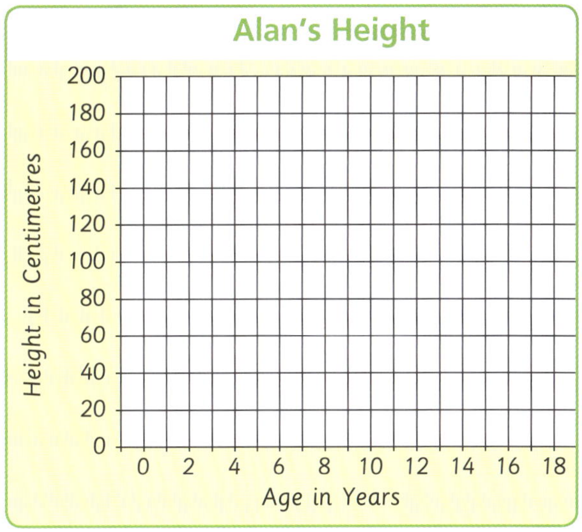

a Draw a line graph for these facts.

b Use the graph to estimate Alan's height at:

i 1 year ☐ ii 5 years ☐ iii 13 years ☐ iv 11 years ☐

c Six months before birth, Alan's height was 10 cm. Show this on the graph.

2 Redraw this column graph as a line graph.

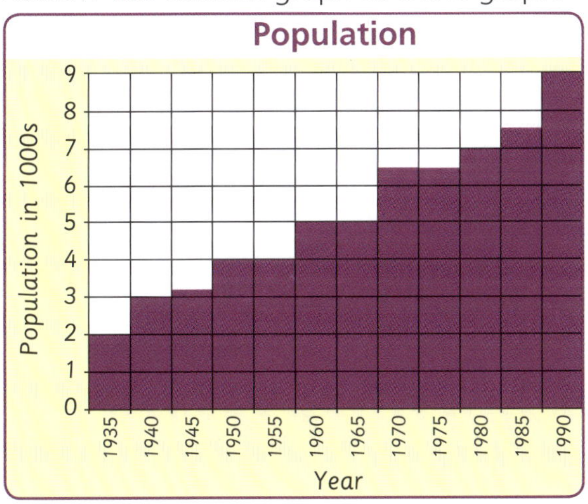

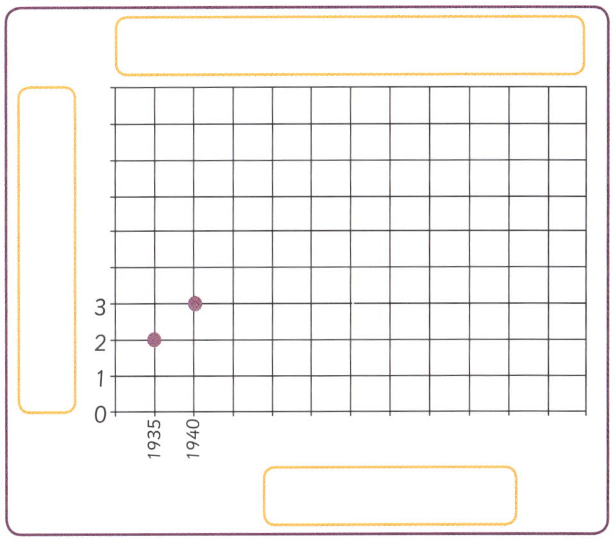

3 To draw a conversion graph we need two points. We then join these with a straight line.

0 acres = 0 hectares	40 acres = 16 hectares

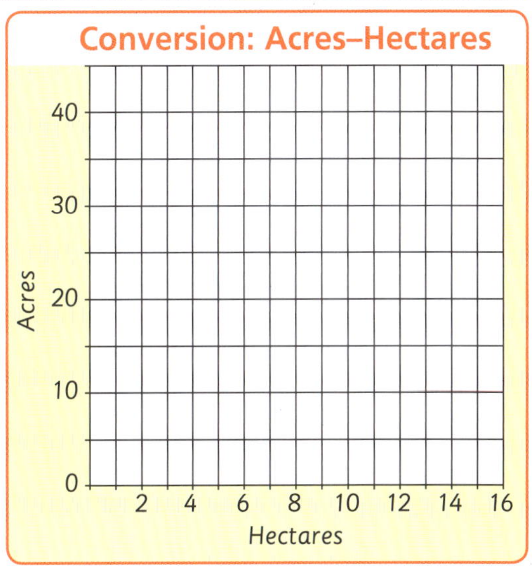

a Use the information above to draw the graph.

b Use your graph to convert these to hectares.

i 20 acres ☐ ii 15 acres ☐

c Use your graph to convert these to acres.

i 12 hectares ☐ ii 2 hectares ☐

d Which is greater:

i 15 acres or 7 hectares? ☐

ii 6 hectares or 16 acres? ☐

5:06 Dot Plots

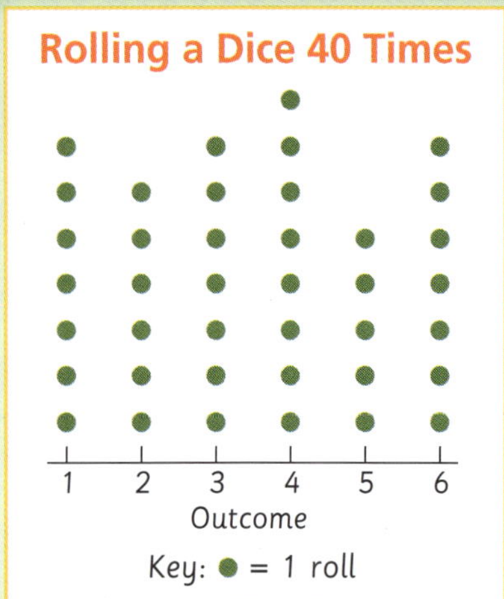

- This is a **dot plot**.
- Here, one dot has been drawn for every roll.
- 4 was rolled 8 times.
- When a dice is rolled, the possible outcomes are 1, 2, 3, 4, 5, and 6.

CONCEPT

1 **a** What question could we try to answer by rolling a dice 40 times?

b Write as much information as you can about these results.

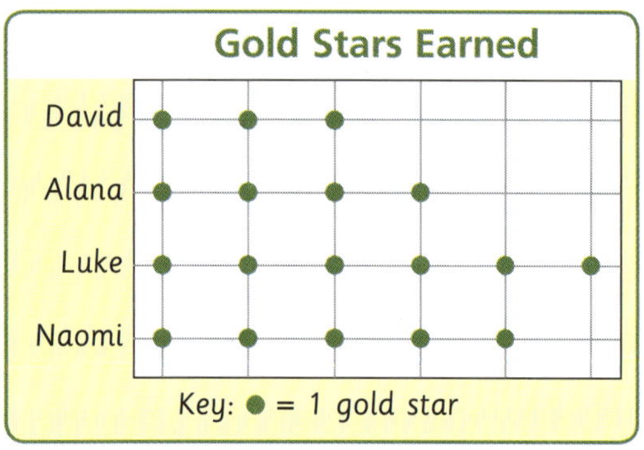

2 **a** Who has the most stars?

b Who has the least?

c How many stars have been earned by:

i Alana? **ii** Luke?

iii Naomi? **iv** David?

INVESTIGATION

3 Roll a dice 40 times. Enter each outcome in the tally below and complete the dot plot.

Rolling a Dice 40 Times

Outcome	1	2	3	4	5	6
Tally						
Totals						

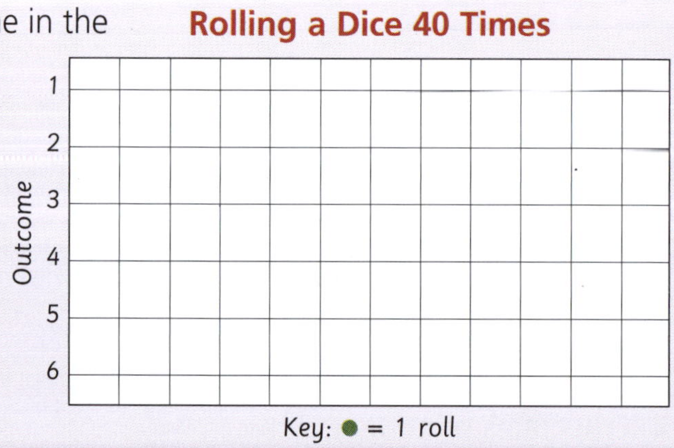

 • *AUSTRALIAN SIGNPOST MATHS 5* • ISBN 978 1 4886 2190 1

5:07 Drawing Dot Plots

Two dice were rolled 25 times. The totals are shown here.

The Sum When Rolling Two Dice

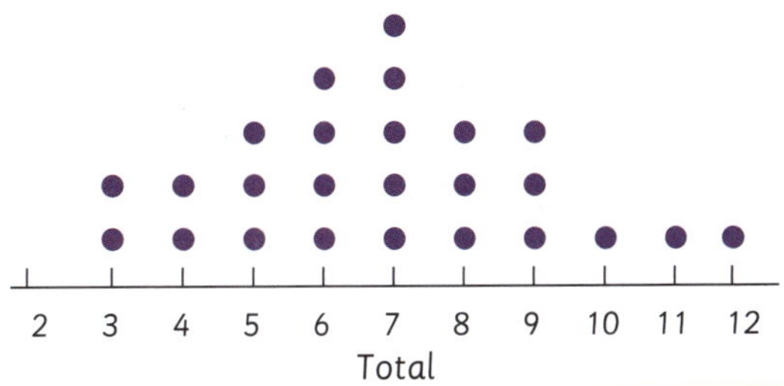

Discussion:

1 Which totals are most likely? Which are least likely?

INVESTIGATION

2 Repeat this experiment. Did you get similar results?

1 **a** How many students were tested?

b The dots of this graph were placed above a ______.

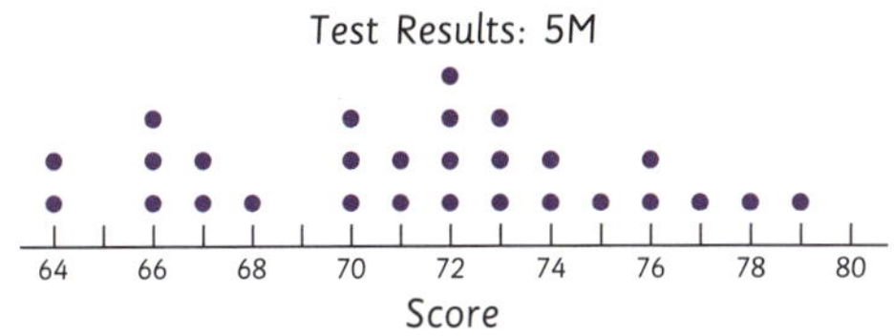

c Write as much information as you can about these results.

d Use the dot plot to complete the table.

Test Results: 5M																
Score	64	65	66	67	68	69	70	71	72	73	74	75	76	77	78	79
Number																

2 **a** Draw a dot plot for these scores.

8 7 5 6 2 9 4 0 3 7
4 7 4 9 3 7 2 7 6 9
3 8 2 4 6 2 6 6 5 7

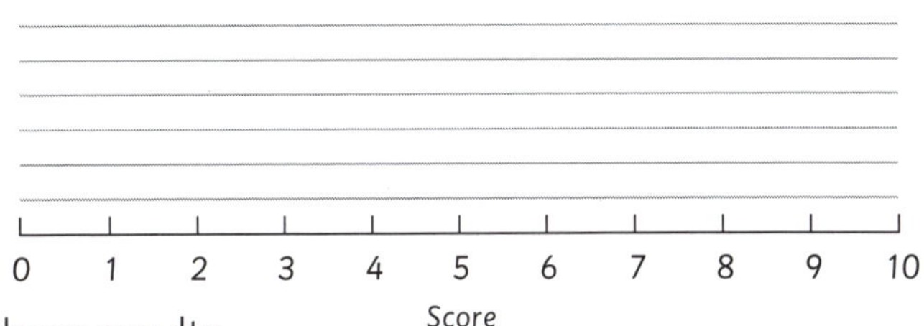

b Write as much information as you can about these results.

3 Show the following data on a dot plot and comment on the result.

2·0 1·8 2·0 1·7 2·1 1·8 2·1 1·7 2·2
1·6 2·0 1·5 1·7 1·6 2·0 1·9 2·1 1·7

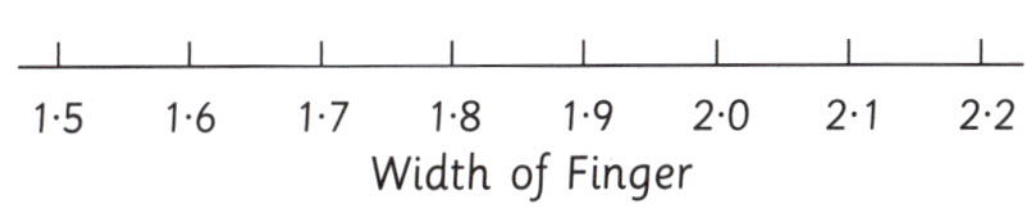

 • *AUSTRALIAN SIGNPOST MATHS 5* • ISBN 978 1 4886 2190 1

5:08 The Likelihood of an Event

The **chance** or **likelihood** of choosing a red ball at random is 1 out of 3 or $\frac{1}{3}$.

CONCEPT

1 What is the chance, as a fraction, that the spinner will stop on:

a 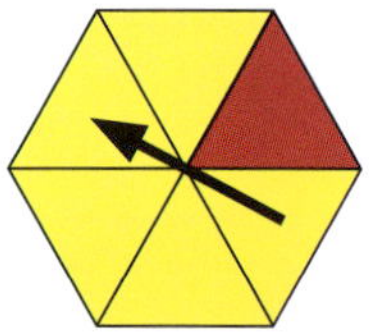 red? ☐ yellow? ☐

b 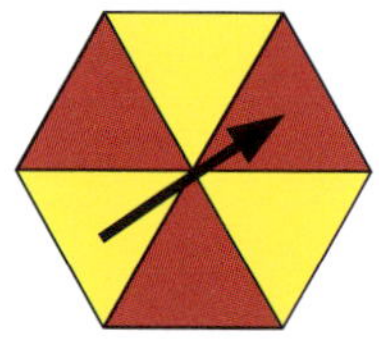 red? ☐ yellow? ☐

c 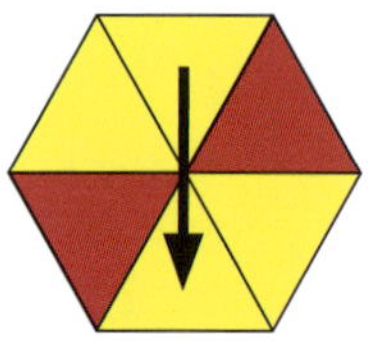 red? ☐ yellow? ☐

d 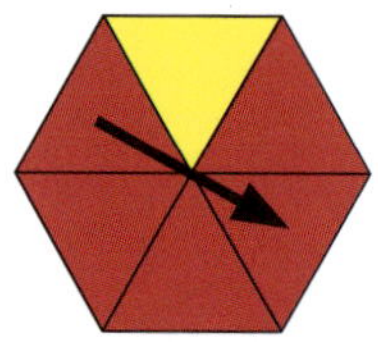 red? ☐ yellow? ☐

In Question 1, the chance of spinning red plus the chance of spinning yellow is equal to ☐.

2 What is the likelihood, as a fraction, that the dice will show:

a 1? ☐ **b** 6? ☐ **c** 5? ☐

d 3? ☐ **e** an even number? ☐

$\frac{1}{6}+\frac{1}{6}+\frac{1}{6}+\frac{1}{6}+\frac{1}{6}+\frac{1}{6}=1$

3 What is the chance, as a fraction, that the spinner will stop on:

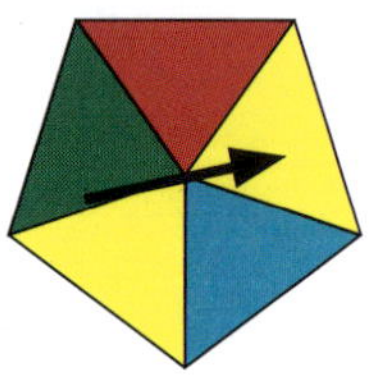

a red? ☐ **b** yellow? ☐ **c** green? ☐

d blue? ☐ **e** yellow or blue? ☐

4 In each part, colour shapes to match the chance given for choosing red at random.

a 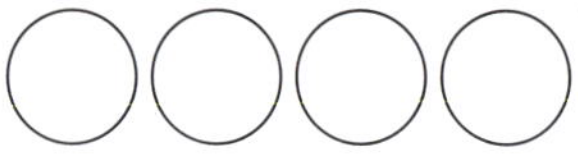 $\frac{1}{2}$

b 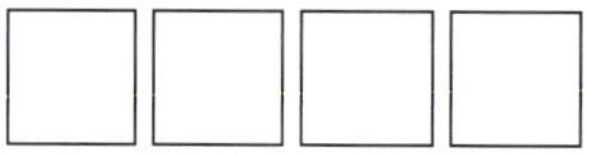 $\frac{3}{4}$

c  $\frac{5}{6}$

d  $\frac{3}{5}$

e  $\frac{1}{2}$

f 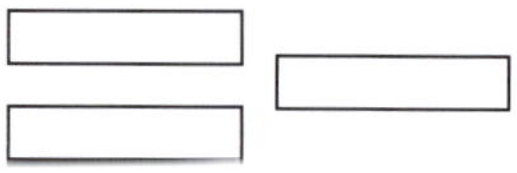 $\frac{2}{3}$

g 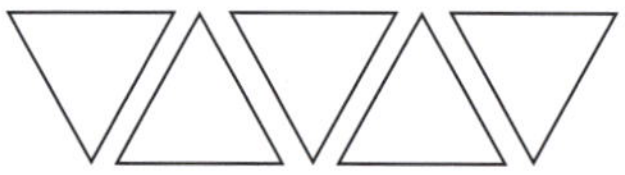 $\frac{2}{5}$

h 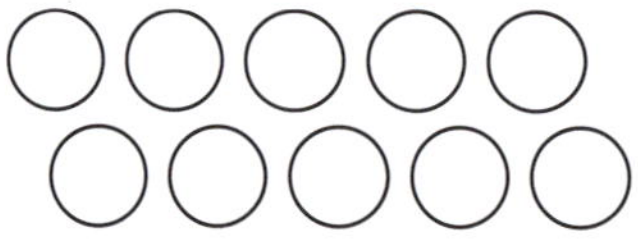 $\frac{1}{2}$

i 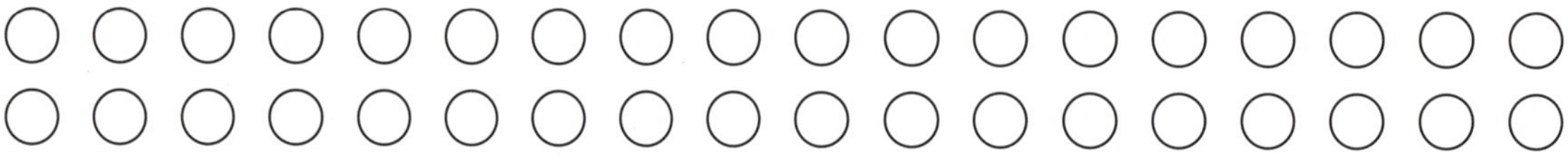 $\frac{1}{2}$

 ISBN 978 1 4886 2190 1

5:09 Chance

CONCEPT

When each item has the same chance of occurring, we can use fractions to give the chance of an event happening.

- The chance of choosing a yellow peg (at random) is $\frac{1}{10}$.
- The chance of choosing a blue peg (at random) is $\frac{2}{10}$.
- The chance of choosing a red peg (at random) is $\frac{5}{10}$.

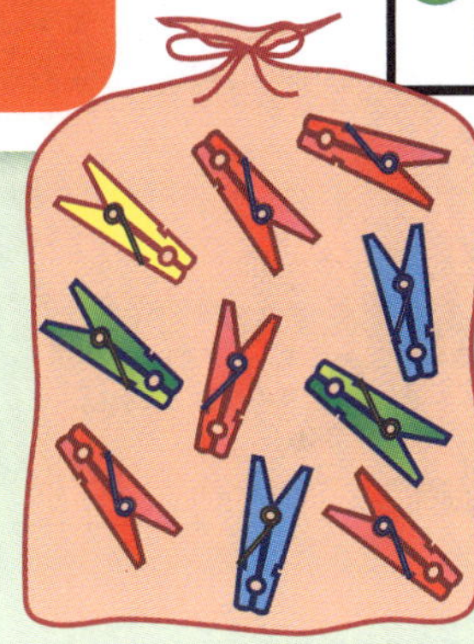

1 20 coloured pegs were placed in a paper bag as shown.

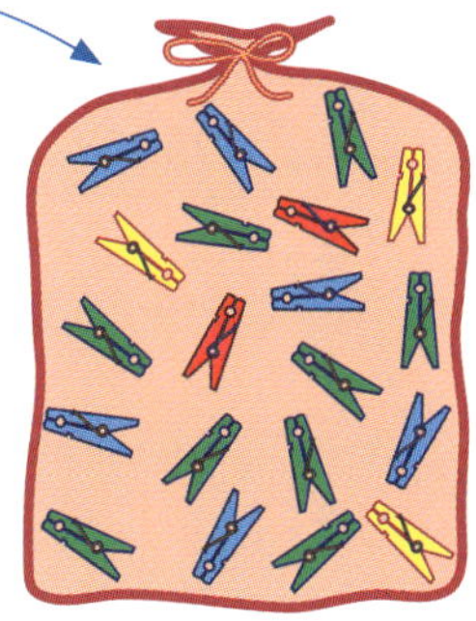

a Which colour is most likely to be drawn from the bag?

b Which colour is least likely to be drawn from the bag?

c Are we more likely to draw a blue peg or a red peg?

d Order the colours from least likely to be selected, to most likely.

Use fractions to give the chance of drawing a peg from the bag that is:

e yellow   **f** green   **g** red   **h** blue

2 Alan owned the 15 toys above. He chose one at random (without looking). What is the chance (written as a fraction) that he chose:

a a standing soldier?   **b** a knight on a horse or a nurse?

c a knight on a horse?   **d** a nurse or a standing soldier?

e a toy?   **f** a ball?   **g** a nurse?

FUN SPOT

3 Describe an event to match each probability.

a $\frac{1}{2}$

b 1

c $\frac{5}{40}$

d 0

e $\frac{10}{40}$

5:10 Divided Bar Graphs

CONCEPT

The Colour of Rachel's Budgerigars

1 The bar graph below is divided into three unequal parts.

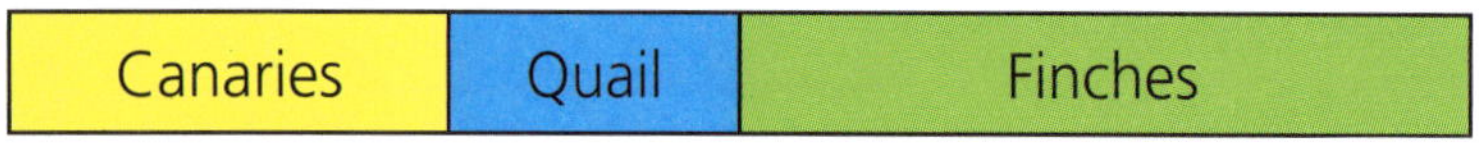

Naomi and Luke have 100 birds, so Luke drew a bar 100 mm long so that each millimetre stood for one bird.

a How long, to the nearest millimetre, is the part of the graph showing:
  i canaries? ___ **ii** quail? ___ **iii** finches? ___

b Of these birds owned by Luke and Naomi, how many were:
  i canaries? ___ **ii** quail? ___ **iii** finches? ___

c Measuring in millimetres, what fraction of the graph shows:
  i canaries? ___ **ii** quail? ___ **iii** finches? ___

d Measuring in centimetres what fraction of the graph shows:
  i canaries? ___ **ii** quail? ___ **iii** finches? ___

2 Twenty Easter eggs were hidden in the backyard and Kate, Sarah, Emma and Ben searched until they found them all. The bar graph below shows the success each had in the search.

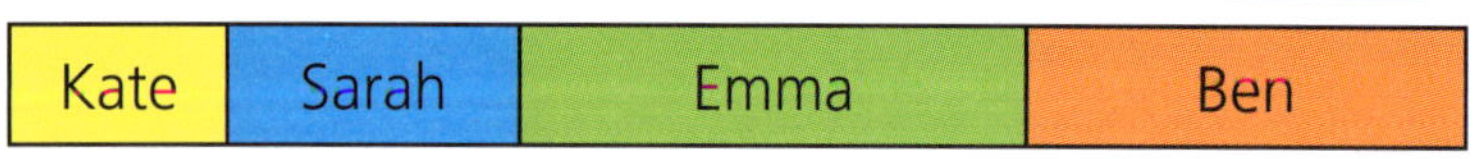

Who found:

a the greatest number of eggs? ___

b the least number? ___

c Write as much information as you can about this graph. ___

3 Draw a divided bar graph to show that our team has 7 males and 3 females.

5:11 Sector (or Pie) Graphs

CONCEPT

If we could bend a divided bar graph, we would get a sector (or pie) graph.

Big Cats at the Zoo

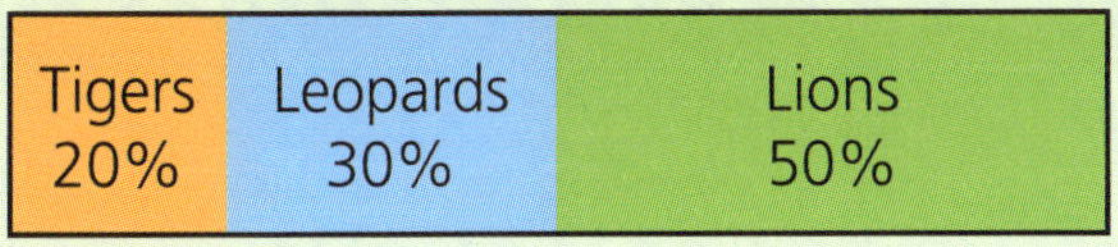

Big Cats at the Zoo

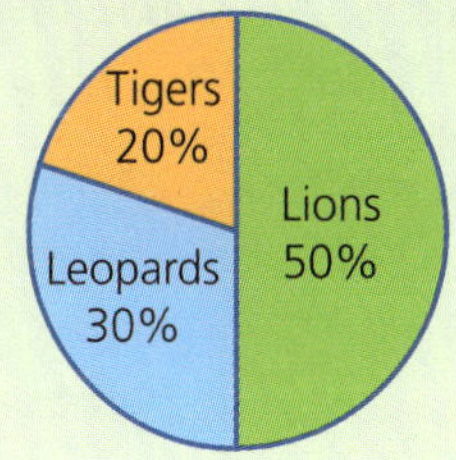

1. Use the graphs above to answer these questions.
   a What kind of big cat is most plentiful at the zoo?
   b What kind is least plentiful?
   c What percentage of big cats in the zoo are leopards?
   d If there are 20 big cats in the zoo, how many are tigers?

Eighty Year 6 students were asked to name their favourite sport. This pie graph was then drawn.

Favourite Sports of Year 6 Students

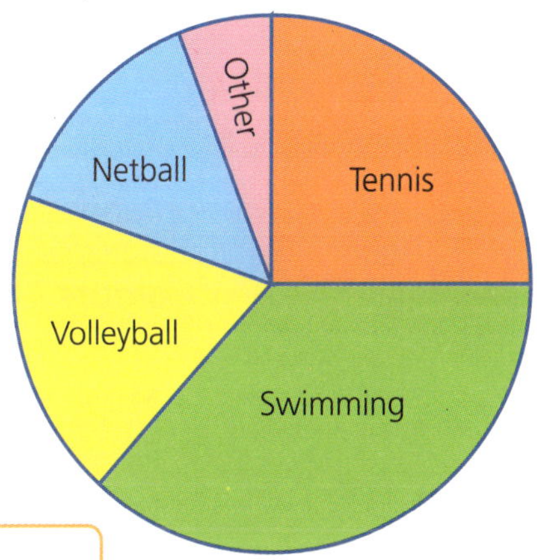

2. Which sport was more popular:
   a volleyball or netball?
   b tennis or volleyball?
   c netball or tennis?

3. a What fraction of the group chose tennis?
   b How many students chose tennis?

4. Of the sports chosen, do we know the least popular?
   Explain your answer.

5. Naomi took a cube at random from a jar containing 5 red, 3 white and 2 blue cubes, noted its colour, then replaced it. She did this 40 times. Luke did the same and they made these graphs.
   a Who drew out more:
      i red? ii white?
   b How often did Naomi choose a:
      i red? ii white?
   c If Naomi and Luke carried out this experiment again, would you expect the same results?

40 draws each

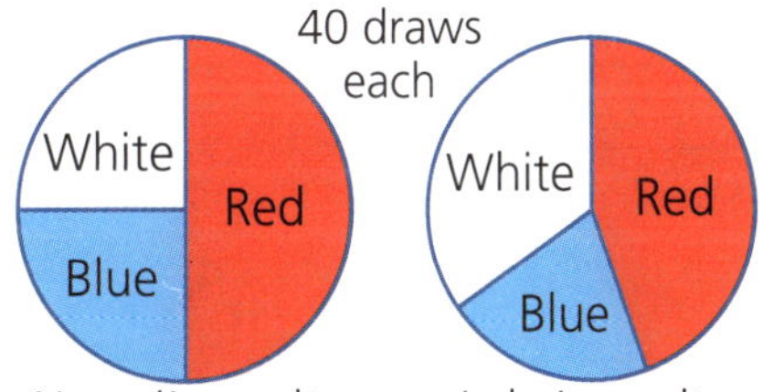

Naomi's results   Luke's results

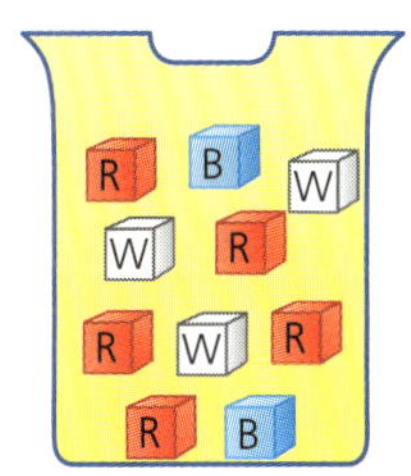

 • *AUSTRALIAN SIGNPOST MATHS 5* • ISBN 978 1 4886 2190 1

5:12 Questionnaires/Surveys

Class 5M was given this survey.
The results were tallied.

Questionnaire/Survey

1 Which colour do you like best: green, blue, red or pink?

2 Which do you like most: playing sport, reading books, walking or watching sport?

<table>
<tr><th>Category</th><th>Tally</th></tr>
<tr><td>1 green
blue
red
pink</td><td>卌
卌 卌 |
卌 |||
|||</td></tr>
<tr><td>2 playing sport
reading books
walking
watching sport</td><td>卌 卌
卌 ||
卌 |
||||</td></tr>
</table>

1 How many students chose these categories?

a green ______ b blue ______ c pink ______

d playing sport ______ e walking ______ f reading books ______

2 a Which was the most popular:

i colour? ______ ii activity? ______

b Which was the least popular:

i colour? ______ ii activity? ______

3 Complete these graphs using the results above.

a **Colours Chosen by 5M**

green
blue
red
pink

0 2 4 6 8 10

b **Activities Chosen by 5M**

playing sport
reading books
walking
watching sport

0 2 4 6 8 10

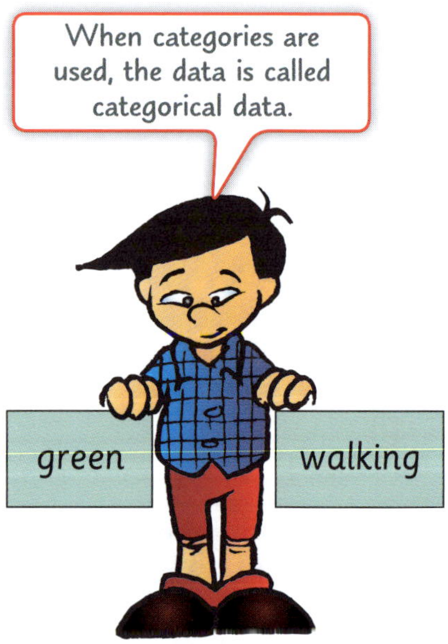

Use the questionnaire on this page to ask people the two questions.
Keep a tally of their answers and report your findings.

ACTIVITY

5:13 Data Investigation

INVESTIGATION

1 Work in small groups to complete this investigation.

Step 1

Choose a topic to investigate.

Topic: How will the temperature change during the day?

Step 2

Decide when and how you will measure the changes.

When will the temperature be measured and how?

Step 3

Decide how you will record your measurements.

Time	Temperature	Comments

What do you expect to discover?

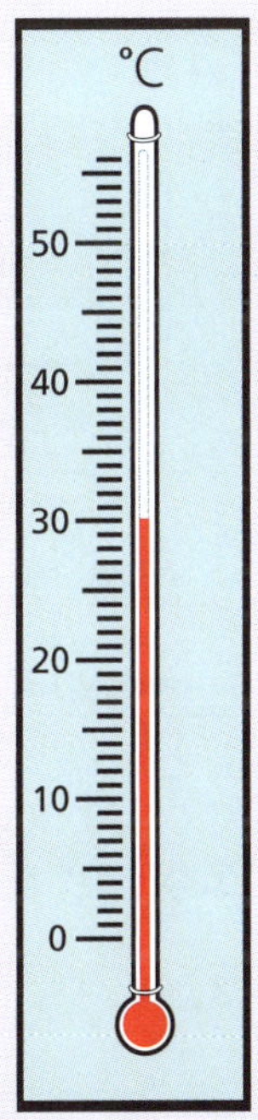

Step 4

Write down what you found and explain your conclusions.

Enter the results in a computer spreadsheet program such as Microsoft® Excel®.
Report on your results by using computer-generated graphs.

ICT

5:14 More Line Graphs

1 Alan planted lawn seeds and graphed the growth at the end of each day for 14 days.

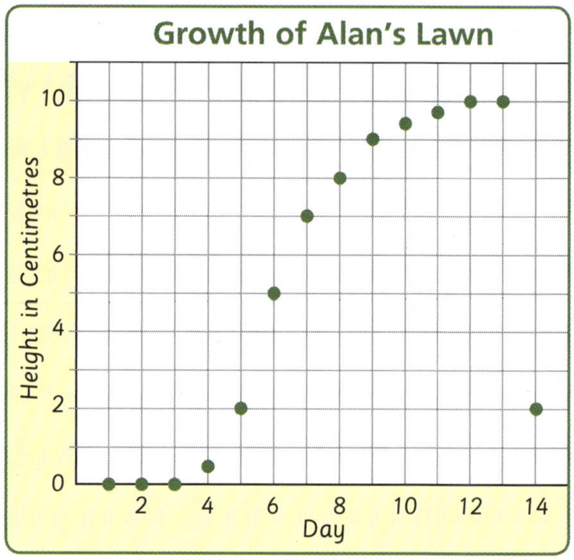

a What was the height of the lawn at the end of:

i day 7? ☐ **ii** day 2? ☐

iii day 14? ☐ **iv** day 13? ☐

b On what day was the height:

i 2 cm? ☐ **ii** 9 cm? ☐

iii 9·5 cm? ☐ **iv** 6 cm? ☐

c When did Alan first see the lawn? ☐

d On which day was the growth the greatest? ☐

e By joining the dots, draw a line graph.

f How many readings did Alan take? ☐

g On which day did he mow the lawn? ☐

h On which day was the height 4 cm? ☐

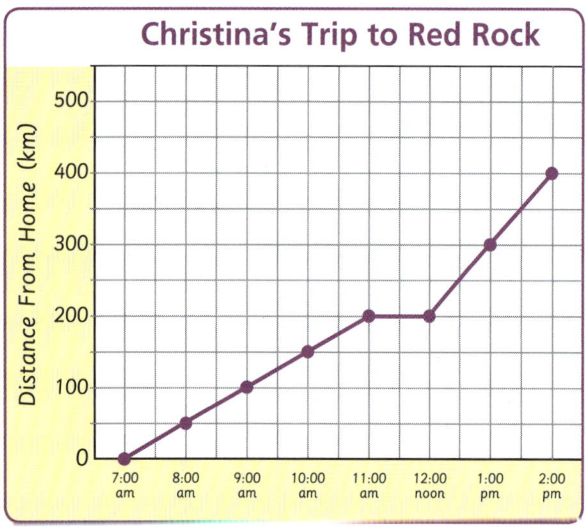

2 Christina left home early to travel on a straight road to Red Rock. This graph shows her progress.

a When did she leave home? ☐

b How far was she from home at:

i 9:00 am? ☐ **ii** 10:00 am? ☐

iii 1:00 pm? ☐ **iv** 2:00 pm? ☐

v 11:00 am? ☐ **vi** 12 noon? ☐

c When did she stop for lunch? ☐

d How long did she take for lunch? ☐

e How far did she travel between 10:00 am and 11:00 am? ☐

f How far did she travel between 12 noon and 1:00 pm? ☐

3 Rhonda needed an electrician to fix her stove. Use the table to decide which quote is cheapest for a 3-hour, 15-min or $1\frac{1}{2}$-hour job.

INVESTIGATION

Quote	Fee to arrive	Cost per 15 min	Cost for parts to be replaced	a 3-hour job	b 15-min job	c $1\frac{1}{2}$-hour job
1	$25	$40	$80			
2	$100	$15	$115			
3	$50	$30	$70			

The cheapest quote would be: **a** ☐ **b** ☐ **c** ☐

5:15 Information Collected Over Time

1 a What was the maximum temperature for:

i Monday? ☐ ii Thursday? ☐

b What was the minimum temperature for:

i Monday? ☐ ii Wednesday? ☐

c On which day was the difference between the maximum and minimum greatest? ☐

Maximum and Minimum Temperatures

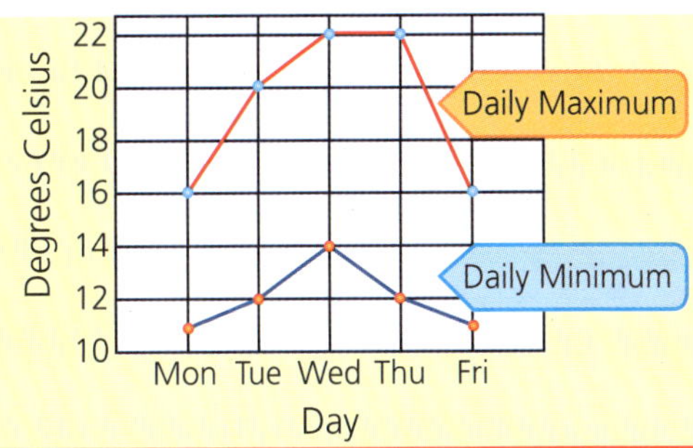

2 Use the internet to find the maximum and minimum temperatures over 5 days.

a Record the temperatures in the table below.

	Mon	Tues	Wed	Thurs	Fri
Highest Temp.					
Lowest Temp.					

b Graph the results, as in Question 1.

c Describe and explain the results.

Maximum and Minimum Temperatures

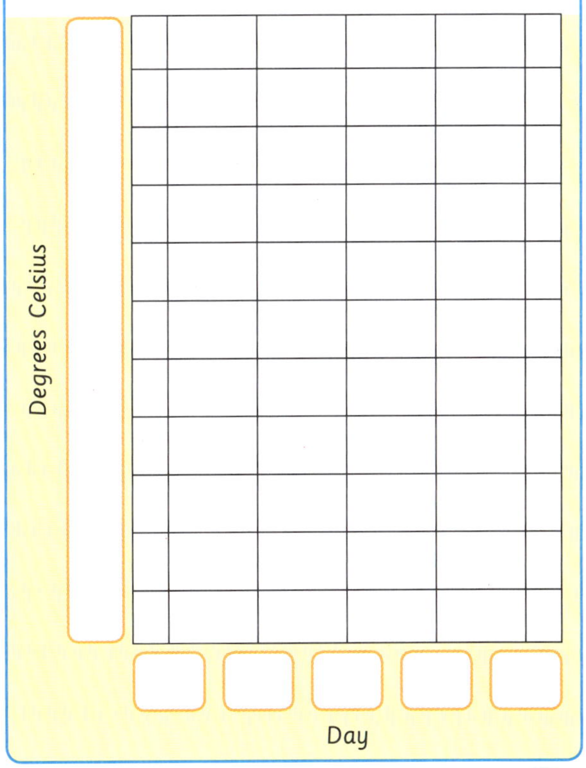

Insect Diversity in the Playground

INVESTIGATION

- Choose an area in the playground that has an insect population.
- Tape a 1 metre square border on the playground.
- Record the type and number of insects in the square at different times.

Time	Number of insects	Description of insects

Conclusions: ☐

5:16 Chance

1 Why would the chance of getting a head when tossing a coin be $\frac{1}{2}$?

2 What is the probability, as a fraction, that the spinner will land on red?

a 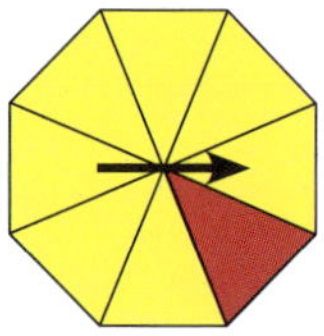 **b** 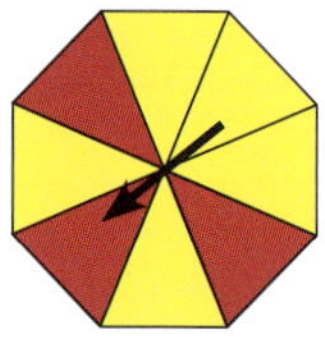 **c** 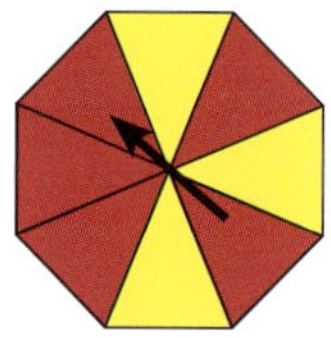 **d** 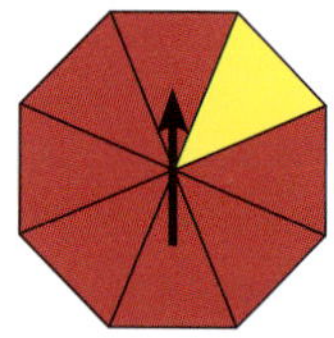

3 The spinner is divided into ten equal sectors.
Use the scale 0 to 1 to rate the chance of spinning each colour shown.

a R **b** B **c** O **d** G

Each part is $\frac{1}{10}$.

4 What is the chance, as a fraction, that a ball drawn at random from the bag will be:

a red? **b** yellow? **c** not blue?

d green? **e** red or green? **f** not green?

g yellow or blue? **h** red or blue?

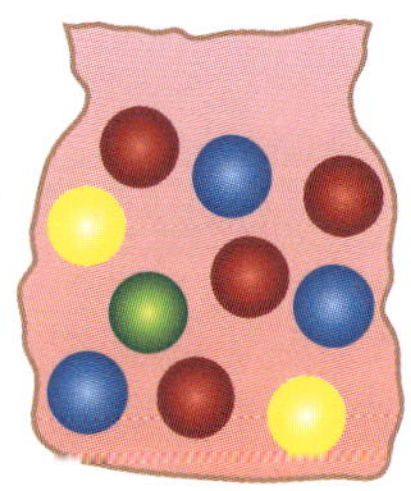

Place everyday expressions related to chance on a number line, e.g. no way (N), sure thing (S), probably (P).

 • *AUSTRALIAN SIGNPOST MATHS 5* • ISBN 978 1 4886 2190 1

5:17 Chance Events

CONCEPT

- The chance of choosing a red triangle is $\frac{1}{8}$.
- The chance of choosing a blue square is $\frac{2}{8}$.
- The chance of choosing a green hexagon is $\frac{5}{8}$.
- The chance of not choosing a green hexagon is $\frac{3}{8}$.

The sum of all probabilities is $\frac{8}{8}$ or 1.

1 One counter is taken at random from a bowl.
What is the chance of choosing:

a a 1? ☐ **b** a 2? ☐ **c** a 3? ☐

d What is the chance of choosing a 1, 2 or 3? ☐

e What is the chance of choosing a 2 or 3? ☐

2 **a** What is the chance of choosing a blue square from the bowl in the Concept box? ☐

b What is the chance of not choosing a blue square from the bowl? ☐

c Do the probabilities in part **a** and part **b** add to 1? ☐

3 How many of the 40 balls in the bag on the right are:

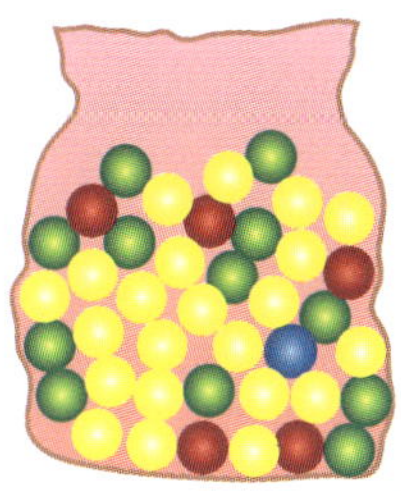

a red? ☐ **b** green? ☐

c yellow? ☐ **d** blue? ☐

e Which colour is least likely to be drawn from the bag? ☐

f The chance of choosing blue = ☐. **g** The chance of not choosing blue = ☐.

h Do the probabilities in part **f** and part **g** add to 1? ☐

i The chance of choosing red = ☐. **j** The chance of not choosing red = ☐.

k Do the probabilities in part **i** and part **j** add to 1? ☐

l The chance of choosing green = ☐. **m** The chance of choosing yellow = ☐.

n Do the chances of choosing blue, red, green and yellow add to 1? ☐

 • *AUSTRALIAN SIGNPOST MATHS 5* • ISBN 978 1 4886 2190 1

5:18 Comparing Types of Graphs

Which of the graphs below (column, sector, picture, divided bar and line) are best suited to show the 'Sales of Books'?

1 Name each graph and match each with its description.

1 ______ graph

2 ______ graph

3 ______ graph

4 ______ graph

5 ______ graph

A A circle is divided into parts.
No axes are necessary.
It shows how the whole is divided into parts.
It usually does not give details.
It is easy to compare the size of categories.
Uncomplicated and takes up little space.

B It is easy to read and understand.
It is easy to draw.
Two axes are used.
It is impressive in appearance.
It allows comparisons to be made at a glance.
It shows more detail than most graphs.

C It is more attractive than most graphs.
It does not give detailed information.
It allows us to compare the sizes of each category easily. It uses a key.
Only one axis is necessary.
It is easy to understand.

D It shows more detail than most graphs.
All points on the line should have some meaning.
Each point has a reading on both axes.
A line is used to show trends and relationships clearly.
Two axes are used.

E No axes are necessary.
It is uncomplicated and takes up little space.
It shows how the whole is divided into parts.
It often does not give details.
It is easy to compare the size of categories.
A rectangle is divided into parts.

2 Make a list of the advantages of using tables.

3 On your own paper, use a column graph to graph the sales of books in this table.

Sales of Books				
Day	Mon	Tues	Wed	Thurs
Sales	35	48	44	63

Selecting the Best Graph to Present Data

We can use the information in the table to create these two graphs.

Water in the Tank

Day	Mon	Tue	Wed	Thurs	Fri
Litres	4	10	8	4	8

Can the information be displayed on a sector graph? Discuss.

CONCEPT

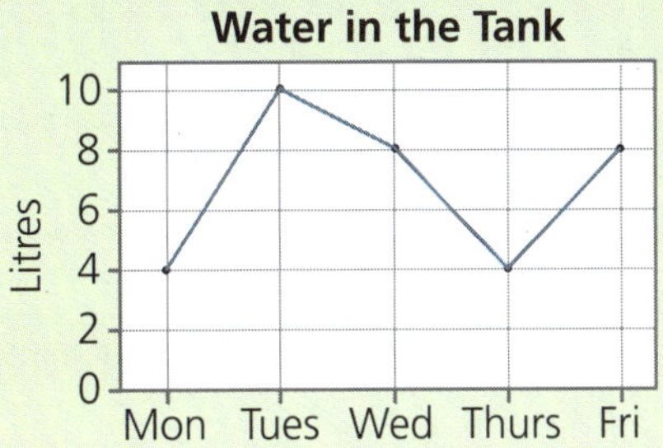

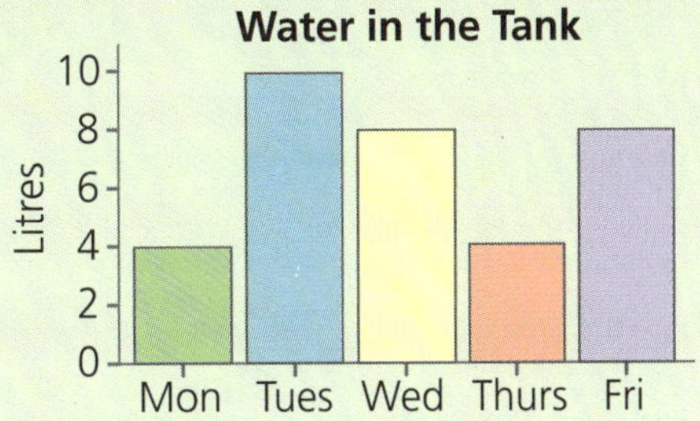

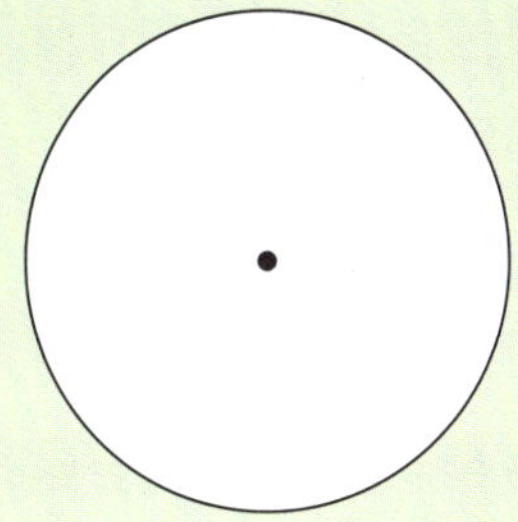

1 Use the information given in the tables to draw these graphs.
Circle the graph you think is the best way to present the data.

a **Have Been to Perth**

Yes	5
No	95

Have Been to Perth
(Line, Picture or Bar Graph)

Have Been to Perth
(Sector Graph)

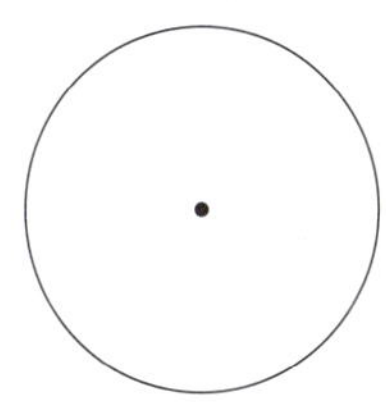

b **Jan's Holiday**

Where	Days spent
Coast	2
Home	5
Country	3

Jan's Holiday
(Divided Bar Graph)

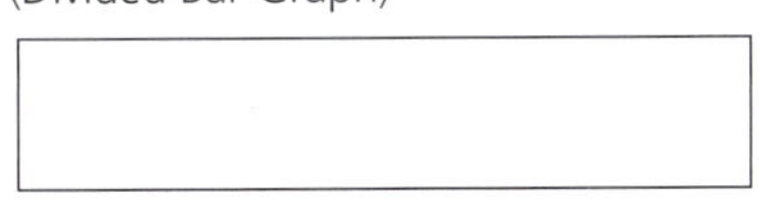

Jan's Holiday
(Column, Picture or Line Graph)

c **Luke's Mass from Birth**

Weeks	0	2	4	6
Mass (kg)	3.4	4	5.2	5.6

Luke's Mass from Birth
(Line Graph)

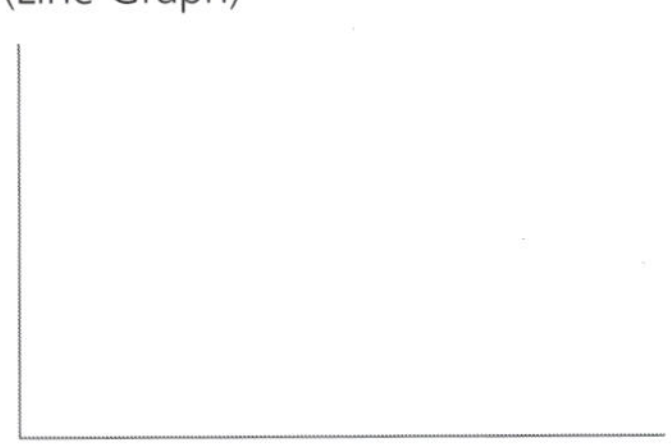

Luke's Mass from Birth
(Column Graph)

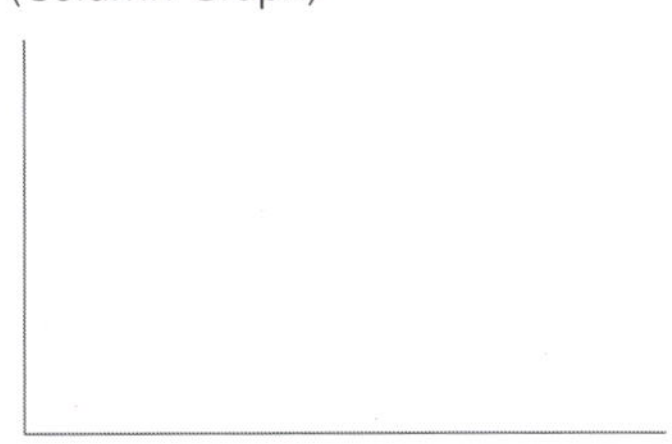

2 Write the best type of graph for the data in each table. Discuss the reasons for your answers.

a **Money in the Bank**

Month	Jan	Feb	Mar	Apr	May	Jun	Jul
Money ($)	48	62	35	7	33	69	54

b **Favourite Pet**

Dog	40
Cat	12

 • *AUSTRALIAN SIGNPOST MATHS 5* • ISBN 978 1 4886 2190 1

5:20 Collecting Data from Experiments

1 Alana tossed a coin many times.

She kept this tally of the results.

	Throwing a Coin	Total
Heads	卌 ||||	
Tails	卌 卌 |	

a How many times did she toss the coin?

b What fraction of the time did she toss:

i a head? ii a tail?

2 If two coins are tossed we could get two heads, two tails, or a head and a tail.

		Total
Two Heads		
Two Tails		
Head and Tail		

a Toss two coins 40 times. Record the results.

b Which result occurred most often?

c What fraction of the time did you toss:

i two heads? ii two tails? iii a head and a tail?

d Are the chances of each happening about the same or very different?

To take a counter **at random** from a bag means to take a counter without looking.

3 5 red counters, 3 blue counters and 1 yellow counter are placed in a bag.

A counter is chosen at random from the bag.

What is the chance (as a fraction) that it will be:

a red? **b** blue? **c** yellow?

d Put 1 yellow, 3 blue and 5 red counters into a container.

- Take a counter.
- Record its colour.
- Return the counter to the container.

Keep a tally of your results.

Red		
Blue		
Yellow		

What have you discovered?

What's My Age?

- Ask a person to write the month in which they were born. (1 for January, 2 for February, etc.)
- Tell them to multiply this number by 5, then add 2, then multiply by 20, then add their age in years, then subtract 40.
- Ask the person for their answer.

Solution: The last two digits of the answer give you the age, and the other digits give you the month of birth, e.g. 714 means the person is 14 and was born in July (the 7th month).

 ISBN 978 1 4886 2190 1

5:21 Collecting Data

1 Stickers were stuck on a dice so that the faces showed three 3s, two 5s and one 1.
Write as a fraction the probability of rolling a:

a 3 ☐ **b** 5 ☐ **c** 1 ☐ **d** 6 ☐

e Which number would you expect to roll most often? ☐

2 Carry out the experiment in Question 1. Record your results in the table as you roll the dice 9 times.

a Which number was rolled most often? ☐

b What fraction of the time did you roll a:

i 3? ☐ **ii** 5? ☐ **iii** 1? ☐

Result	Tally	Total
1		
3		
5		

3 If you threw three coins 60 times, how often would you expect to throw three heads? ☐

Check your guess by carrying out the experiment.

Result	Tally	Total
No Heads		
1 Head		
2 Heads		
3 Heads		

4 A bag contains 20 lollies. They are wrapped in identical wrappings.
There are 10 chocolate, 6 caramel, 3 candy and 1 licorice.

a If one lolly is taken at random from the bag, which type is:

i least likely to be chosen? ☐ **ii** most likely to be chosen? ☐

b Write as a fraction the chance of choosing a:

i chocolate ☐ **ii** caramel ☐ **iii** candy ☐ **iv** licorice ☐

5 Instead of using lollies, use counters (or cards) to carry out the experiment in Question 4.

- Mark 10 counters with an A, 6 with a B, 3 with a C and 1 with a D.
- Put these in a hat, choose one, record the result and then return it to the hat. Do this 60 times.
- Do you think this experiment is a good model of the one in Question 4? ☐

	Result	Tally	Total
A	Chocolate		
B	Caramel		
C	Candy		
D	Licorice		

6 In this game of dice soccer, two dice are rolled. The ball (a counter) is placed on **Start** and is moved one place towards Goal A if the total is less than 7, and one place towards Goal B if 7 or more.
Who do you think should win? ☐ Play some games to find out.

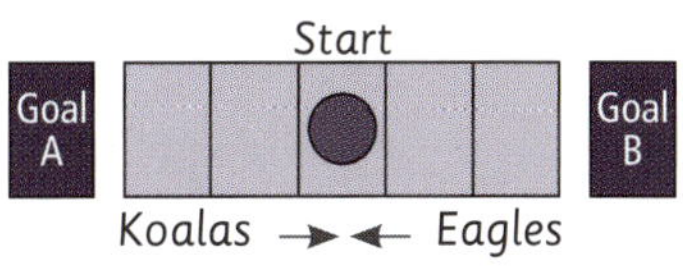

5:22 Reasoning with Graphs

1 In the USA, distance is measured in miles. Use the graph to convert:

a 40 kilometres into miles

b 60 kilometres into miles

c 12·5 miles into kilometres.

Which is the greater distance:

d 30 miles or 50 kilometres?

e 20 miles or 30 kilometres?

f Howler monkeys can be heard up to 16 km away. What is this distance in miles?

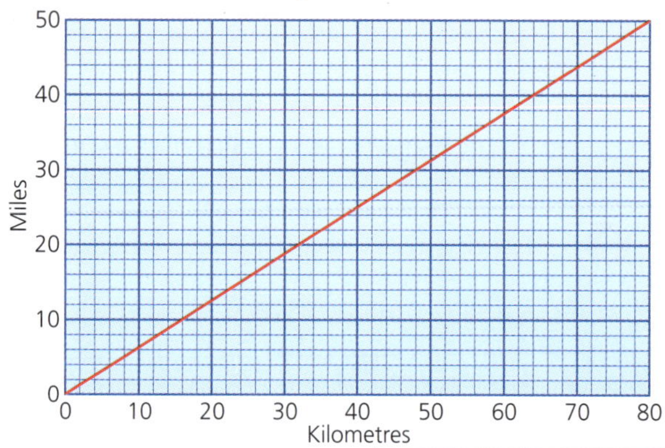

Discuss why each graph has been used to show the data.

2 Luke drew a graph of how he passed the time on his birthday.

a What activity took up most time?

b Was more time spent working or playing?

c Was less time spent working than sleeping?

d Which two categories used the same time?

e Which category used $\frac{1}{4}$ of the day?

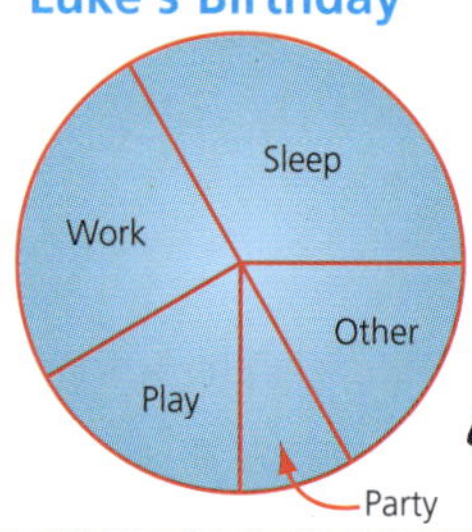

3 Heights of Year 6 students were measured to the nearest centimetre.

a How many had heights from 121 to 130 cm?

b How many had heights from 171 to 180 cm?

c How many had heights from 141 to 180 cm?

d Which category had the greatest number of people in it?

e Is it possible to tell how many people were 100 cm tall?

f How many Year 6 students were measured?

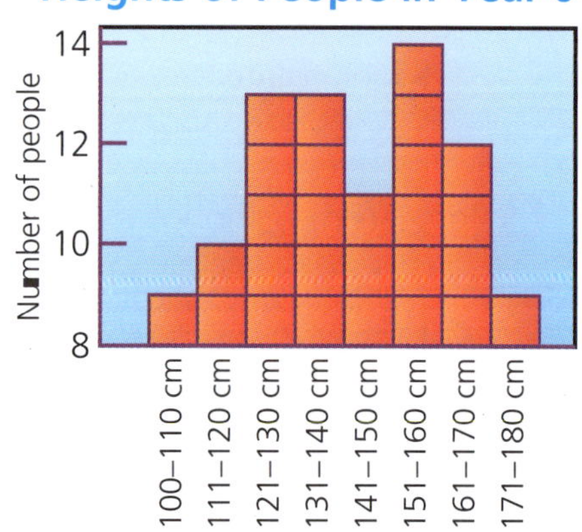

Tally of Year 5 Test Results

Mark Group					
1 to 10	𝍸				
11 to 20	𝍸	𝍸	||||		
21 to 30	𝍸	𝍸	𝍸	𝍸	||
31 to 40	𝍸	𝍸	|||		

4 a How many students scored 11 to 20 marks?

b How many students scored less than 21?

c How many scored more than 20?

d How many students were in Year 5?

e Do we know how many people scored 40?

5:23 Comparing Mobile Phone Plans

Three Different Phone Plan Costs

Plan	Average Cost per Call	Cost per GB of Data	Cost per SMS
1	$3.35	$5	$0.25
2	$1.44	$10	$0.15
3	$2.62	$12	$0.10

Data refers to internet download usage. GB stands for gigabytes. An SMS is a text message.

1 Use the information in the table above to find the cheapest plan for each person.
Use a calculator to complete the tables.

a Jake made 12 calls, sent 10 messages and used no data each month.

Plan	Total Cost of Calls	Total Cost of Data	Total Cost of Messages	Total Cost per Month
1				
2				
3				

The cheapest plan for Jake is Plan ______.

Realistically, the average time per call is different for different people. Usually, the longer the call, the more expensive it will be.

b Simon used 1 GB of data, sent 35 messages and made no calls each month.

Plan	Total Cost of Calls	Total Cost of Data	Total Cost of Messages	Total Cost per Month
1				
2				
3				

The cheapest plan for Simon is Plan ______.

There are many things to consider when deciding on a plan.

c Katie used 3 GB of data, made 10 calls and sent 12 messages each month.

Plan	Total Cost of Calls	Total Cost of Data	Total Cost of Messages	Total Cost per Month
1				
2				
3				

The cheapest plan for Katie is Plan ______.

Some mobile phone plans offer special deals, e.g. $30 for $300 worth of calls and messages.

d Jasmine used 1 GB of data, made 3 calls and sent 200 messages each month.

Plan	Total Cost of Calls	Total Cost of Data	Total Cost of Messages	Total Cost per Month
1				
2				
3				

The cheapest plan for Jasmine is Plan ______.

Some mobile phone plans come with a free phone if you commit to the plan for 2 or more years. This will appeal to some people.

5 Answers

1:01

1

	HThous	TThous	Thous	Hunds	Tens	Ones
a	3	4	6	9	7	5
b	5	2	3	4	8	2
c	7	1	6	8	5	9
d	6	7	8	9	6	0
e	2	5	1	6	0	3

2 **a** 4 thousands **b** 6 thousands **c** 7 tens
d 3 tens of thousands **e** 3 thousands **f** 5 ones
g 4 hundreds of thousands **h** 3 thousands

3 **a** 147 624 **b** 496 539 **c** 869 376
d 658 192 **e** 275 406

1:02

1 **a** 2 386 531 **b** 7 843 266 **c** 3 521 653

2 **a** 659 874 **b** 763 759 **c** 3 848 631 **d** 5 916 592

3 **a** 9 tens **b** 3 tens of thousands
c 8 hundreds of thousands **d** 5 thousands
e 6 millions **f** 2 hundreds
g 1 hundreds of thousands **h** 4 thousands

4 **a** 3 654 761; 4 607 519; 5 814 903
b 7 135 976; 7 323 916; 7 651 411
c 4 238 175; 4 572 391; 4 962 345

5 **a** 2 000 000 **b** 6 000 000 **c** 1 000 000
d 5 000 000 **e** 5 000 000 **f** 8 000 000

1:03

1 **a** 4287 **b** 8695 **c** 7454 **d** 5138
e 9375 **f** 2761 **g** 6529 **h** 1916

2 **a** $(3 \times 10^3) + (7 \times 10^2) + (4 \times 10) + 2$
b $(8 \times 10^3) + (4 \times 10^2) + (6 \times 10) + 3$
c $(9 \times 10^3) + (5 \times 10^2) + (2 \times 10) + 9$
d $(7 \times 10^3) + (3 \times 10^2) + (8 \times 10) + 5$
e $(6 \times 10^3) + (9 \times 10^2) + (5 \times 10) + 6$

3 **a** 3×10^1 **b** 5×10^2 **c** 6×10^3 **d** 7×10^1
e 2×10^2 **f** 4×10^3 **g** 8×10^2 **h** 7×10^3

4 **a** 7583 **b** 3195 **c** 6639 **d** 9743
e 4837 **f** 9638 **g** 5749 **h** 4768

1:04

1 **a** $\frac{9}{10}$ **b** $\frac{5}{12}$ **c** $\frac{5}{6}$ **d** $\frac{4}{6}$ **e** $\frac{5}{8}$ **f** $\frac{3}{8}$

2 **a** $\frac{78}{100}$ or 0·78 **b** $\frac{56}{100}$ or 0·56 **c** $\frac{63}{100}$ or 0·63
d $\frac{39}{100}$ or 0·39 **e** $\frac{28}{100}$ or 0·28 **f** $\frac{43}{100}$ or 0·43
g $\frac{92}{100}$ or 0·92 **h** $\frac{87}{100}$ or 0·87

3 **a**
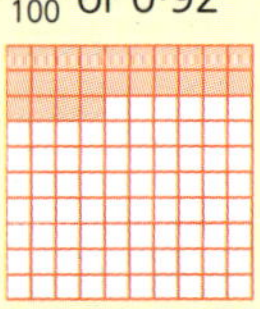
b

c
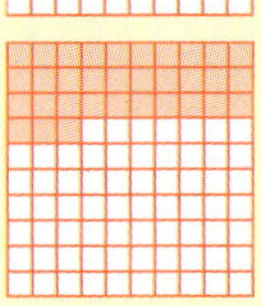
d
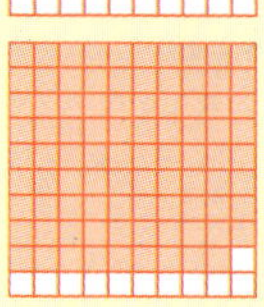

4 **a** $\frac{22}{100}$ **b** $\frac{44}{100}$ **c** $\frac{37}{100}$ **d** $\frac{61}{100}$

1:05

1 **a** $\frac{3}{4}$, $\frac{1}{4}$. $\frac{3}{4}$ and $\frac{1}{4}$ makes 1 whole.
b $\frac{3}{8}$, $\frac{5}{8}$. $\frac{3}{8}$ and $\frac{5}{8}$ makes 1 whole.

2 **a** 1 **b** $\frac{3}{4}$ **c** $\frac{1}{3}$ **d** 1 **e** $\frac{5}{8}$ **f** $\frac{6}{10}$

3 **a** $\frac{1}{4}$ **b** $\frac{3}{8}$ **c** $\frac{2}{5}$ **d** $\frac{1}{10}$ **e** $\frac{5}{8}$ **f** $\frac{3}{10}$

4 watermelons $\frac{1}{20}$; apples $\frac{4}{20}$; pears $\frac{5}{20}$; bananas $\frac{5}{20}$; oranges $\frac{3}{20}$; cherries $\frac{2}{20}$

1:06

1 **a**
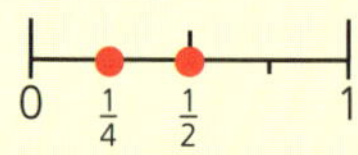

$\frac{1}{2} > \frac{1}{4}$

b
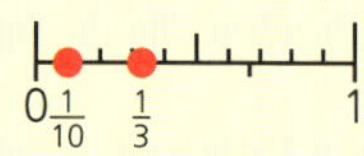

$\frac{1}{10} < \frac{1}{3}$

c
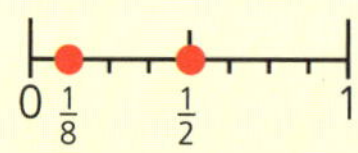

$\frac{1}{2} > \frac{1}{8}$

d
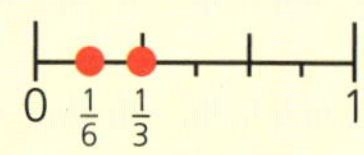

$\frac{1}{6} < \frac{1}{3}$

2 **a** $\frac{1}{5}$, $\frac{1}{4}$, $\frac{1}{2}$ **b** $\frac{1}{100}$, $\frac{1}{20}$, $\frac{1}{10}$
c $\frac{1}{8}$, $\frac{1}{3}$, $\frac{1}{2}$ **d** $\frac{1}{8}$, $\frac{1}{4}$, $\frac{1}{2}$

3
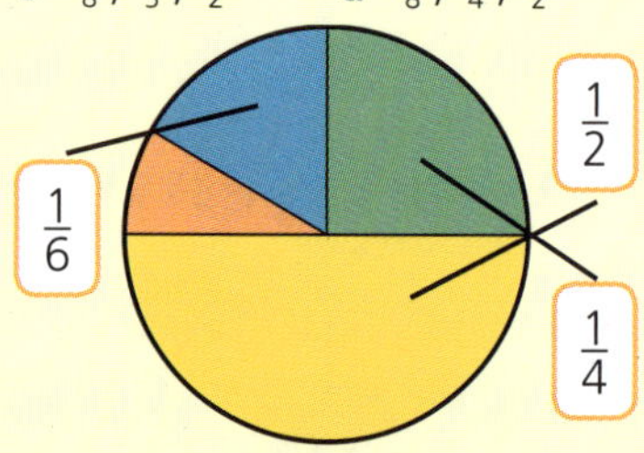

4
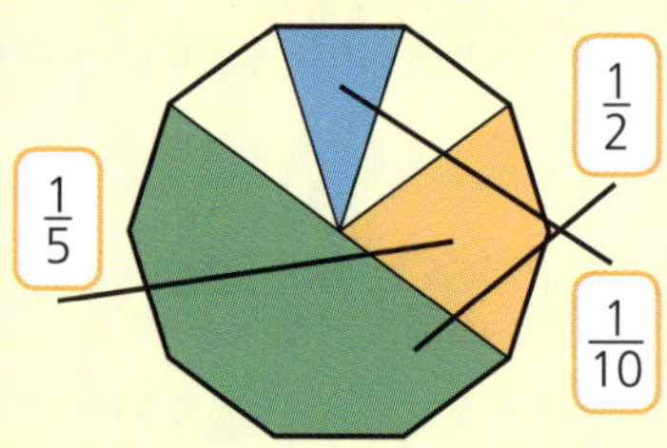

5 smaller

1:07

1 **a** 0·9 **b** 0·5 **c** 0·3 **d** 0·7 **e** 0·4 **f** 1·6
g 2·8 **h** 1·1 **i** 2·8 **j** 3·2 **k** 6·9 **l** 2·8

2

3 **a** 0·8 **b** 0·3 **c** 0·9 **d** 0·4 **e** 0·5
f 0·1 **g** 0·2 **h** 0·6 **i** 0·8 **j** 0·3
k 1·0 **l** 1·5 **m** 4·2 **n** 8·9

4 **a**
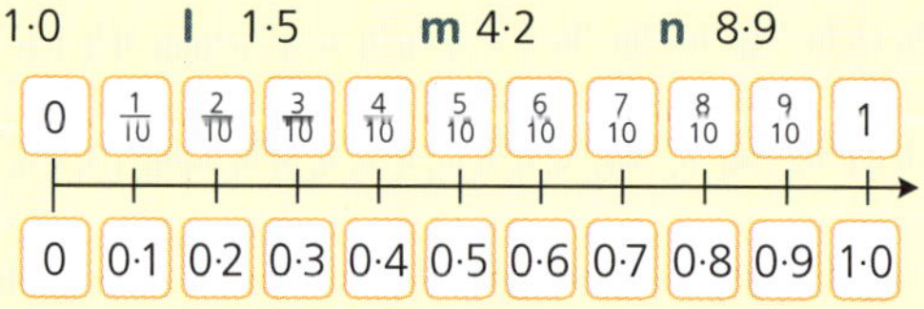

b
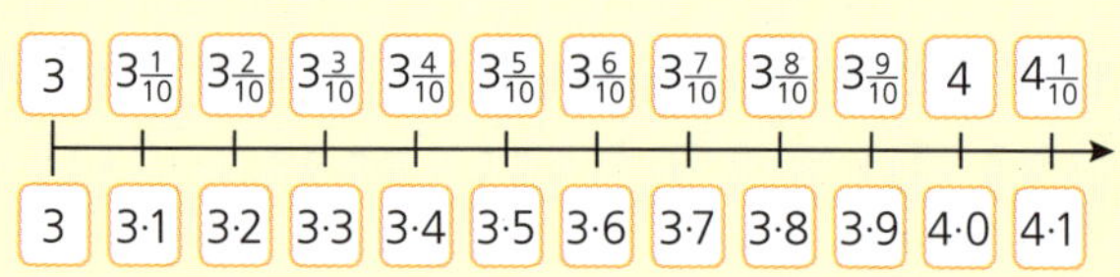

 • *AUSTRALIAN SIGNPOST MATHS 5* • ISBN 978 1 4886 2190 1

1:08

1. 100
2. a 40 out of 100, $\frac{40}{100}$, 0·40   b 18 out of 100, $\frac{18}{100}$, 0·18
   c 18 out of 100, $\frac{18}{100}$, 0·18
3. 0·24
4. a 0·49, $\frac{49}{100}$ b 0·94, $\frac{94}{100}$ c 0·71, $\frac{71}{100}$ d 0·89, $\frac{89}{100}$
   e 0·40, $\frac{40}{100}$ f 0·35, $\frac{35}{100}$ g 0·74, $\frac{74}{100}$ h 0·53, $\frac{53}{100}$

1:09

1. a

Ones		Tenths	Hundredths
0	·	4	6

b

Ones		Tenths	Hundredths
0	·	5	5

c

Ones		Tenths	Hundredths
1	·	2	8

2. a $1\frac{73}{100}$ — 1·73; $\frac{17}{100}$ — 0·17; $1\frac{53}{100}$ — 1·53
   b $2\frac{51}{100}$ — 2·51; $1\frac{35}{100}$ — 1·35; $2\frac{15}{100}$ — 2·15
   c $\frac{10}{100}$ — 0·1; $2\frac{49}{100}$ — 2·49; $3\frac{49}{100}$ — 3·49
3. 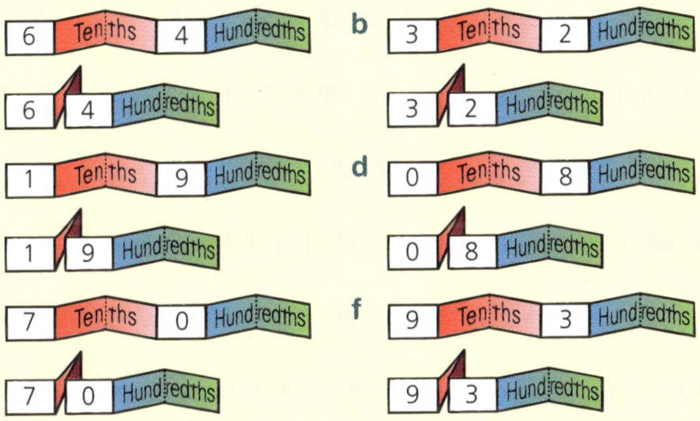

1:10

1. a 5·753   b 4·957   c 5·748   d 9·358
   e 8·254   f 2·574   g 3·495   h 4·935
2.

	Units	·	10ths	100ths	1000ths
a	3	·	1	9	7
b	5	·	6	3	8
c	9	·	2	4	9
d	6	·	5	4	8
e	8	·	3	5	2
f	2	·	7	1	9

1:11

1. a 53·64   b 75·46   c 95·48   d 37·90
2. a 15·75   b 80·36   c 27·92   d 90·08
   e 62·73   f 71·99
3. a 9 tenths and 1 hundredth
   b 4 tenths and 0 hundredths
   c 0 tenths and 9 hundredths
4. a 583·27   b 603·02   c 563·8   d 410·39
   e 6·9   f 4·2   g 5·3   h 9·1
5. a 3   b $\frac{9}{10}$ c 10   d 800
   e 20   f $\frac{4}{100}$ g 0   h $\frac{7}{10}$
   i $\frac{6}{100}$ j 500   k 30   l $\frac{9}{100}$

1:12

1. a 3·842   b 7·473   c 4·349   d 4·258
   e 5·348   f 4·738   g 5·745   h 2·558
2. a 3·619 < 4·207   b 9·138 < 9·56   c 4·196 < 5·231
   d 8·953 > 7·641   e 1·403 < 1·936   f 6·925 > 5·863
   g 5·306 < 5·638   h 7·612 > 7·59
3. a false   b true   c true
4. a 49·357   b 45·750   c 53·254   d 25·740
   e 34·957   f 73·848

1:13

1. a 34·956   b 25·748   c 45·753   d 73·848
2. a 6 tenths   b 8 thousandths   c 3 hundredths
   d 5 hundredths   e 4 hundredths   f 9 tenths
   g 4 ones   h 7 hundredths   i 2 thousandths
3. a 0·47   b 0·359   c 0·7   d 0·206
   e 0·6   f 0·008
4. a $\frac{36}{100}$ b $\frac{417}{1000}$ c $\frac{5}{10}$ d $\frac{3}{100}$ e $\frac{48}{100}$ f $\frac{745}{1000}$
5. a 0·6 because $\frac{6}{10}$ is greater than $\frac{2}{10}$
   b 0·75 because $\frac{75}{100}$ is greater than $\frac{72}{100}$
   c 0·999 because $\frac{999}{1000}$ is greater than $\frac{998}{1000}$
   d 3·72 because $\frac{372}{100}$ is greater than $\frac{370}{100}$ or $\frac{72}{100}$ is greater than $\frac{70}{100}$

1:14

1. 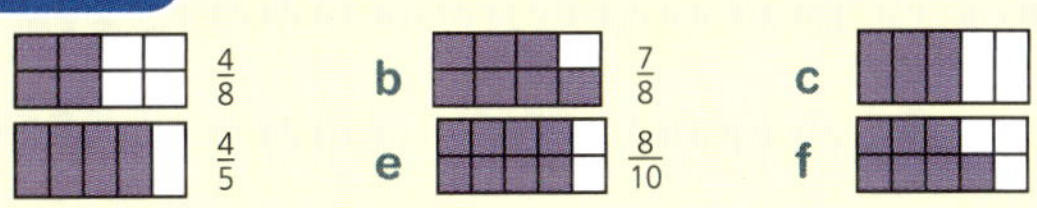
   a $\frac{4}{8}$ b $\frac{7}{8}$ c $\frac{3}{5}$
   d $\frac{4}{5}$ e $\frac{8}{10}$ f $\frac{7}{10}$
2. a $\frac{6}{10}$ b $\frac{7}{10}$ c $\frac{9}{10}$ d $\frac{9}{10}$
   e $\frac{7}{10}$ f $\frac{9}{10}$ g $\frac{8}{10}$ h $\frac{9}{10}$
3. a $\frac{7}{8}$ b $\frac{6}{8}$ c $\frac{6}{8}$ d $\frac{5}{8}$
4. a true   b false   c true   d false
   e true   f true   g true   h true

1:15

1. 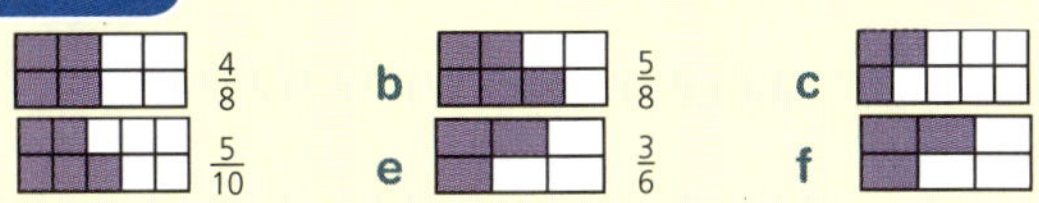
   a $\frac{4}{8}$ b $\frac{5}{8}$ c $\frac{3}{10}$
   d $\frac{5}{10}$ e $\frac{3}{6}$ f $\frac{3}{6}$
2. a $\frac{2}{10}$ b $\frac{2}{10}$ c $\frac{5}{10}$ d $\frac{4}{10}$
   e $\frac{1}{10}$ f $\frac{5}{10}$ g $\frac{1}{10}$ h $\frac{3}{10}$
3. a $\frac{4}{8}$ b $\frac{2}{8}$ c $\frac{4}{8}$ d $\frac{3}{8}$
4. a true   b false   c true   d true
   e true   f false   g true   h true

1:16

1. 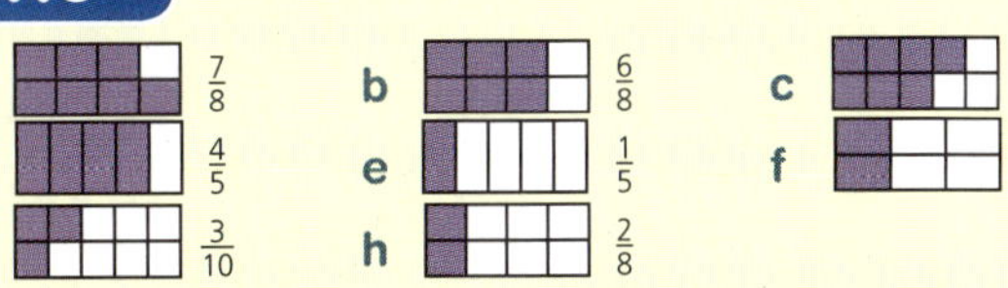
   a $\frac{7}{8}$ b $\frac{6}{8}$ c $\frac{7}{10}$
   d $\frac{4}{5}$ e $\frac{1}{5}$ f $\frac{2}{6}$
   g $\frac{3}{10}$ h $\frac{2}{8}$
2. a $\frac{1}{10}$ b $\frac{3}{10}$ c $\frac{5}{10}$ d $\frac{2}{10}$
   e $\frac{2}{10}$ f $\frac{1}{10}$ g $\frac{6}{10}$ h $\frac{2}{10}$
3. a $\frac{4}{8}$ b $\frac{2}{8}$ c $\frac{4}{8}$ d $\frac{3}{8}$
   e $\frac{3}{8}$ f $\frac{1}{8}$ g $\frac{3}{8}$ h $\frac{1}{8}$

4 a $\frac{3}{8} > \frac{1}{8}$ b $\frac{7}{8} > \frac{3}{8}$ c $\frac{2}{8} < \frac{5}{8}$ d $\frac{4}{8} < \frac{5}{8}$
e $\frac{1}{10} < \frac{3}{10}$ f $\frac{9}{10} > \frac{4}{10}$ g $\frac{7}{10} > \frac{2}{10}$ h $\frac{3}{10} < \frac{4}{10}$

1:17

1 a $\frac{4}{8}$ b $\frac{3}{4}$ c $\frac{9}{10}$ d $\frac{4}{5}$
e $\frac{4}{5}$ f $\frac{7}{8}$ g $\frac{3}{4}$ h $\frac{7}{10}$

2 a $\frac{2}{4}$ b $\frac{2}{5}$ c $\frac{3}{10}$ d $\frac{2}{8}$
e $\frac{2}{5}$ f $\frac{3}{8}$ g $\frac{4}{10}$ h $\frac{5}{10}$
i $\frac{1}{10}$ j $\frac{1}{5}$ k $\frac{1}{4}$ l $\frac{5}{10}$

3 a $\frac{6}{8}$ b $\frac{4}{5}$ c $\frac{5}{10}$ d $\frac{9}{10}$

4 a $\frac{10}{8}$ b $\frac{9}{5}$ c $\frac{7}{4}$ d $\frac{7}{2}$ e $\frac{13}{5}$ f $\frac{19}{10}$

5 a $1\frac{3}{5}$ b $1\frac{1}{3}$ c $2\frac{1}{4}$ d $1\frac{1}{8}$

1:18

1 a $\frac{5}{6}$ b $\frac{9}{10}$ c $\frac{7}{8}$ d $\frac{11}{12}$ e $\frac{4}{5}$ f $\frac{1}{4}$
g $\frac{3}{10}$ h $\frac{3}{5}$ i $\frac{1}{3}$ j $\frac{1}{6}$ k $\frac{5}{8}$ l $\frac{7}{12}$

2 a $\frac{4}{8}$ b $\frac{3}{6}$ c $\frac{1}{3}$ d $\frac{2}{4}$ e $\frac{2}{6}$ f $\frac{3}{8}$
g $\frac{6}{10}$ h $\frac{2}{5}$ i $\frac{5}{12}$ j $\frac{3}{10}$ k $\frac{2}{5}$ l $\frac{4}{12}$

3 a $2\frac{1}{2}$ b $1\frac{2}{3}$ c $1\frac{5}{6}$ d $2\frac{4}{5}$ e $3\frac{9}{10}$
f $3\frac{11}{12}$ g $2\frac{7}{8}$ h $3\frac{3}{4}$ i $1\frac{1}{4}$

4 a $\frac{1}{6}$ b $\frac{5}{6}$ c $\frac{2}{3}$ or $\frac{8}{12}$
d $\frac{4}{6}$ or $\frac{8}{12}$ e $\frac{4}{6}$ or $\frac{2}{3}$ f $\frac{6}{12}$

5 a true b true c false d true

1:19

1 a Yes b Yes c Yes

2
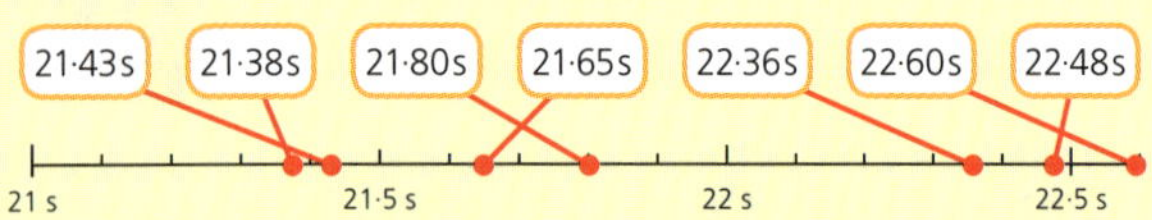

3 a 4·75 b 31·46 c **39·74** d 59·60
4·70 **30·40** 40·06 **59·50**
e **29·56** f **19·14** g 36·43 h 19·63
29·83 26·99 **27·86** **19·62**

4 Answers will vary.

1:20

1 a 0·9 b 1·3 c 1·4 d 2·0
e 4·6 f 0·6 g 1·1 h 2·3

2 a 0·5 b 0·8 c 0·6 d 1·5 e 0·6

3 a 0·7, 0·9 b 2·3, 2·8 c 0·6, 0·5
d 2·5, 2·3 e 0·21, 0·23 f 1·52, 1·62
g 0·76, 0·71 h 3·24, 3·30

4 Answers will vary.

1:21

1 a $\frac{13}{10}, \frac{15}{10}, \frac{17}{10}, \frac{19}{10}$ Rule: Add $\frac{2}{10}$
b $3\frac{4}{10}, 3\frac{2}{10}, 3, 2\frac{8}{10}$ Rule: Subtract $\frac{2}{10}$
c 0·91, 0·93, 0·95 Rule: Add 0·02
d 1·3, 1·2, 1·1 Rule: Subtract 0·1

2 a $\frac{1}{8}, \frac{5}{8}, \frac{9}{8}, \frac{13}{8}, \frac{17}{8}, \frac{21}{8}, \frac{25}{8}$; add $\frac{4}{8}$
b $\frac{16}{3}, \frac{14}{3}, \frac{12}{3}, \frac{10}{3}, \frac{8}{3}, \frac{6}{3}$; subtract $\frac{2}{3}$

3 Answers will vary.

4 a $\frac{2}{4}, \frac{3}{4}, \frac{4}{4}, \frac{5}{4}, \frac{6}{4}, \frac{7}{4}$; add $\frac{1}{4}$
b $\frac{2}{6}, \frac{3}{6}, \frac{4}{6}, \frac{5}{6}, \frac{6}{6}, \frac{7}{6}$; add $\frac{1}{6}$
c 0·3, 0·5, 0·7, 0·9, 1·1, 1·3; add 0·2
d $\frac{2}{3}, \frac{3}{3}, \frac{4}{3}, \frac{5}{3}, \frac{6}{3}, \frac{7}{3}$; add $\frac{1}{3}$

1:22

1 a $\frac{2}{6}$ or $\frac{3}{9}$ b $\frac{3}{3}$ or $\frac{6}{6}$ or 1
c $\frac{1}{3}$ or $\frac{2}{6}$ d $\frac{4}{6}$ or $\frac{6}{9}$
e $\frac{1}{3}$ or $\frac{3}{9}$ f $\frac{2}{3}$ or $\frac{6}{9}$
g $\frac{3}{3}$ or $\frac{9}{9}$ or 1 h $\frac{2}{3}$ or $\frac{4}{6}$

2 a false b true c true d true
e true f false g true h false

3 a
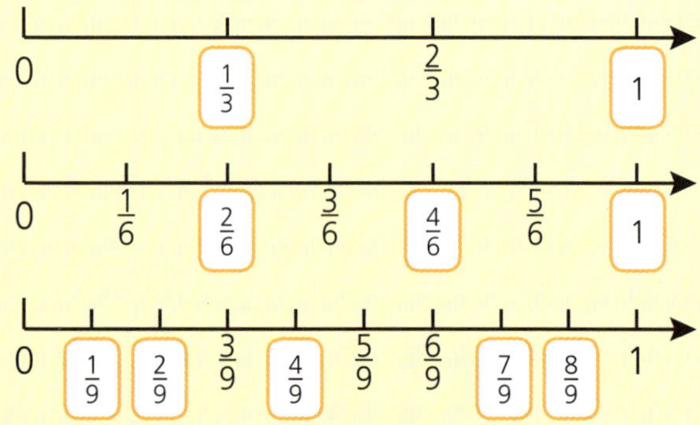

b
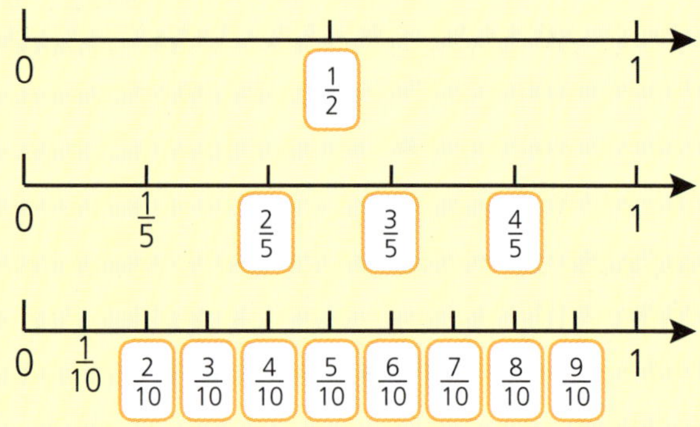

4 a true b true c false d true
e true f false g false h false
i false j true k true l false

5 a $\frac{2}{6}$ or $\frac{3}{9}$ b $\frac{2}{3}$ or $\frac{6}{9}$ c $\frac{5}{10}$ or $\frac{3}{6}$
d $\frac{2}{3}$ or $\frac{4}{6}$ e $\frac{4}{5}$ f $\frac{1}{3}$ or $\frac{2}{6}$
g $\frac{4}{6}$ or $\frac{6}{9}$ h $\frac{2}{10}$

1:23

1 a $\frac{2}{8}$ b $\frac{4}{10}$ c $\frac{6}{8}$ d $\frac{2}{10}$ e $\frac{2}{4}$
f $\frac{6}{10}$ g $\frac{4}{8}$ h $\frac{8}{10}$ i $\frac{4}{8}$ j $\frac{10}{10}$

2 a $\frac{2}{8}$ b $\frac{4}{8}$ c $\frac{2}{10}$ d $\frac{6}{10}$ e $\frac{6}{8}$
f $\frac{8}{10}$ g $\frac{2}{4}$ h $\frac{4}{10}$ i $\frac{2}{6}$ j $\frac{2}{12}$
k $\frac{8}{12}$ l $\frac{8}{12}$

3 a $\frac{1}{3}\,\frac{(\times 4)}{(\times 4)} = \frac{4}{12}$ b $\frac{3}{6}\,\frac{(\times 2)}{(\times 2)} = \frac{6}{12}$ c $\frac{2}{3}\,\frac{(\times 4)}{(\times 4)} = \frac{8}{12}$
d $\frac{2}{6}\,\frac{(\times 2)}{(\times 2)} = \frac{4}{12}$ e $\frac{4}{10}\,\frac{(\times 2)}{(\times 2)} = \frac{8}{20}$ f $\frac{3}{5}\,\frac{(\times 3)}{(\times 3)} = \frac{9}{15}$
g $\frac{3}{5}\,\frac{(\times 2)}{(\times 2)} = \frac{6}{10}$ h $\frac{1}{4}\,\frac{(\times 5)}{(\times 5)} = \frac{5}{20}$ i $\frac{2}{7}\,\frac{(\times 3)}{(\times 3)} = \frac{6}{21}$

1:24

1 a $\frac{2}{6}$ b $\frac{3}{12}$ c $\frac{5}{10}$ d $\frac{3}{15}$
e $\frac{6}{15}$ f $\frac{8}{12}$ g $\frac{6}{8}$ h $\frac{9}{15}$

2 a $\frac{4}{8}$ b $\frac{2}{10}$ c $\frac{3}{9}$ d $\frac{2}{12}$
e $\frac{4}{6}$ f $\frac{3}{6}$ g $\frac{4}{20}$ h $\frac{4}{12}$

3 a $\frac{2}{8}$ b $\frac{2}{12}$ c $\frac{2}{4}$ d $\frac{2}{10}$ e $\frac{2}{6}$

4 a $\frac{3}{18}$ b $\frac{6}{9}$ c $\frac{3}{9}$ d $\frac{9}{12}$ e $\frac{12}{15}$

5 a $\frac{4}{16}$ b $\frac{4}{20}$ c $\frac{4}{12}$ d $\frac{12}{16}$ e $\frac{8}{12}$

6 a 4 b 3 c 3 d 2 e 2
f 2 g 2 h 2 i 5 j 4

7 a $\frac{1}{2} = \frac{2}{4} = \frac{3}{6} = \frac{4}{8} = \frac{5}{10} = \frac{6}{12}$
b $\frac{1}{3} = \frac{2}{6} = \frac{3}{9} = \frac{4}{12} = \frac{5}{15} = \frac{6}{18}$

1:25

1 a 80% b 65% c 75% d 40% e 30%
f 45% g 95% h 90%

2 a 20% b 35% c 25% d 60% e 70%
f 55% g 5% h 10%

3 a $\frac{25}{100}$, 25% b $\frac{65}{100}$, 65% c $\frac{45}{100}$, 45%
d $\frac{80}{100}$, 80% e $\frac{50}{100}$, 50% f $\frac{20}{100}$, 20%
g $\frac{60}{100}$, 60% h $\frac{35}{100}$, 35% i $\frac{75}{100}$, 75%
j $\frac{55}{100}$, 55% k $\frac{95}{100}$, 95% l $\frac{40}{100}$, 40%

1:26

1 a 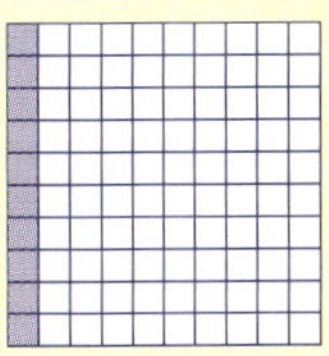 b 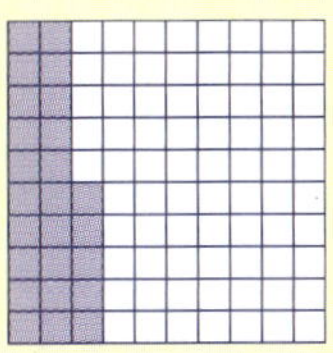 c 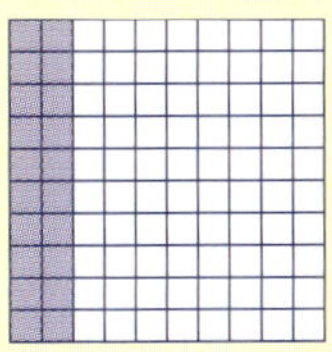
d 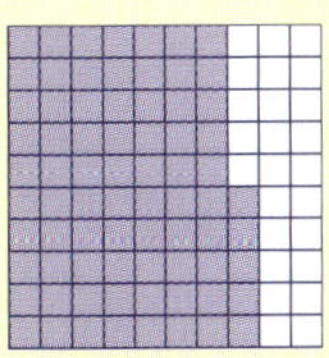 e 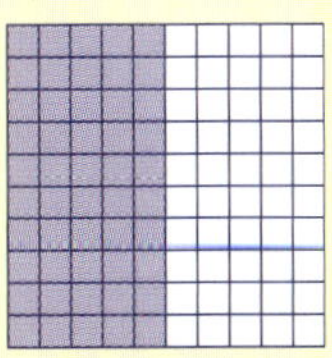 f 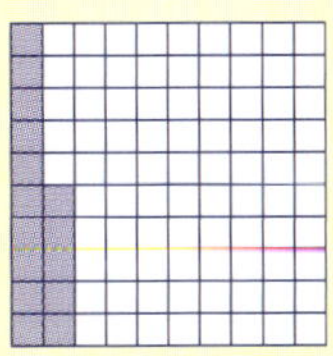
g 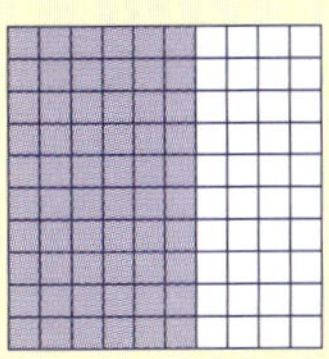 h 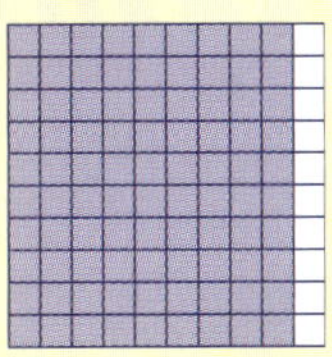

2 a 90% b 75% c 80% d 25%
e 50% f 85% g 40% h 10%

3 a 0·15, 15% b 0·75, 75% c 0·25, 25%
d 0·45, 45% e 0·95, 95% f 0·85, 85%
g 0·35, 35% h 0·1, 10% i 0·3, 30%
j 0·2, 20% k 0·6, 60% l 0·9, 90%

4 a 0·65 b 0·15 c 0·95 d 0·45 e 0·75
f 0·25 g 0·05 h 0·6 i 0·8 j 0·4
k 0·1 l 0·37 m 0·91 n 0·2 o 1

2:01

1 a 27 b 40 c 32 d 20 e 35
f 24 g 14 h 28 i 15 j 30
k 18 l 24

2 a 25 b 36 c 9 d 1 e 10
f 45 g 40 h 24

3 a
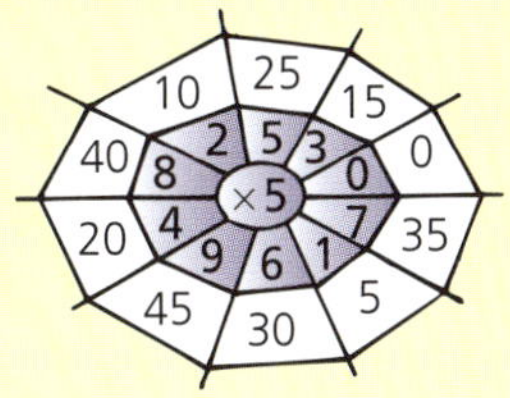

b
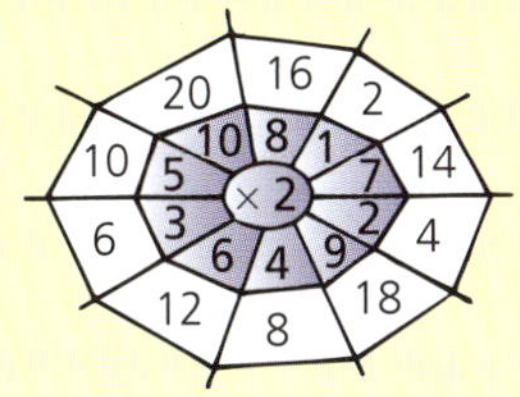

c
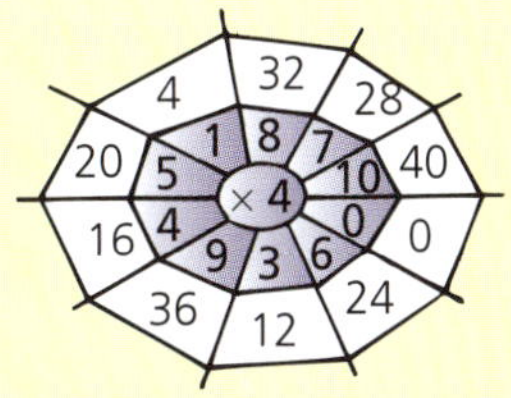

d
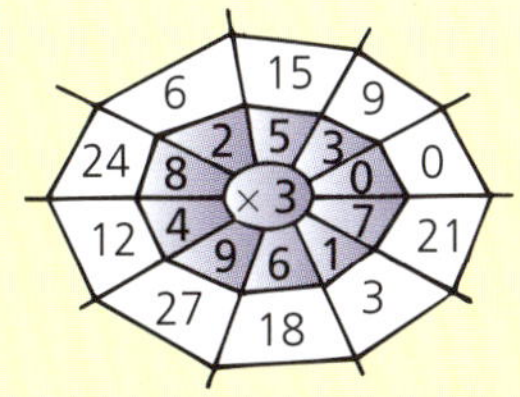

e
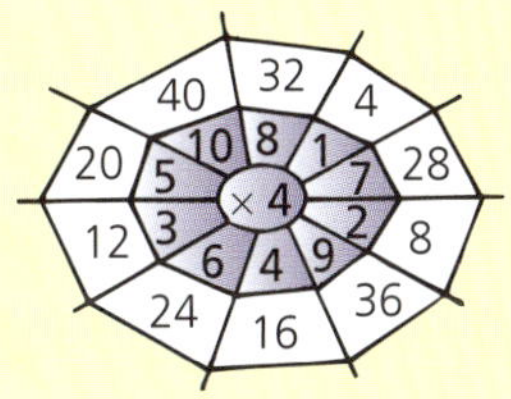

f
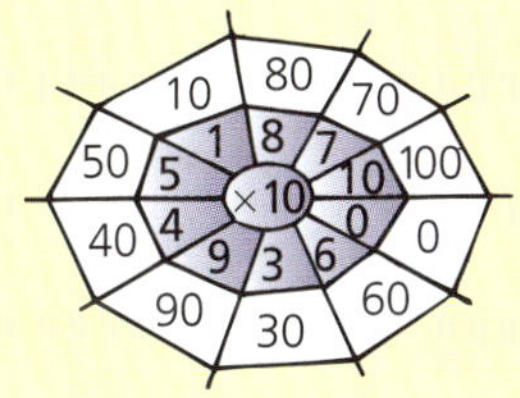

4 a 3, 6, 9, 12, 15, 18, 21, 24, 27, 30
b 4, 8, 12, 16, 20, 24, 28, 32, 36, 40
c 5, 10, 15, 20, 25, 30, 35, 40, 45, 50

2:02

1 a 30 b 24 c 21 d 36 e 40
f 63 g 36 h 49 i 54 j 72
k 48 l 48 m 42 n 56 o 72
p 81 q 24 r 64 s 48

2 a
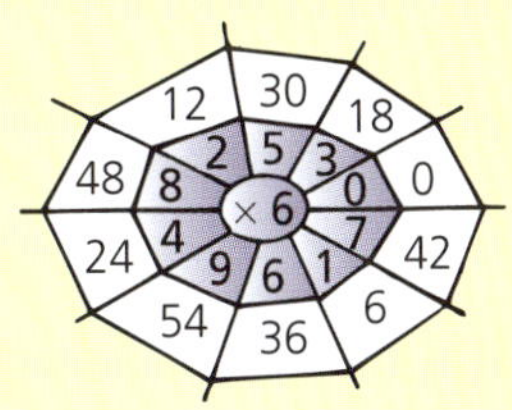

b
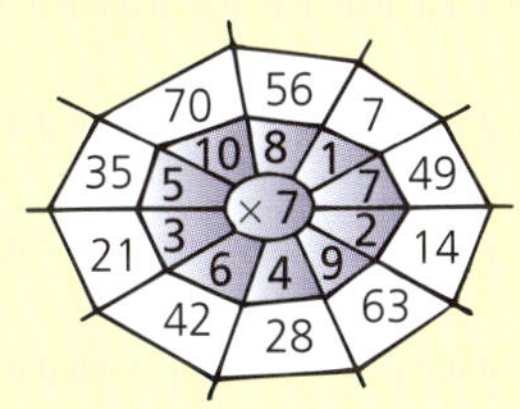

c
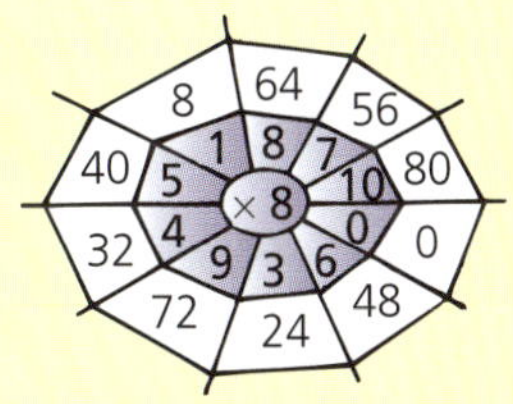

d
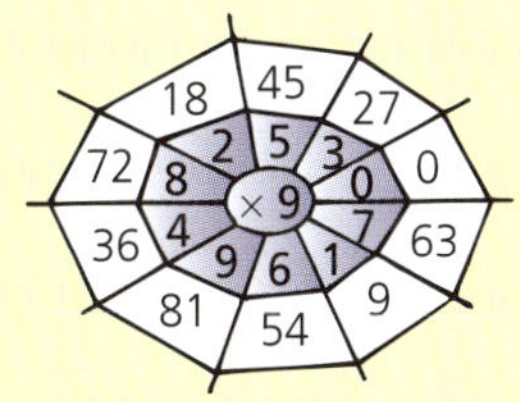

e

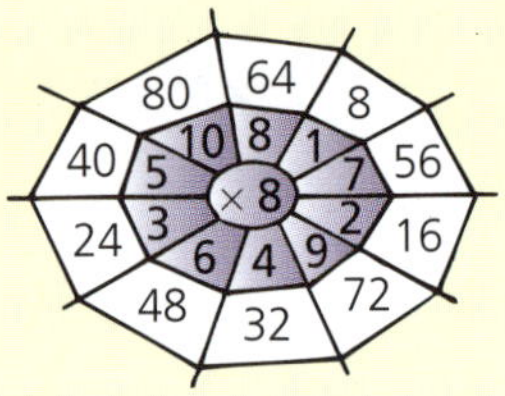

f

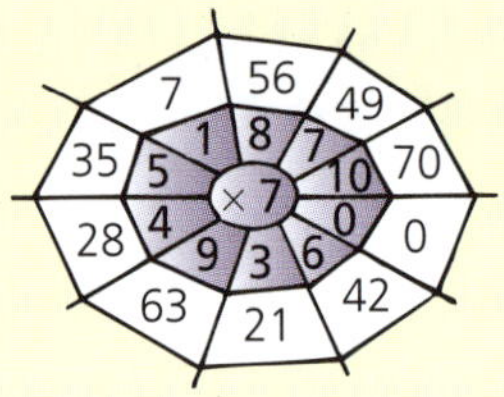

3 a 6, 12, 18, 24, 30, 36, 42, 48, 54, 60
b 7, 14, 21, 28, 35, 42, 49, 56, 63, 70
c 9, 18, 27, 36, 45, 54, 63, 72, 81, 90

2:03

1 a 9 b 12 c 12 d 18 e 24 f 25
g 24 h 36 i 16 j 14 k 27 l 16
m 15 n 30 o 64 p 18 q 54 r 21
s 60 t 48 u 49 v 80 w 42 x 81

2 a

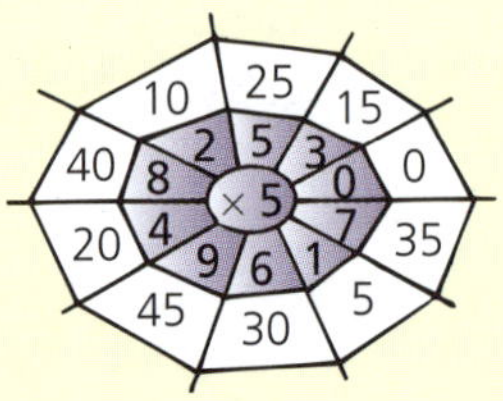

b

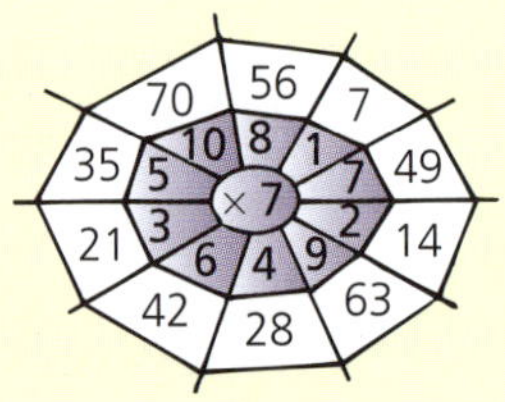

c

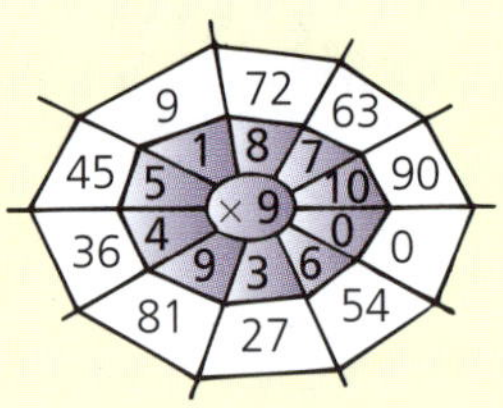

3 a 4, 8, 12, 16, 20, 24, 28, 32, 36, 40
b 8, 16, 24, 32, 40, 48, 56, 64, 72, 80
The multiples of 8 are **twice** the size of the multiples of 4.

2:04

1 a 10, 4 b 5, 8 c 3, 7
d 9, 8 e 8, 9 f 4, 9

2 a 11 b 10 c 7 d 2
e 6 f 9 g 1 h 0
i 7 j 8 k 7 l 9

3 a 6, 6 b 7, 7 c 8, 8
d 10, 10 e 7, 7 f 6, 6

4 a 9 b 5 c 3 d 4
e 10 f 1 g 5 h 3
i 0 j 8 k 2 l 6
m 9 n 5 o 7 p 9

5 a 5, 5 b 9, 9 c 7, 7
d 9, 9 e 5, 5 f 4, 4
g 9 h 6 i 9

2:05

1 a 40 b 80 c 110 d 130
e 310 f 930 g 790 h 330
i 1890 j 1900 k 1900 l 2070

2 a 700 b 400 c 200 d 900
e 900 f 700 g 400 h 200
i 3500 j 9000 k 4100 l 2100

3 a 3000 b 4000 c 8000 d 4000
e 8000 f 7000 g 1000 h 5000

4 a 3672 3196 2878
3945 2764 3285
3812 2246 3349
b 6249 5862 6591
5396 6478 6487
5723 6823 5734

5 a 3275 3227 3270
3249 3185 3175
3159 3129 3207
b 75247 75672 74503
74820 75981 75199
74498 74981 74500

6 a true b true c false d false

2:06

1 a 154 b 117 c 108 d 283 e 790
f 388 g 516 h 510 i 428 j 331
k 514 l 733

2:07

1 a 328 b 601 c 815 d 713 e 561
f 456 g 342 h 463 i 570 j 293

2 a 365 b 605 c 628 d 779 e 414
f 819

2:08

1 a 2 hundreds 4 tens 5 ones
b 6 hundreds 4 tens 5 ones
c 131 d 733 e 533 f 331
g 674 h 934 i 682 j 743
k 362 l 129 m 123 n 443
o 178 p 429 q 70 r 205

2:09

1 a 334 b 662 c 361 d 432 e 615
f 531 g 792 h 657 i 727 j 762
k 321 l 731 m 805 n 640 o 962

2 a 432 minutes b 374 people
c 377 bags d 921 m^2

2:10

1 a 528 b 315 c 229 d 421 e 264
f 108 g 836 h 908 i 711 j 315
k 248 l 628 m 727 n 55 o 105
p 608 q 6 r 734 s 16

2:11

1. a 464 b 255 c 194 d 691 e 376
f 255 g 594 h 40 i 311 j 291
k 312 l 638 m 625 n 80 o 193
p 453 q 206 r 434 s 491

2:12

1. 1 × 5 = 5 2 × 5 = 10
3 × 5 = 15 4 × 5 = 20
5 × 5 = 25 6 × 5 = 30
7 × 5 = 35 8 × 5 = 40
9 × 5 = 45

The first ten multiples of 5 are 5, 10, 15, 20, 25, 30, 35, 40, 45 and 50.

2. a 2, 4, 6, 8, 10, 12, 14, 16, 18 and 20
b 4, 8, 12, 16, 20, 24, 28, 32, 36 and 40
c 5, 10, 15, 20, 25, 30, 35, 40, 45 and 50
d 6, 12, 18, 24, 30, 36, 42, 48, 54 and 60
e 8, 16, 24, 32, 40, 48, 56, 64, 72 and 80
f 10, 20, 30, 40, 50, 60, 70, 80, 90 and 100
3. a 10 b 30 c 30 d 40 e 10 f 20
g 24 h 20 i 40 j 8 k 8 l 10
4. The first 10 common multiples are: 10, 20, 30, 40, 50, 60, 70, 80, 90, 100.

2:13

1. a–d Answers will include 1 × 24, 2 × 12, 3 × 8 and 4 × 6.
e 1, 24, 2, 12, 3, 8, 4, and 6
2. a 1, 6, 2, 3 b 1, 12, 2, 6, 3, 4
c 1, 5 d 1, 20, 2, 10, 4, 5
e 1, 15, 3, 5 f 1, 28, 2, 14, 4, 7
g 1, 30, 2, 15, 3, 10, 5, 6
3. a 1, 80, 2, 40, 4, 20, 5, 16, 8 and 10
4. a 1, 16, 2, 8 or 4 b 1, 21, 3 or 7
c 1, 12, 2, 6, 3 or 4

2:14

1. a $0.78 b $1.52 c $4.73
d $5.31 e $6.69 f $8.25
g $4.70 h $8.51 i $6.43
j $7.11 k $6.08 l $6.97
m $5.40 n $6.29 o $7.14
2. a $3.79 b $5.77 c $5.68
d $8.68 e $6.49 f $8.48
3. a $9.55 b $8.20 c $8.85
d $8.70

2:15

1. a $5.22 b $5.17 c $6.34
d $2.39 e $4.17 f $1.84
g $2.77 h $6.08 i $3.34
j $4.08 k $1.70 l $2.58
m $3.89 n $5.45 o $2.04
2. a $2.11 b $4.98 c $5.35
d $0.05 e $5.50 f $1.51
3. a $1.05 b $6.20 c $5.45
d $1.20

2:16

1. a Yes b Yes c Yes d Yes
e No f Yes g No h Yes
2. a $4.60 b $7.40 c $7.70 d $9.10
e $7.80 f $9.40
3. a $4.40 b $2.80 c $1.20 d $4.10 e $1.60
4. a $9.71 b $9.02 c $9.18 d $7.81 e $7.49
f $3.27 g $6.45 h $3.39 i $6.33 j $3.26
5. a $9.70 b $9.00 c $9.20 d $7.80 e $7.50
f $3.25 g $6.45 h $3.40 i $6.35 j $3.25

2:17

1. 15 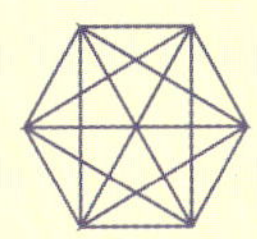
2. 28 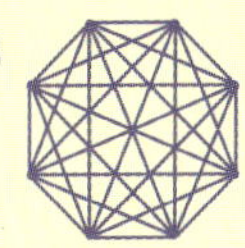
3. a 8 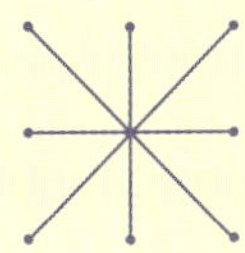 b 5 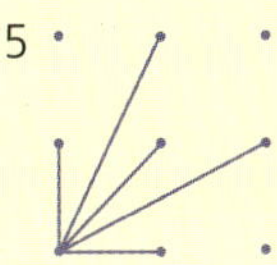 c 28 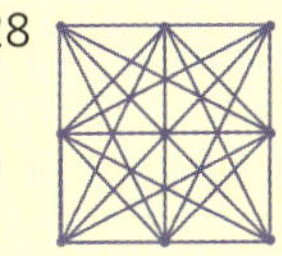
4. 6 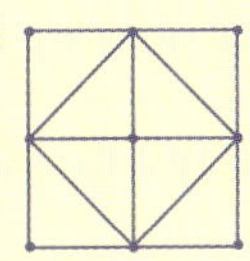
5. 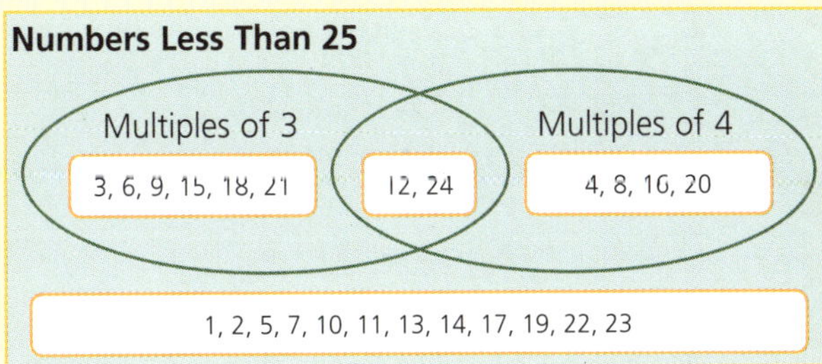

6. Not necessarily. Answers will vary.

2:18

1. a 2 r 2 b 2 r 1 c 2 r 2 d 2 r 4
e 2 r 3 f 2 r 3 g 3 r 3 h 2 r 3
2. a 48 ÷ 7 = 6 packets and 6 remaining
b 48 ÷ 10 = 4 groups and 8 remaining
c 48 ÷ 8 = 6 rows and 0 remaining
d 80 ÷ 8 = 10 lots and 0 remaining
e 80 ÷ 10 = 8 lots and 0 remaining
f 80 ÷ 9 = 8 groups and 8 remaining
3. a 17 ÷ 3 = 5 groups and 2 remaining
b 17 ÷ 6 = 2 groups and 5 remaining

4 a 3 tins 35 ÷ 10 = 3, (r 5)
b 8 sets 33 ÷ 4 = 8, (r 1)
c 5 pages 42 ÷ 8 = 5, (r 2)
d 8 boxes 50 ÷ 6 = 8, (r 2)

2:19

1 a 4 r 1 b 2 r 1 c 7 r 1 d 1 r 3 e 2 r 3
f 5 r 2 g 1 r 3 h 2 r 4 i 8 r 2 j 7 r 8
k 8 r 2 l 6 r 5 m 3 r 4 n 9 r 2 o 2 r 4
p 2 r 1 q 8 r 4 r 5 r 2 s 4 r 3 t 5 r 5
2 a 10 b 30 c 10 d 10 e 20
f 20 g 10 h 30 i 10 j 10
3 a 7 r 4, 7 people b 6 r 1, 6 packets

2:20

1 a 6 r 2 b 4 r 1 c 7 r 2
2 a 2 r 2 b 2 r 3 c 4 r 4 d 5 r 5
e 10 r 1 f 1 r 7 g 8 r 2 h 7 r 1
i 7 r 4 j 3 r 2 k 8 r 4 l 7 r 4
3 a 17 ÷ 3 = 5 r 2 b 27 ÷ 7 = 3 r 6
c 30 ÷ 8 = 3 r 6 d 25 ÷ 6 = 4 r 1
e 53 ÷ 10 = 5 r 3 f 33 ÷ 5 = 6 r 3
g 60 ÷ 9 = 6 r 6 h 36 ÷ 8 = 4 r 4
i 47 ÷ 6 = 7 r 5 j 32 ÷ 8 = 4 r 0
k 29 ÷ 2 = 14 r 1 l 53 ÷ 7 = 7 r 4
m 25 ÷ 4 = 6 r 1 n 27 ÷ 2 = 13 r 1
o 46 ÷ 5 = 9 r 1
4 9

2:21

1 a 16, 19, 22, 25 Rule: add 3
b 43, 55, 67, 79 Rule: add 12
c 60, 51, 42, 33 Rule: subtract 9
d 32, 25, 18, 11 Rule: subtract 7
2 a 120, 135, 150, 165, 180; add 15
b 135, 125, 115, 105, 95; subtract 10
3 Answers will vary.
4 a 6, 9, 12, 15, 18, 21, 24; add 3
b 5, 10, 15, 20, 25, 30, 35; add 5
c 7, 11, 15, 19, 23, 27, 31; add 4
d 6, 11, 16, 21, 26, 31, 36; add 5

2:22

1 a 258 b 179 c 34 d 646 e 274
f 63 g 489 h 589 i 138 j 177
k 659 l 567
2 a $0.29 b $0.99 c $0.87
d $4.97 e $4.97 f $2.98

2:23

1 a 73 b 162 c 839 d 545 e 398
f 107 g 294 h 47
2 a $0.66 b $1.01 c $1.26 d $0.32
3 a 113 points b 419 albums c 253 books
d $6.35

2:24

1 a 5468 b 8975 c 7588 d 3798 e 7967
f 6324 g 8600 h 7251 i 3573 j 8240
2 a 7195 b 4885 c 4343 d 8925
e 8019 f 7505 g 9281 h 4339
i 7617 j 7932
3 a $9149 b $8291 c $7909
d $6297 e $9430

2:25

1 a 7561 b 6935 c 9818 d 7482
2 a 3904 b 5108 c 7837 d 7088
3 a $86.14 b $52.53 c $90.10 d $86.80
4 a your estimate, 5495 b your estimate, 8456
c your estimate, $8273

2:26

1 a 4353 b 7100 c 5420 d 8000 e 6213
f 6070 g 442 h 5320 i 6303 j 1824
2 a 2776 b 1193 c 706 d 6990 e 3068
f 369 g 2278 h 1602 i 6749 j 3489
3 a 1875 b 6195 c $3659
4 909

2:27

1 a 564 b 1699 c 103 d 2275
e 5168 f 1140 g 1100 h 4573
2 a 4550 b 3012 c 775 d 1645
e 2814 f 1305 g 28 h 6633
3 a 3320 km b 1675 m c 3658 m
d 88 years old e 91 years old

2:28

1 a 617 + 1 = 618 b 1308 + 1 = 1309
c 16 + 1 = 17 d 1075 + 1 = 1076
e 3518 + 1 = 3519 f 1024 + 1 = 1025
g $792 + 1 = $793 h $1917 + 1 = $1918
i $4208 + 1 = $4209
2 a 1027 b 3097 c 6248
d 3754 e $1269

2:29

1 a 865 b 758 c 929 d 733
e 853 f 731
2 a 625 b 405 c 354 d 634
e 218 f 349
3 a 606 + 1 = 607 b 3653 + 1 = 3654 c 318
d 635 e 482 f 2828

 ISBN 978 1 4886 2190 1

g 4564 h 8864 i 6112
j 1086 k 3585

2:30

1. a 35, 7 × 5 = 35
   b 15, 3 × 5 = 15
   c 25, 5 × 5 = 25
   d 45, 9 × 5 = 45
   e 30, 6 × 5 = 30
   f 40, 8 × 5 = 40
2. a 50, 55, 60, 65, 70, 75
   b 22, 24, 26, 28, 30, 32
   c 110, 120, 130, 140, 150, 160
   d 66, 68, 70, 72, 74, 76
3. a 14 and 28
   b 35 and 70
   c 12 and 24
   d 21 and 42
4. a 0 b 5, 0 c 2, 4, 6 ,8, 0
5. a 54, 60, 66, 72, 78, 84, 90, 96, 102
   b 90, 99, 108, 117, 126, 135, 144, 153, 162

2:31

1. (3), (7), 8, 13, 20, (21)
2. 2, 4, 6, 8, 10, 12, 14
3. 1, 32, 2, 16, 4, 8
4. 1, 24, 2, 12, 3, 8, 4, 6
5. 1, 100, 2, 50, 4, 25, 5, 20, 10
6. a 2, 4, 6, 8, 10, 12, 14, 16, 18, 20
   b 4, 8, 12, 16, 20, 24, 28, 32, 36, 40
   c 8, 16, 24, 32, 40, 48, 56, 64, 72, 80
   d 10, 20, 30, 40, 50, 60, 70, 80, 90, 100
7. a 16, even b 12, even c 7, odd
   d 10, even e 14, even f 35, odd
8. a even b even c odd
9. Answers will vary.

2:32

1. a 22 b 34 c 44 d 12
   e 11 f 22 g 13 h 12
   i 11 j 12 k 21 l 22
2. a 17 b 15 c 23
   d 18 e 13 f 17

2:33

1. a 23 b 11 c 21 d 41 e 30
   f 12 g 34 h 11
2. a 25 b 48 c 13 d 24 e 13
   f 15 g 12 h 12
3. a 18 b 14 c 14 d 17 e 17
   f 19 g 16 h 29
4. a 13 brushes b 17 ladders c 13 globes
   d 18 plants e 22 caps f 15 spades

2:34

1. a $12\frac{1}{2}$ b $11\frac{2}{3}$ c $22\frac{1}{4}$ d $10\frac{3}{5}$
   e $11\frac{3}{6}$ or $11\frac{1}{2}$ f $12\frac{1}{3}$
2. a $17\frac{1}{2}$ b $24\frac{2}{3}$ c $16\frac{1}{2}$ d $12\frac{3}{8}$
   e $13\frac{1}{6}$ f $14\frac{3}{4}$
3. a 15 b 17 c 14 d 19
   e 16 f 28 g 14 h 13
4. a 17 b 16 c 19 d 18 e 48
   f 37 g 16 h 27 i 18 j 15
5. a 18 km b $12 c 25°C d 23 cm

2:35

1. a 134 b 285 c 144 d 122 e 133
   f 174 g 135 h 145 i 119 j 127
2. a 147 r 5 b 141 r 1 c 234 r 1 d 117 r 5
   e 257 r 1 f 117 r 3 g 196 r 2 h 135 r 4
   i 183 r 3 j 118 r 7
3. a 81 b 50 c 72 d 63
   e 33 f 73
4. a 42 r 1 b 31 c 81 r 1 d 114 r 2 e 75
   f 112 r 4 g 44 r 4 h 28 r 4 i 38 j 49
   k 67 l 94 m 84
5. a 21 m b 97 m

2:36

1. a 80 b 90 c 60 d 40
   e 110 f 140 g 120 h 130
   i 320 j 210 k 480 l 850
   m 900 n 770 o 600 p 990
2. a 80 b 70 c 100 d 90
   e 130 f 180 g 270 h 220
   i 350 j 610 k 840 l 750
3. a 18 tens = 180 b 21 tens = 210
   c 20 tens = 200 d 18 tens = 180
   e 320 f 400 g 360 h 420
4. a 180 b 400 c 280 d 200
5. a 120 b 180 c 320 d 350
   e $140 f $540 g $640 h $360

2:37

1. a 70 b 210 c 900 d 1200 e 1000
   f 1100 g 2500 h 3000 i 4900
2. a 90 b 270 c 560 d 600 e 1300
   f 2500
3. a 24 tens, 240 b 21 hundreds, 2100
   c 32 hundreds, 3200 d 35 tens, 350
   e 15 hundreds, 1500 f 36 hundreds, 3600
4. a 240 b 350 c 540 d 800 e 2100
   f 3000
5. a 120 b 320 c 300 d 1800 e 2800
   f 7200

 • *AUSTRALIAN SIGNPOST MATHS 5* • ISBN 978 1 4886 2190 1

2:38

1. a 80 b 60 c 50 d 30 e 90
   f 37 g 86 h 54 i 29 j 18
   k 11 l 20 m 99 n 73 o 55
2. a 15 r 6 b 37 r 5 c 52 r 1 d 21 r 8 e 86 r 6
   f 68 r 4 g 56 h 73 r 7 i 90 r 8 j 36 r 7
   k 71 r 1 l 40 r 7 m 89 r 9 n 40 o 60 r 1
3. a $95 b 51 runs c 47 mL d 20 cups
4. a 750 b 831 c 600 d 909

2:39

1. a 204 b 208 c 408 d 307 e 205
   f 104 g 105 h 102 i 204 j 209
2. a 101 b 102 c 102 d 403
   e 203 f 100 r 2 g 101 r 3 h 304 r 1
   i 101 r 1 j 100 r 5
3. a 108 sheep b 102 groups c 203 fish
   d 104 people e 106 buckets
4. a 843 b 912 c 815 d 403

2:40

1. a 24, 30, 46, 76, 88, 100
   b 30, 45, 60, 75, 95, 120, 135
   c 21, 63, 81, 102, 162
   d 36, 72, 684, 693, 126, 234
2.

Divisible by 2	Divisible by 5	Divisible by 3	Divisible by 9
64, 96, 80, 54, 92, 134, 168, 190, 162, 348, 460, 378	80, 75, 145, 135, 190, 460, 645, 385	96, 54, 75, 168, 135, 162, 348, 645, 378	54, 135, 162, 378

3. a 3 × 14 = 42; 42 ÷ 3 = 14; 42 ÷ 14 = 3
   b 5 × 16 = 80; 80 ÷ 5 = 16; 80 ÷ 16 = 5
   c 9 × 13 = 117; 117 ÷ 9 = 13; 117 ÷ 13 = 9
4. a 1, 8, 2, 4 b 1, 16, 2, 8, 4
   c 1, 35, 5, 7 d 1, 28, 2, 14, 4, 7
   e 1, 49, 7 f 1, 99, 3, 33, 9, 11
5. Not necessarily.

2:41

1. a 1, 15, 3, 5 b 1, 6, 2, 3
   c 1, 36, 2, 18, 3, 12, 4, 9, 6
2. a 6, 12, 18, 24, 30, 36, 42, 48, 54, 60
   b 9, 18, 27, 36, 45, 54, 63, 72, 81, 90
   c 7, 14, 21, 28, 35, 42, 49, 56, 63, 70
3. a 1, 80, 2, 40, 4, 20, 5, 16, 8, 10
   b 1, 92, 2, 46, 4, 23
   c 1, 63, 3, 21, 7, 9
   d 1, 56, 2, 28, 4, 14, 7, 8
   e 1, 160, 2, 80, 4, 40, 5, 32, 8, 20, 10, 16
   f 1, 108, 2, 54, 3, 36, 4, 27, 6, 18, 9, 12
   g 1, 48, 2, 24, 3, 16, 4, 12, 6, 8
   h 1, 60, 2, 30, 3, 20, 4, 15, 5, 12, 6, 10
   i 1, 144, 2, 72, 3, 48, 4, 36, 6, 24, 8, 18, 9, 16, 12

2:42

1. a 4 cm b Yes, because there were no 4 cm nails.
2. a A: 9, B: 18, C: 11, D: 7 b 4 units
   c 4 units
3. a $34 b $30 c 22 tickets
4. a $2\frac{1}{2}$ or 2·5 tickets b $2\frac{1}{2}$ or 2·5 tickets

2:43

1. a 45 cm b 4·5 kg or $4\frac{1}{2}$ kg c 90 cm
   d 53 mL e 160 mL
   f Each receives 160 mL.
2. a 11 chocolates b 11 chocolates
3. a 12 mm b $23\frac{3}{6}$ mm (or $23\frac{1}{2}$ mm)
   c $30\frac{3}{6}$ (or $30\frac{1}{2}$ squares)

2:44

1. a 16 × 5 = 8 × 2 × 5 = 80
   b 12 × 5 = 6 × 2 × 5 = 60
   c 18 × 5 = 9 × 2 × 5 = 90
   d 14 × 5 = 7 × 2 × 5 = 70
   e 8 × 25 = 2 × 4 × 25 = 200
   f 16 × 25 = 4 × 4 × 25 = 400
   g 28 × 25 = 7 × 4 × 25 = 700
   h 36 × 25 = 9 × 4 × 25 = 900
   i 20 × 5 = 10 × 2 × 5 = 100
   j 12 × 25 = 3 × 4 × 25 = 300
   k 20 × 25 = 5 × 4 × 25 = 500
   l 32 × 25 = 8 × 4 × 25 = 800
2. a 12 × 6 = 2 × 6 × 6 = 2 × 36 = 72
   b 14 × 4 = 2 × 7 × 4 = 2 × 28 = 56
   c 8 × 14 = 8 × 7 × 2 = 56 × 2 = 112
   d 6 × 16 = 6 × 8 × 2 = 48 × 2 = 96
   e 4 × 18 = 2 × 2 × 18 = 2 × 36 = 72
   f 20 × 9 = 2 × 10 × 9 = 2 × 90 = 180
   g 16 × 4 = 2 × 8 × 4 = 2 × 32 = 64
   h 14 × 6 = 2 × 7 × 6 = 2 × 42 = 84

2:45

1. a 80 + 32 = 112 b 40 + 8 = 48 c 60 + 30 = 90
   d 50 + 45 = 95 e 90 + 63 = 153 f 80 + 48 = 128
2. a 40 + 24 = 64 b 160 + 24 = 184 c 200 + 15 = 215
3. a 92
   b 51 × 2 = 102
   51 × 4 = 2 × 102
   = 204
   c 2 × 21 = 42
   4 × 21 = 2 × 42
   = 84

2:46

1 a

First number	1	2	3	4	5	6
Second number	15	16	17	18	19	20

b Add 14 to the 1st number.

c 21

2 a

First number	1	2	3	4	5	6
Second number	4	8	12	16	20	24

b Multiply the 1st number by 4.

c 40

3 a

First number	90	80	70	60	50	40
Second number	82	72	62	52	42	32

b Subtract 8 from the 1st number.

c 22

4 a

Input number	36	46	56	66	76	86
Output number	43	53	63	73	83	93

b Add 7 to the input number.

c 133

5 a

Input number	29	39	49	59	69	79
Output number	12	22	32	42	52	62

b Subtract 17 from the input number.

c 102

6 Answers will vary.

2:47

1 a

First number	1	2	3	4	5	6
Second number	7	14	21	28	35	42

Multiply the first number by 7.

b

First number	1	2	3	4	5	6
Second number	9	18	27	36	45	54

Multiply the first number by 9.

c

First number	1	2	3	4	5	6
Second number	24	25	26	27	28	29

Add 23 to the first number.

d

First number	1	2	3	4	5	6
Second number	36	37	38	39	40	41

Add 35 to the first number.

e

First number	35	45	55	65	75	85
Second number	27	37	47	57	67	77

Subtract 8 from the first number.

f

First number	92	82	72	62	52	42
Second number	104	94	84	74	64	54

Add 12 to the first number.

g

First number	37	47	57	67	77	87
Second number	24	34	44	54	64	74

Subtract 13 from the first number.

h

First number	78	68	58	48	38	28
Second number	53	43	33	23	13	3

Subtract 25 from the first number.

2 a 15, 20, 25, 30, 35, 40, 45, 50

b 5

c Add 5 to the first number.

2:48

1 a 80 + 6 = 86
b 50 + 25 = 75
c 180 + 48 = 228
d 80 + 36 = 116

2 a 160 + 10 = 170
b 80 + 32 = 112
c 180 + 12 = 192
d 70 + 42 = 112
e 270 + 9 = 279
f 160 + 24 = 184
g 200 + 20 = 220
h 240 + 6 = 246
i 360 + 12 = 372

2:49

1 120 + 30 = 150 blocks,

	H	T	U		H	T	U		H	T	U
		2	0				5		1	2	0
×			6	×			6	+		3	0
	1	2	0	+		3	0	=	1	5	0

2 210 + 56 = 266 dollars,

	H	T	U		H	T	U		H	T	U
		3	0				8		2	1	0
×			7	×			7	+		5	6
	2	1	0	+		5	6	=	2	6	6

3 90 + 54 = 144 people,

	H	T	U		H	T	U		H	T	U
		1	0				6			9	0
×			9	×			9	+		5	4
		9	0	+		5	4	=	1	4	4

4 560 + 24 = 584 dollars,

	H	T	U		H	T	U		H	T	U
		7	0				3		5	6	0
×			8	×			8	+		2	4
	5	6	0	+		2	4	=	5	8	4

2:50

1 a 215 b 292 c 216 d 231
e 474 f 182 g 226 h 152

2 a 147 b 288 c 324 d 375
e 384 f 308 g 296 h 189

2:51

1 a 842 b 957 c 903 d 790 e 804
f 920 g 868 h 875 i 944 j 924
k 987 l 1253

2 a 192 horseshoes b 768 pages
c 576 eggs d $891

2:52

1 a 190 b 0
c 60 + 90 = 150 d 170 – 50 = 120

2 a 188 b 7 c 151 d 122

3 a 500 b 600 – 500 = 100
c 800 + 400 = 1200 d 800 – 700 = 100

4 a 514 b 90 c 1 178 d 97

5 a 5 000 b 7 000 – 4 000 = 3 000
c 6 000 + 4 000 = 10 000

6 a 4 425 b 3 193 c 10 241

7 Estimates will vary.
a 2199 + 693 b 2199 + 4904 c 693 + 4904
d 2199 + 9251 e 4904 + 7057 f 7057 + 9251
g No. We can estimate to decide whether the two numbers would be close to the answer.

2:53

1 a (47 × 10) 47 × 5 (470) 235
47 × 10 (47 × 5) 470 (235)
b (58 × 10) 58 × 5 290 (580)
58 × 10 (58 × 5) (290) 580

2 a 73 × 9 < 73 × 10 but > 73 × 5 so 73 × 9 is between 730 and 365.
b 62 × 6 < 62 × 10 but > 62 × 5 so 62 × 6 is between 620 and 310.
c 95 × 7 < 95 × 10 but > 95 × 5 so 95 × 7 is between 950 and 475.
d 81 × 8 < 81 × 10 but > 81 × 5 so 81 × 8 is between 810 and 405.
e 79 × 9 < 79 × 10 but > 79 × 5 so 79 × 9 is between 790 and 395.
f 86 × 6 < 86 × 10 but > 86 × 5 so 86 × 6 is between 860 and 430.

3 a 340 b 464 c 675 d 2 409

4 a CCCXLVI b DCCCLXII c CMXXVIII
d MCCLXIII

2:54

1 a 160 b 217 c 249 d 78 e 90 f 81
g 315 h 342 i 460 j 600 k 228 l 810
m 120 n 158 o 56 p 588

2 a 384 b 692 c 585 d 836 e 975
f 780 g 968 h 879 i 968 j 680
k $9.90 l $8.89 m $9.18 n $8.48 o $9.45

2:55

1 a 1 890 b 4 866 c 5 600 d 1 648 e 2 799
f 1 166 g 1 676 h 3 663 i 2 496 j 5 826
k 2 730 l 7 680 m 2 583 n 2 313 o 4 980
p 4 263 q 7 650 r 5 994

2 a 9 063 b 8 448 c 9 099 d 6 408 e 3 507
f 9 050 g 6 558 h 9 948 i 9 296 j 9 900
k 6 516 l $96.95 m $89.95 n $74.82 o $93.28
p $413.50

2:56

1 a

Deaths from Traffic Accidents

Type of road user	NSW	Vic	Qld	SA	WA	Tas	Total
Drivers of motor vehicles	284	188	138	71	78	16	775
Motor cyclists	55	56	53	22	23	9	218
Pedal cyclists	14	7	5	3	6	0	35
Passengers	124	89	88	40	48	4	393
Pedestrians	95	58	37	18	24	6	238
Other	-	-	-	-	1	-	1
Total	**572**	**398**	**321**	**154**	**180**	**35**	**1 660**

b Tasmania c New South Wales
d 393 e 35

2 537 3 99 4 183 5 69

6 Qld and WA 7 SA and Tas

8 Both horizontal and vertical totals should add up to 1 660.

2:57

1 a 1 kg 250 g b 3 L
c 250 g d 72 ads

2 a 5 kg b 7 000 L c 5 pkts, 9 250 pkts
d 6·1 m

3 a 900 m b 1 008 L

2:58

1 a 300 b 300 c 500 d 200 e 700
f 400 g 900 h 700 i 900 j 600

2 a 50 × 5 b 90 × 5 c 300 × 10

3 a 210 b 400 c 600 d 790
e 180 f 30 g 400 h 40

4 Estimates will vary. The actual answers are:
a $114 b $205.80 c $48.51

2:59

1. a 30 b 50 c 80 d 50
   e 30 f 90 g 70 h 50
   i 70 j 60
2. a 700 b 300 c 300 d 900
   e 900 f 500 g 500 h 700
   i 300 j 400
3. a 800 b 60 × 30 = 1 800 c 50 × 70 = 3 500
   d 90 × 30 = 2 700 e 40 × 10 = 400 f 70 × 50 = 3 500
   g 50 × 20 = 1 000 h 60 × 40 = 2 400
4. a 24 000 b 40 × 200 = 8 000
   c 80 × 700 = 56 000 d 60 × 700 = 42 000
5. a true b true
6. a E = 3 600, A = 3 478 b E = 2 100, A = 2 015
   c E = 18 000, A = 19 779 d E = 30 000, A = 29 375

2:60

Answers will vary.

2:61

1. a Answers will vary. b $87.65
   c deposit $7.05 d $78.65
2. a Answers will vary. b $91.35
   c withdraw $3.15 d $66.95

2:62

1. $233
2. $536
3. $129
4. 144 tonnes, 256 tonnes
5. $267
6. 555 bricks
7. 72 DVDs
8. 8 L, $12.80
9. 380 km, $5\frac{1}{2}$ hours
10. $1 945
11. a

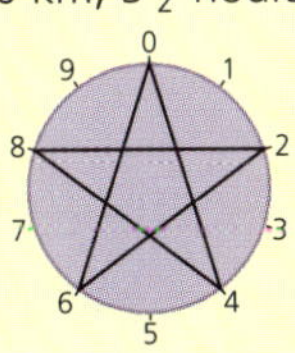

b

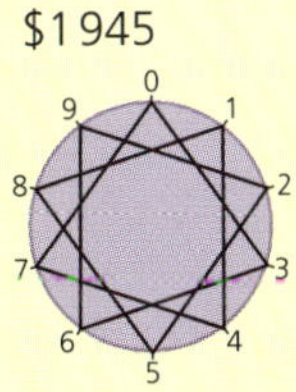

c

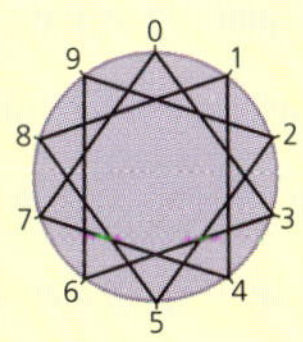

2:63

1. a (30 × 50) + (30 × 7) + (4 × 50) + (4 × 7)
   = 1 500 + 210 + 200 + 28
   = 1 938
   b (60 × 20) + (60 × 8) + (2 × 20) + (2 × 8)
   = 1 200 + 480 + 40 + 16
   = 1 736
   c (50 × 10) + (50 × 6) + (7 × 10) + (7 × 6)
   = 500 + 300 + 70 + 42
   = 912
2. a 61 × 36 = 2 196

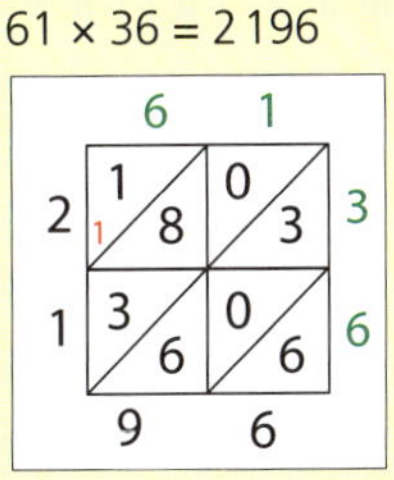

   b 62 × 28 = 1 736

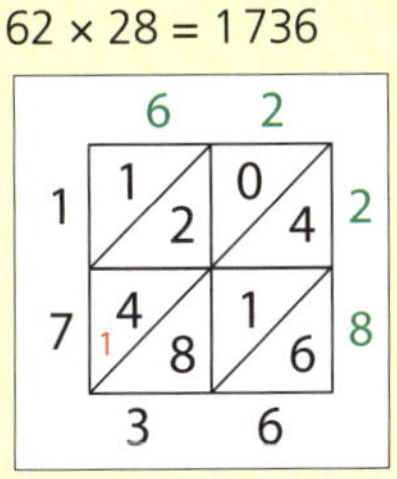

   c 18 × 73 = 1 314

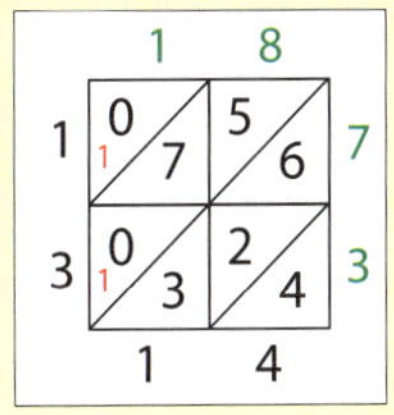

2:64

1. a 3 b 6 c 6 d 87 e 9
   f 6 g 7 h 8 i 11
2. a 78 coins b 5 jars
   c 125 soldiers d 26 counters
3. a 3 b 37 c 110 d 220 e 209
   f 271 g 179 h 478 i 125
4. a 15 b 19 c 11 d 9 e 5
   f 8 g 60 h 120 i 69
5. a 30 b 32 c 42 d 81 e 108
   f 90 g 5 h 6 i 9
6. a 109 b 22 c 31 d 93 e 8
   f 4 g 350 h 90 i 74

2:65

1. a 2 b 4 c 7 d 3 e 9
   f 0 g 4 h 5 i 0 j 6
   k 5 l 2 m 11 n 10 o 29
2. a 7 b 9 c 5 d 0 e 10
   f 20 g 15 h 9 i 7 j 4
   k 8 l 6 m 12 n 12 o 20
   p 31
3. a 3 b 7 c 6 d 9 e 5
   f 3 g 10 h 5 i 8 j 6
   k 7 l 7 m 8 n 9 o 9
   p 7
4. a 16 b 12 c 40 d 8 e 6
   f 3 g 3 h 8 i 10 j 7
5. a 2 b 10 c 5 d 2 e 4
   f 6 g 4 h 5 i 3
6. a 4 b 2 c 4 d 8

2:66

1. a 211 matches b 142 matches
2. a 91 matches b 47 matches
3. a 12 faces b 102 faces c 266 faces
4. a 19 faces b 121 faces c 223 faces
5. a 20 b 9 c 20

2:67

1 2   2 6   3 24

4 a 20c + 5c, 10c + 10c + 5c, 10c + 5c + 5c + 5c
b 20c + 20c, 20c + 10c + 10c, 20c + 10c + 5c + 5c, 10c + 10c + 10c + 5c + 5c
c 50c + 10c + 5c, 50c + 5c + 5c + 5c, 20c + 20c + 20c + 5c, 20c + 20c + 10c + 10c + 5c, 20c + 20c + 10c + 5c + 5c + 5c, 20c + 10c + 10c + 10c + 5c + 5c + 5c

5 a 3   b 2   c 3   d 6

6

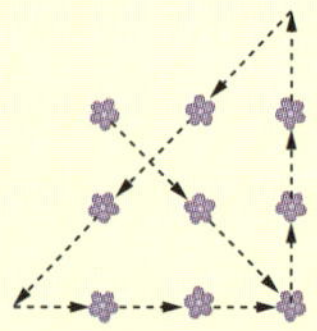

2:68

1 a

In	1	2	3	4	5	6
Out	4	8	12	16	20	24

Rule: Out = In × 4

b

In	1	2	3	4	5	6
Out	4	7	10	13	16	19

Rule: Out = In × 3 + 1

c

In	1	2	3	4	5	6
Out	17	27	37	47	57	67

Rule: Out = In × 10 + 7

d

In	1	2	3	4	5	6
Out	3	8	13	18	23	28

Rule: Out = In × 5 – 2

e

In	1	2	3	4	5	6
Out	12	14	16	18	20	22

Rule: Out = In × 2 + 10

f

In	1	2	3	4	5	6
Out	0	9	18	27	36	45

Rule: Out = In × 9 – 9

3:01

1 a 21   b 48   c 420   d 3   e 240
f 12   g 48   h 5   i 2   j 8
k 100   l 14   m 6   n 156   o 10

2 a 35 min   b 26 min   c 1 h 6 min
d 1 h 49 min   e 2 h 44 min   f 2 h 6 min

3 a 43 min   b 2 h 13 min   c 4 h 48 min
d 6 h 48 min   e 7 h 58 min   f 11 h 3 min

4 a 12 h 30 min   b 12·5 h
c 750 min   d 45 000 s

3:02

1 a metres   b metres   c kilometres
d kilometres   e kilometres   f metres
g metres   h metres

2 a 2000 m   b 5000 m   c 4000 m
d 10 000 m   e 3000 m

3 a 6 km   b 10 km   c 7 km   d 9 km   e 8 km

4 a 4 km   b 6 km   c 3 km
d 5 km   e 4 km   f 5 km

3:03

1 a 4000 m   b 7000 m   c 9000 m
d 6000 m   e 2000 m   f 3000 m
g 5000 m   h 10 000 m   i 14 000 m

2 a 3 km   b 9 km   c 5 km   d 7 km
e 12 km   f 16 km

3 a km   b m   c cm   d mm
e m   f km   g m   h km

4 a 60 km/h   b 95 km/h   c 70 km/h
d 50 km/h   e 190 km/h

5 a 3 km 850 m   b 1 km 56 m   c 9 km 5 m
d 5 km 600 m   e 3 km 275 m   f 8 km 241 m

6 a 5200 m   b 1341 m   c 4750 m
d 8005 m   e 10 030 m   f 7038 m

7 Answers will vary.

3:04

1

	Length	Width	Perimeter
a	5 cm	2 cm	14 cm
b	4 cm	3 cm	14 cm
c	3 cm	3 cm	12 cm
d	6 cm	2 cm	16 cm

2 a 14 cm   b 16 cm   c 18 cm   d 20 cm
e 20 cm   f 24 cm

3 a 24 cm   b 24 cm   c 24 cm   d 26 m
e 32 m   f 28 m

4 98 cm

3:05

1 a 24 cm   b 24 cm   c 30 cm
d 20 cm   e 30 cm   f 32 cm

2 a 26 cm   b 34 cm   c 44 cm   d 44 cm   e 72 cm
f 32 m   g 40 m   h 42 m   i 56 m   j 100 m

3 a 36 cm   b 28 cm   c 28 cm   d 36 cm

4 Answers will vary.

3:06

1 a 12 cm^2 b 12 cm^2 c 18 cm^2 d 15 cm^2 e 10 cm^2

2 a 24 cm^2 b 45 cm^2 c 35 cm^2 d 60 cm^2

3 588 cm^2

3:07

1 a 20 m^2 b 10 m^2 c 6 m^2 d 14 m^2 e 4 m^2

2 a 36 m^2 b 42 m^2 c 80 m^2 d 56 m^2 e 54 m^2
f 42 cm^2 g 30 cm^2 h 63 cm^2 i 40 cm^2 j 80 cm^2

3 a 63 m^2 b 48 m^2

4 a 24 m^2 b 72 m^2 c 72 m^2 d 36 m^2
5 Floor measurements will vary.

3:08

1 a 20 000 m^2 b 50 000 m^2 c 70 000 m^2 d 40 000 m^2
e 60 000 m^2 f 90 000 m^2 g 30 000 m^2 h 80 000 m^2
2 a 1 ha b 5 ha c 3 ha d 6 ha e 4 ha
f 2 ha g 9 ha h 7 ha
3 a ha b m^2 c m^2 d ha e ha
f m^2 g ha h ha i m^2
4 a 10 000 m^2, equal to 1 hectare
b 7 000 m^2, less than 1 hectare
c 10 000 m^2, equal to 1 hectare
d 13 500 m^2, more than 1 hectare
e 10 000 m^2, equal to 1 hectare

3:09

1 a 30 m^2 b 38 m^2 c 34 m^2 d 44 cm^2
e 43 cm^2 f 38 cm^2
2 a 1 cm × 12 cm, 2 cm × 6 cm, 3 cm × 4 cm
b 1 cm × 20 cm, 2 cm × 10 cm, 4 cm × 5 cm
3 a m^2 b ha c ha d cm^2 e m^2 f m^2
4 a 8 ha b 56 ha

3:10

1 a 90 cm^3 b 70 cm^3 c 126 cm^3 d 72 cm^3 e 36 cm^3
2 a 60 cm^3 b 72 cm^3 c 135 cm^3
3 a 270 cm^3 b 84 cm^3 c 125 cm^3 d 216 cm^3
e 192 cm^3

3:11

1 a 7 cm , 2 cm , 2 cm , 28 cm^3
b 5 cm, 4 cm, 2 cm, 40 cm^3
c 10 cm, 3 cm, 2 cm, 60 cm^3
d 7 cm, 3 cm, 3 cm, 63 cm^3
e 6 cm, 4 cm, 3 cm, 72 cm^3
f 12 cm, 2 cm, 2 cm, 48 cm^3
2 12 cm^3
3 Answers will vary.
e.g. 10 cm, 2 cm, 2 cm
5 cm, 2 cm, 4 cm
10 cm, 1 cm, 4 cm
4 12 cm^3

3:12

1 a 1:00 pm b 3:00 pm c 8:00 pm d 9:00 am
e 6:00 pm f 7:00 am g 11:00 pm h 4:00 pm
i 2:00 pm j 9:00 pm
2 a 17:00 b 11:00 c 20:00 d 14:00 e 04:00
3 a 16:00 b 08:00 c 19:00 d 22:00 e 02:00
f 21:00 g 12:00 h 24:00 or 00:00 i 17:00

3:13

1 Answers may vary. Some are:
a 5 pm = 17:00 or 1700 b 0800 = 08:00 or 8 am
c 0300 = 03:00 or 3 am d 2 am = 02:00 or 0200
e 7 pm = 19:00 or 1900 f 0100 = 01:00 or 1 am
g 1100 = 11:00 or 11 am h 0400 = 04:00 or 4 am
i 10 pm = 22:00 or 2200 j 6 pm = 18:00 or 1800
k 0900 = 09:00 or 9 am l 12 am = 24:00 or 2400
2 a 1 pm or 13:00 b 15:00 or 3 pm
c 4 am or 04:00 d 9 am or 09:00
e 1 pm or 13:00 f 19:00 or 7 pm
g 3 pm or 15:00 h 06:00 or 6 am
i 6 pm or 18:00 j 6 am or 06:00
3

	07:00	09:00	04:00	09:00	10:00	11:00	12:00	14:00	20:00	21:00	23:00	
	06:00	02:00	03:00	11:00	09:00	10:00	13:00	15:00	19:00	22:00	20:00	
Start	**00:00**	01:00	04:00	07:00	08:00	11:00	14:00	17:00	18:00	23:00	**24:00**	Finish
	02:00	08:00	05:00	06:00	07:00	13:00	15:00	16:00	19:00	20:00	21:00	

3:14

1 a 3 m^3 b 7 m^3 c 2 m^3 d 10 m^3 e 15 m^3
f 9 m^3 g 4 m^3 h 50 m^3 i 12 m^3
2 a m^3 b cm^3 c m^3 d m^3 e m^3 f cm^3
3 a m^3 b m^3 c cm^3 d m^3 e m^3 f cm^3
g m^3 h m^3 i cm^3

3:15

1 a 6 cm, 2 cm, 2 cm = 24 cm^3
b 5 cm, 2 cm, 3 cm = 30 cm^3
c 6 cm, 5 cm, 3 cm = 90 cm^3
d 5 cm, 3 cm, 2 cm = 30 cm^3
e 4 cm, 3 cm, 2 cm = 24 cm^3
f 5 cm, 4 cm, 3 cm = 60 cm^3
g 12 cm, 3 cm, 2 cm = 72 cm^3
2 a 18 m^3 b 16 m^3 c 24 m^3 d 20 m^3

3:16

1 a 75 mm b 53 mm c 105 mm d 17 mm
e 26 mm f 98 mm
2 d, e, b, a, f, c
3 a cm b km c m or cm d cm
e cm f mm g m h m
4 a 6·4 cm b 8·3 cm c 3·6 cm d 4·1 cm
e 9·2 cm f 5·9 cm
5 a 3·9 cm b 6·5 cm c 4·3 cm d 8·2 cm
e 7·6 cm f 9·1 cm g 16·2 cm h 15·8 cm
6 a 148 mm b 120 mm c 108 mm d 162 mm
e 160 mm
7 a 10 cm b 9 cm c 13 cm

3:17

1. a 4·8 cm b 7·4 cm c 9·5 cm d 2·7 cm
   e 6·0 cm f 6·7 cm
2. a 13·4 cm b 12·2 cm c 22·2 cm d 27·4 cm
3. The shapes will be drawn.

3:18

1. a 22:31 b 07:57 c 15:37 d 04:09 e 21:02
   f 20:19 g 01:23 h 11:43
2. a 7:51 pm b 2:37 am c 6:13 am d 4:24 pm
   e 5:49 am f 1:06 pm g 6:27 pm h 8:59 am
   i 10:33 am j 5:42 pm k 9:16 am l 3:01 am
3. a 15:54 b 09:32 c 12:38 d 22:09
   e 06:18 f 17:47 g 08:14 h 01:26
   i 16:23 j 14:31 k 23:56 l 07:28
4. a 1 h 45 min b 4 h 45 min c 4 h 35 min d 3 h 11 min

3:19

1. a 1 pm b 4 pm c 8 pm
2. a 19:30 b 22:30 c 21:30
3. a 8:30 am b 3:30 pm c 11:30 pm
4. a 18:30 b 17:30 c 23:30
5. a 8:00 am b 3:00 pm c 7:00 pm d 10:00 pm
6. a 05:00 b 08:00 c 20:00 d 22:00
7. a 8:00 pm b 18:30
8. Answers will vary.
9.

	3.00 am	4.00 am	7.00 am	8.00 am	9.00 am	6.00 pm	2.00 pm	8.00 pm	9.00 pm	10.00 pm	9.00 pm	
	4.00 am	7.00 am	6.00 am	4.00 am	10.00 am	7.00 pm	3.00 pm	7.00 pm	2.00 pm	11.00 pm	**12.00 am**	**Finish**
	1.00 am	3.00 am	5.00 am	7.00 am	11.00 am	12.00 pm	8.00 pm	6.00 pm	3.00 pm	4.00 pm	7.00 pm	
Start	**12.00 am**	1.00 am	4.00 am	9.00 am	6.00 am	1.00 pm	7.00 pm	5.00 pm	2.00 pm	5.00 pm	8.00 pm	
	3.00 am	2.00 am	3.00 am	10.00 am	5.00 am	2.00 pm	3.00 pm	4.00 pm	1.00 pm	6.00 pm	9.00 pm	

3:20

1. a kg b t c kg d kg e kg f t
2. a 3000 kg b 6000 kg c 9000 kg d 5000 kg
   e 7000 kg f 10 000 kg g 4000 kg h 40 000 kg
   i 15 000 kg j 25 000 kg k 60 000 kg l 70 000 kg
3. a 1 t b 5 t c 2 t d 9 t
   e 4 t f 6 t g 10 t h 8 t
   i 20 t j 17 t k 30 t l 45 t
4.

	kilograms	tonnes
a	3500	3·5
b	4500	4·5
c	7500	7·5

	kilograms	tonnes
d	8500	8·5
e	6500	6·5
f	9500	9·5

5. a 15·3 t b 425 t

3:21

1. a 7000 g b 4000 g c 8000 g
   d 1625 g e 3750 g f 2125 g
   g 3500 g h 9500 g i 7500 g
2. a 2 kg 475 g b 1 kg 295 g c 6 kg 456 g
   d 3 kg 865 g e 5 kg 125 g f 4 kg 796 g
   g 9 kg 350 g h 7 kg 642 g i 8 kg 923 g
3. a g b t c kg d kg
   e t f g g t h g
4. a false b true c false d false e true f true

3:22

1. a 3 cm, 28 mm b 4 cm, 44 mm
2. a 8 kg b 7 kg c 14 kg d 3 mL
   e 10 mL f 6 mL g 8 mL h 80 mL
   i 30 mL j 100 mL k 37°C l 40°C
   m 39°C n 36°C

3:23

1. (A) 26 mm, 2·6 cm (B) 39 mm, 3·9 cm
   (C) 64 mm, 6·4 cm (D) 91 mm, 9·1 cm
   (E) 128 mm, 12·8 cm
2. a 4·9 cm b 6·4 cm c 8·3 cm d 5·1 cm
   e 9·2 cm f 7·5 cm g 10·8 cm h 12·3 cm
3. a 2·51 m b 8·29 m c 3·75 m d 6·42 m
   e 5·63 m f 9·25 m g 10·21 m h 11·65 m
4. a 316 cm b 831 cm c 465 cm d 954 cm
   e 527 cm f 295 cm g 745 cm h 1075 cm
5. a m b m c km d m
   e mm f km g cm h km

3:24

1. a 00:04·70 b 00:30·40 c 01:39·74
   d 03:59·50 e 06:29·56 f 13:19·14
   g 25:27·86 h 37:19·62
2. a 4·7 seconds b 30·4 seconds
   c 1 minute 39·74 seconds d 3 minutes 59·5 seconds
3. a 0:00·65 b 00:03·04 c 02:46·33 d 0:01·21
   e 0:00·07 f 0:02·03 g 0:04·84 h 0:01·99

3:25

1. a

Rectangle	Perimeter	Length	Height	Area
	24 mm	9 mm	3 mm	27 mm^2
	24 mm	7 mm	5 mm	35 mm^2
	24 mm	10 mm	2 mm	20 mm^2
	24 mm	6 mm	6 mm	36 mm^2

ISBN 978 1 4886 2190 1

b Answers will vary.
Two are given below.

Perimeter	Length	Height	Area
30 mm	8 mm	7 mm	56 mm^2
30 mm	9 mm	6 mm	54 mm^2

2 Area of smaller rectangle = 6 cm^2
Area of larger rectangle = 24 cm^2
a 4
b 4

3 **a** 1 cm^3, 8 cm^3 **b** 10 m^3, 80 m^3 **c** 8

4 Volume = Length × Width × Height

4:01

1 **a** false **b** true **c** true **d** true

2 **a** square pyramid **b** rectangular prism

3 Two prisms and two pyramids will be drawn.

4 **a** 6 **b** 8 **c** 12

5 **a**

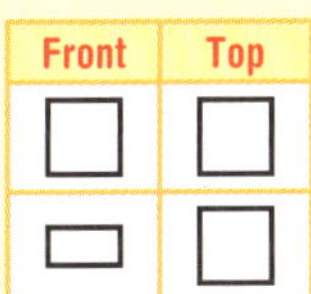

b

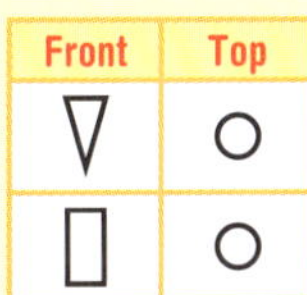

4:02

1 **a** F **b** B **c** C **d** A **e** D **f** E

2 **a** rectangle **b** hexagon **c** square **d** pentagon

3

Object	Number of Corners	Number of Edges	Number of Faces
A	5	8	5
B	7	12	7
C	8	12	6
D	8	12	6
E	5	8	5
F	6	9	5

4 **a** square pyramid **b** pentagonal prism

4:03

1

Transformation	Figure Changes Position	Figure Changes Shape	Figure Changes Size	Figure Changes Orientation
translation	yes	no	no	no
reflection	yes	no	no	yes
rotation	yes	no	no	yes

2 **a** B, D **b** C, E, F

3

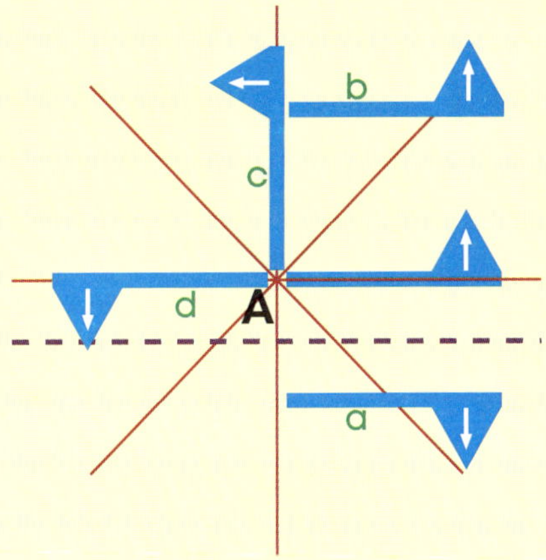

4:04

1 **a** B **b** A, D **c** C **d** D **e** E **f** F

2 **a** 0 **b** 1 **c** 0 **d** 5 **e** 8

3 **a**

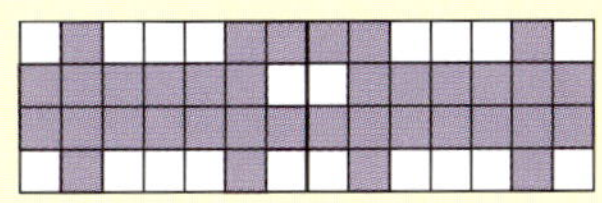

b

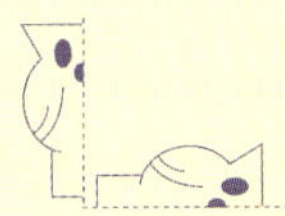

c

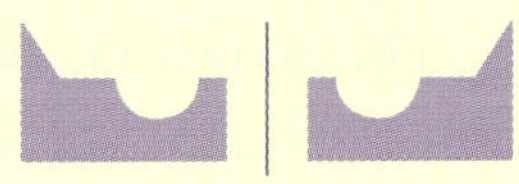

d

4 **a**

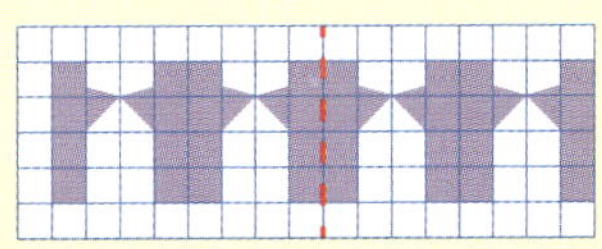

b

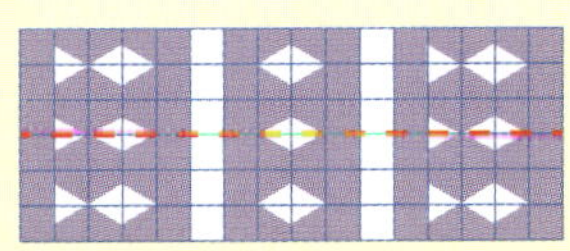

c–d Answers will vary.

4:05

1

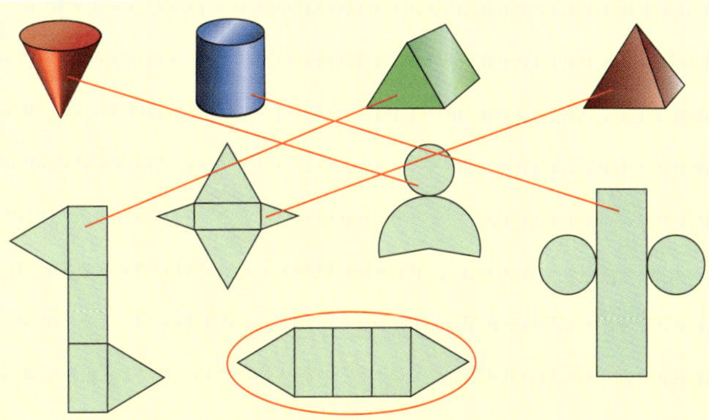

2 **a** cube

3

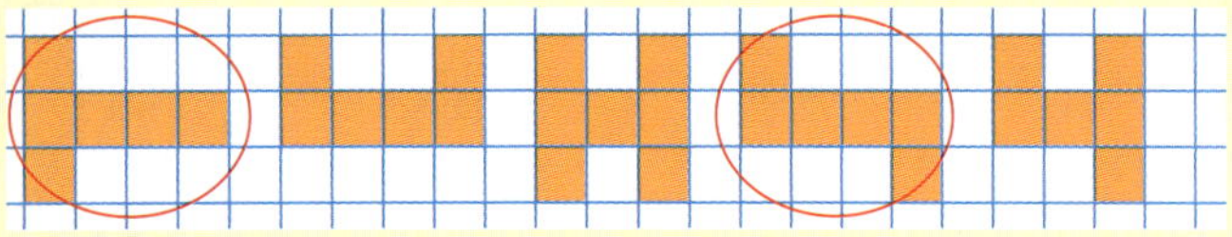

4:06

1 a 2C b 4B c 1D d 1A e 4A f 2B g 1C h 3C i 3A j 4D k 5A l 4C m 3D n 2A (or 2C) o 5B p 1C, 2B, 3C, 5C q 3B

2 a triangle b pentagon c square d left upper, middle upper e right lower, non-rigid

4:07

1 a 30°, acute b 110°, obtuse c 120°, obtuse d 140°, obtuse e 80°, acute f 65°, acute g 25°, acute h 165°, obtuse i 95°, obtuse

2 a 90 b 180 c 360

3 Answers will vary. The baseline of the protractor must be aligned with one of the arms of the angle with the centre point of the protractor over the vertex of the angle. The angle can be read off the outer scale. The second arm of the angle may need to be extended to aid this process.

4:08

1 a 90° b 40° c 120° d 80° e 110° f 140° g 60° h 100°

2

Picture	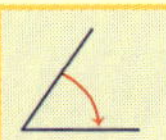	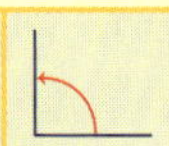	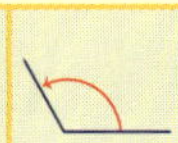	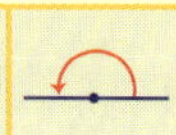		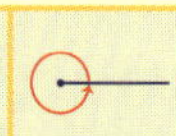
Type	acute	right	obtuse	straight	reflex	revolution (full turn)
Size	between 0° and 90°	90°	between 90° and 180°	180°	between 180° and 360°	360°

3 a acute b straight c revolution (full turn) d obtuse e acute f reflex g right h obtuse i acute j reflex k reflex l obtuse m acute n acute o reflex p obtuse

4 One of each angle type in Question 2 will be drawn.

4:09

1 a 60° b 40° c 100° d 120° e 50° f 20° g 120° h 40° i 140° j 70° k 80° l 130°

4:10

1 a acute b straight c right d reflex e revolution f obtuse g reflex h acute

2 Yes

3 a acute b revolution c straight d obtuse e right f reflex g obtuse h obtuse

4 a square or rectangle b regular pentagon, hexagon, octagon, etc. c triangle

5 Answers will vary.

4:11

1 a W b S c E d N e NW f SE g NE h SW i SE j SW k E l NW

2 a E b N c W d S e SE f NW g SW h NE i NW j NE k W l SE

3 a Karumba b Ceduna, Sydney, Canberra or Adelaide c Rockhampton, Brisbane or Longreach d Coral Bay e Derby or Coral Bay f Derby, Fitzroy Crossing or Coral Bay g Melbourne h Rockhampton or Brisbane

4 a south b west c north-east d north-west e south-west f south-east

4:12

1 a Springbank Island b Vernon Circle c Lennox Gardens (or Lotus Bay)

2 a 5A (or 4A) b 5E c 2E d 3D e 6C f 4E

3 a Perth Avenue b Coronation Drive c Moonah Place

4:13

1 a Yes, No b No, Yes

2 A, B, C, D, G, H, I, J

3 A, B, C, D, E, G, H, J

4 a Yes, 4 b No, 1 c Yes, 2 d Yes, 8 e Yes, 6 f Yes, 2 g Yes, 3 h No, 1

4:14

1 a 150° b 60° c 240° d 160° e 100° f 220°

2 Estimates will vary. a 60° b 90° c 120° d 30° e 70° f 180°

4:15

1 a 3 b 5

4:16

1

Number of:	A	B	C	D
surfaces	2	3	1	5
edges	1	2	0	8
vertices (corners)	1	0	0	5
curved surfaces	1	1	1	0
faces (flat surfaces)	1	2	0	5

2 **a** A **b** B, C **c** D

3 **a** B **b** C **c** A, D

4 B, E

5 A, B, D

6 **a** triangular prism **b** 5 + 6 – 9 = 2
c cube = 2, rectangular prism = 2

4:17

1 **a** Circular Quay **b** Opera House
c Farm Cove **d** Conservatorium of Music
e Royal Botanic Gardens **f** Darling Harbour

2 **a** F6 **b** G6 **c** H7
d B3 or C3 **e** G2 **f** C6

3 **a** A8, B8 **b** F8, F9, F10 or F7

4 400 m

5 **a** Opera House
b Answers will vary. Proceed along Circular Quay East, veer right into Alfred Street, then turn left into George Street.

4:18

1 **a** Colombia **b** Panama **c** Brazil **d** Argentina
e Paraguay **f** Uruguay **g** Peru **h** Ecuador

2 G14

3 **a** G7 **b** B13 **c** G14 **d** F9

4 three of C13, C14, C15 or D14

5 D10, E10, F10, D11, E11

6 **a** 600 km **b** 1 200 km **c** 4 200 km **d** 1 200 km

7 **a** Chile (or Peru) **b** Venezuela (or Brazil)

8 A map of Victoria will be drawn.

4:19

1

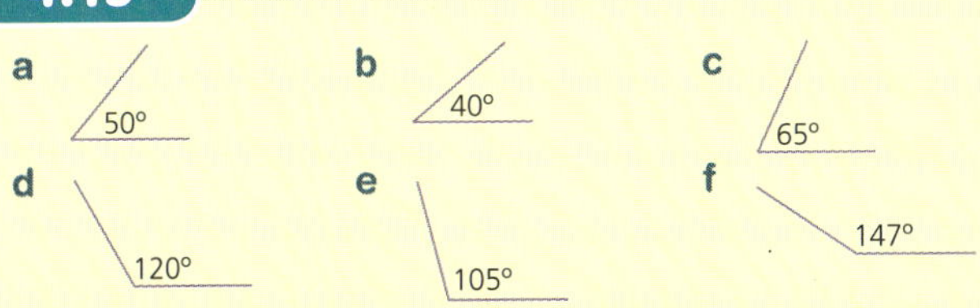

2

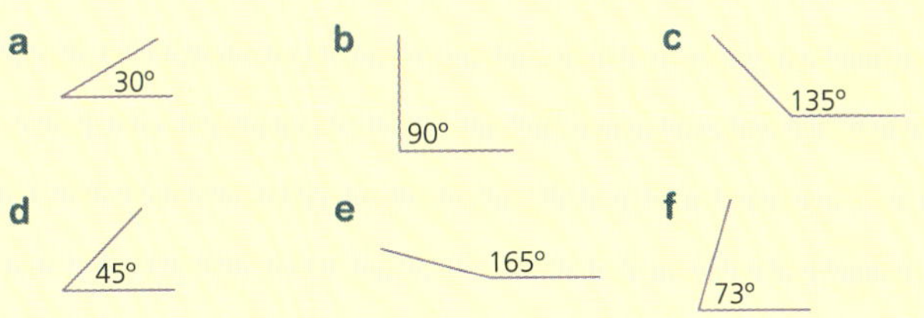

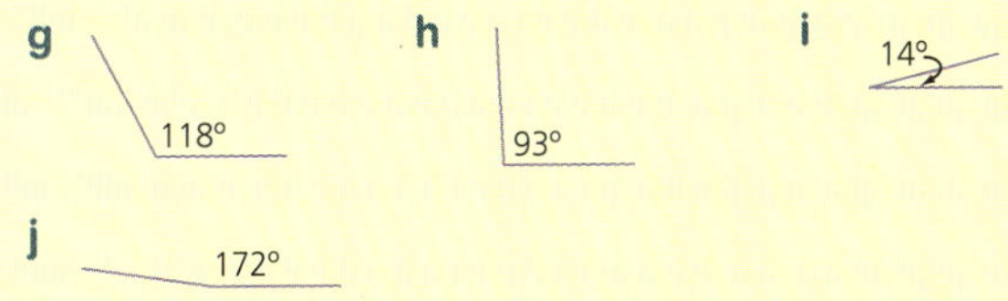

3

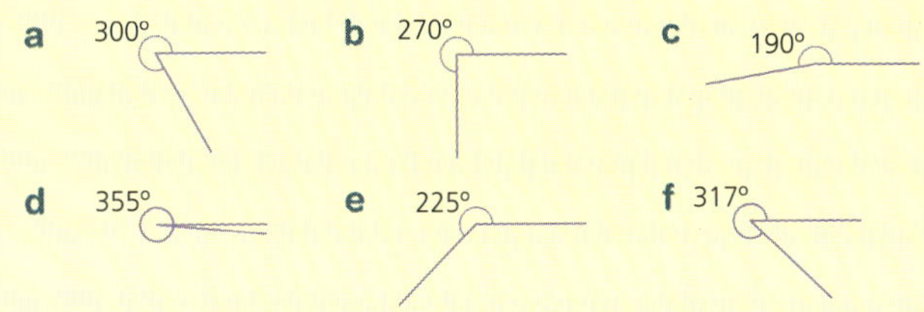

4:20

1 **a** 200° **b** 270° **c** 340°

2 **a** 160° **b** 90° **c** 20°

3 **a** 300° **b** 270° **c** 230°
d 200° **e** 280°

4 **a** 200° or 210° **b** 230° **c** 260°
d 290° **e** 340° **f** 350°

4:21

1

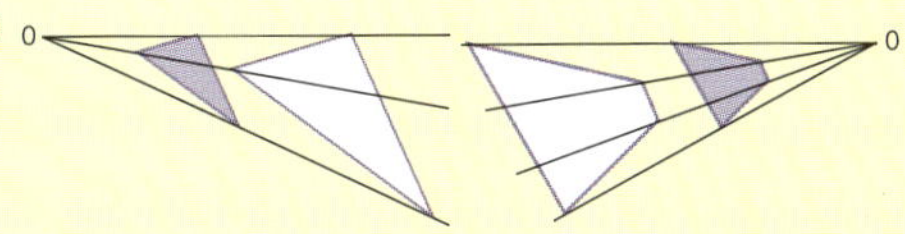

The enlargement factor is 2 or 200%.

2

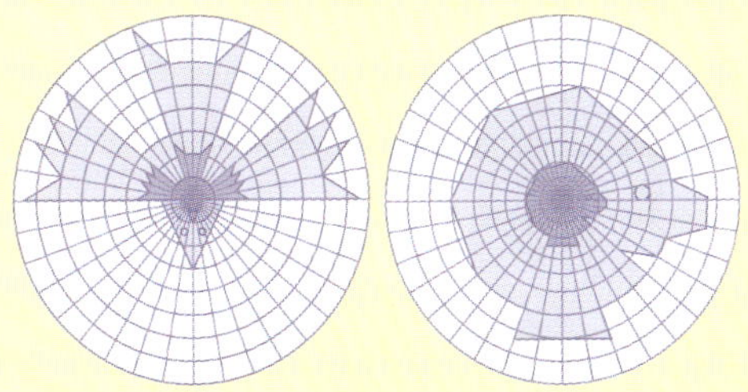

The enlargement factor is 3 or 300%.

3 **a** 2 **b** 3 **c** $1\frac{1}{2}$ **d** $\frac{1}{2}$

4:22

1 2

2 1

3 The camel will be drawn on the smaller grid.

4 Answers will vary.

4:23

Answers will vary.

1 **a** Four times longer **b** 2 cm 2 mm **c** 8 cm 8 mm

2 Approximately:
a 140 cm **b** 160 cm **c** 180 cm **d** 120 cm

3 **a** 75:1 **b** about 1 mm

4 **a** about 180 cm (or 1·8 m) **b** about 150 cm (or 1·5 m)
c about 290 cm (or 2·9 m) **d** about 3 600 cm (or 36 m)

5:01

1. **a** 12 **b** Monday
   c Tuesday and Wednesday **d** 4, 4 books
   e 76 **f** borrowed for the weekend
2. **a i** October **ii** February
   b i 65 **ii** 85 **iii** 80 **iv** 105 **v** 130 **vi** 205
   c 335
3. **a** No. The bars do not end on the lines of the horizontal scale.
   b i 40 **ii** 110 **iii** 100 **iv** 10
   c about 330
   d They are small, cheap to feed and easy to care for.
4. 24

5:02

1. 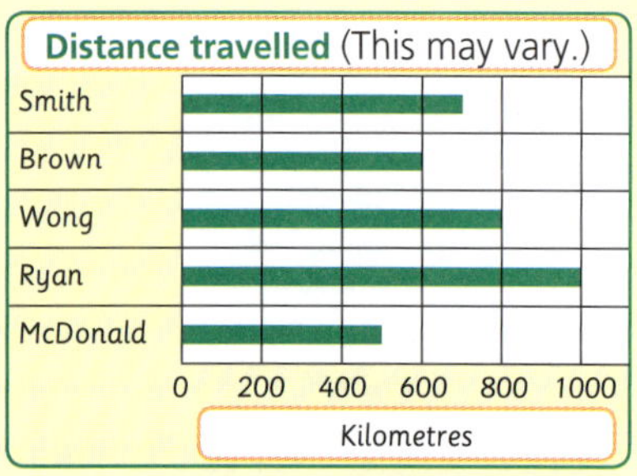

2. **a** 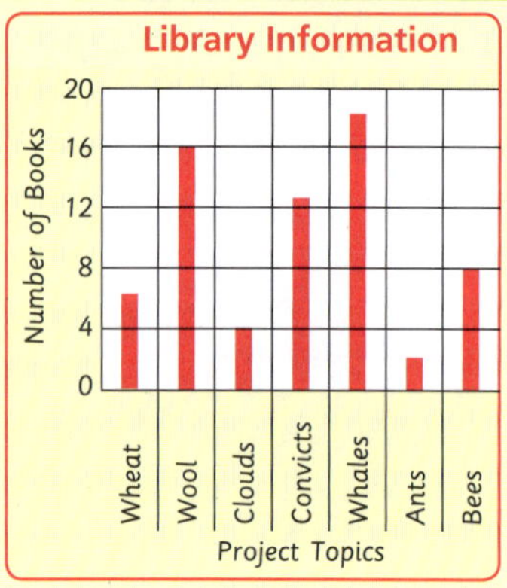

   b Wool, convicts and whales. There are more books on these.
3. **a** 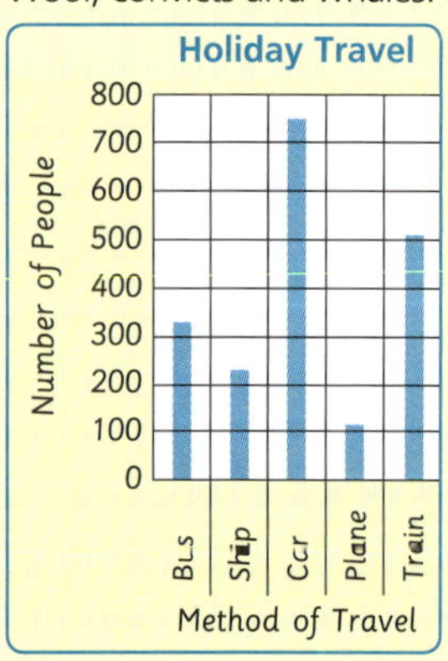

   b 40
4. **a** $2.10 **b** $2.80

5:03

1. 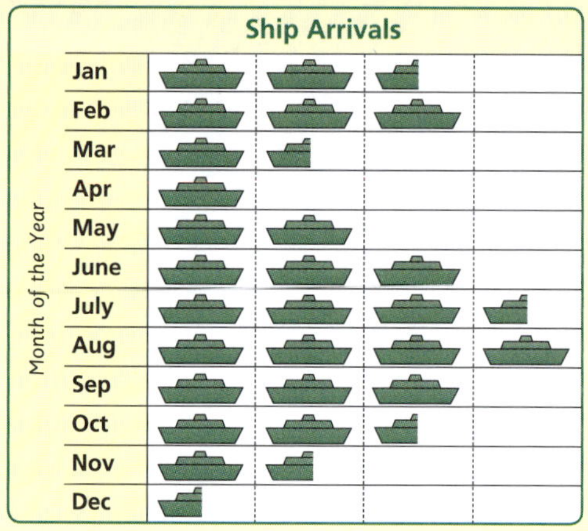

2. 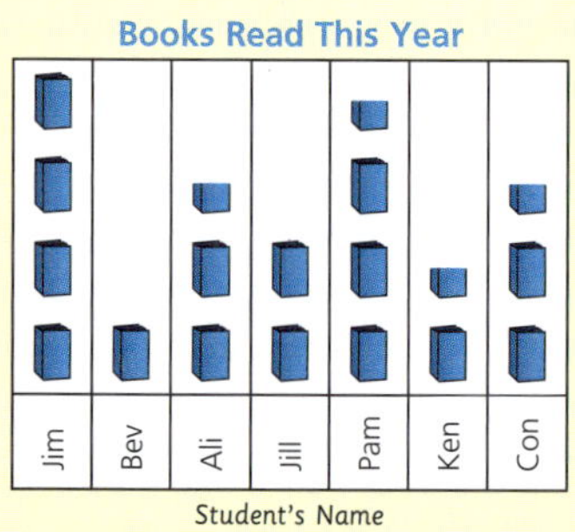

3.

Name	Tally	Total
Alana	卌 卌 卌 卌 卌 |||	28
Rachel	卌 卌 卌 卌 ||	22
Naomi	卌 卌 卌 卌 卌 卌 卌 卌	40
Luke	卌 卌 卌 卌 卌 卌	30
Heather	卌 卌 卌 卌 卌 卌 卌		36
Sandy	卌 卌 卌 卌 卌 卌 ||	32

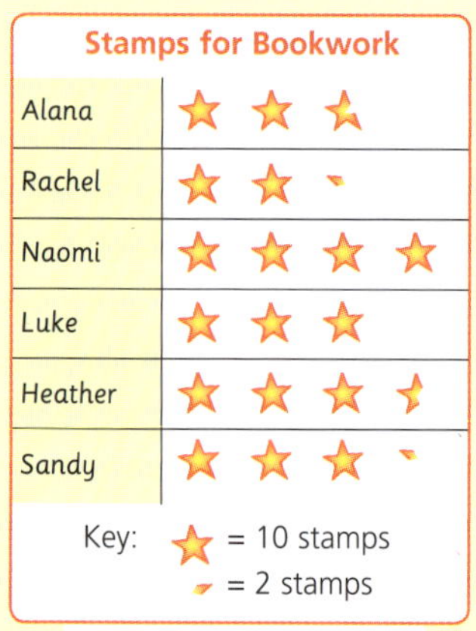

5:04

1. **a i** 200 g **ii** 400 g **iii** 150 g **iv** 350 g
   b i 3 cups **ii** 5 cups **iii** 1 cup **iv** $\frac{1}{2}$ cup
   c 600 g
2. **a i** 18°C **ii** 3°C **b** 21°C at 1 pm
   c i 7 pm **ii** noon and 2 pm
   d i 17°C **ii** 8°C **iii** 3°C **iv** 3·5°C

3 a i 300 min ii 180 min iii 90 min iv 270 min
b i 3 h ii 4 h iii $2\frac{1}{2}$ h iv $4\frac{1}{2}$ h

5:05

1 a

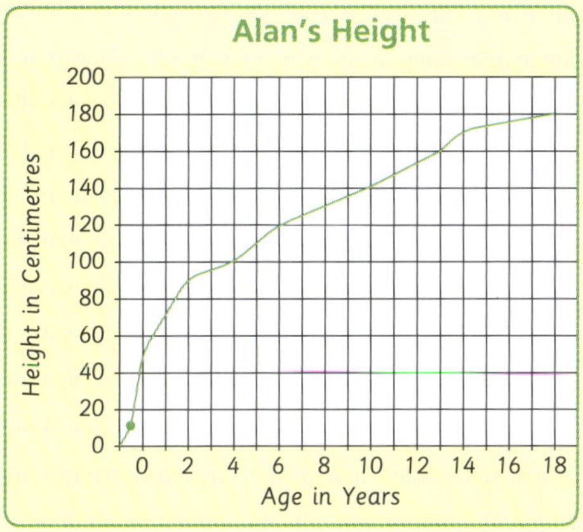

b i 70 cm ii 110 cm iii 160 cm iv 145 cm
c Alan's height 6 months before birth is marked on the graph.

2

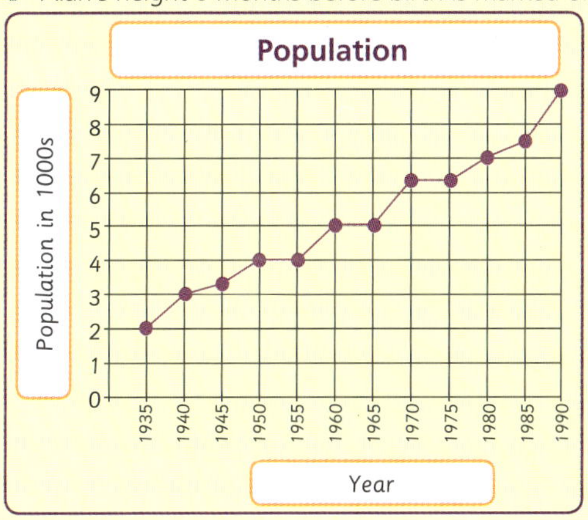

3 a

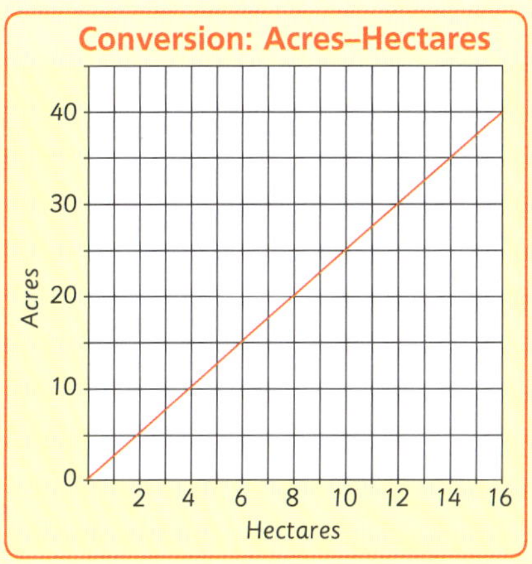

b i 8 ha ii 6 ha
c i 30 acres ii 5 acres
d i 7 hectares ii 16 acres

5:06

1 a Answers will vary. (e.g. Is there more chance of rolling a 6 than of other numbers?)
b Answers will vary. (e.g. All numbers have about the same chance of being rolled.)

2 a Luke b David
c i 4 ii 6 iii 5 iv 3

3 Answers will vary.

5:07

1 a 28 b number (line)
c Answers will vary. Two students got a score of 64. 64 was the lowest score. 79 was the highest score. No students received a score of 69.
d

Test Results: 5M																
Score	64	65	66	67	68	69	70	71	72	73	74	75	76	77	78	79
Number	2	0	3	2	1	0	3	2	4	3	2	1	2	1	1	1

2 a

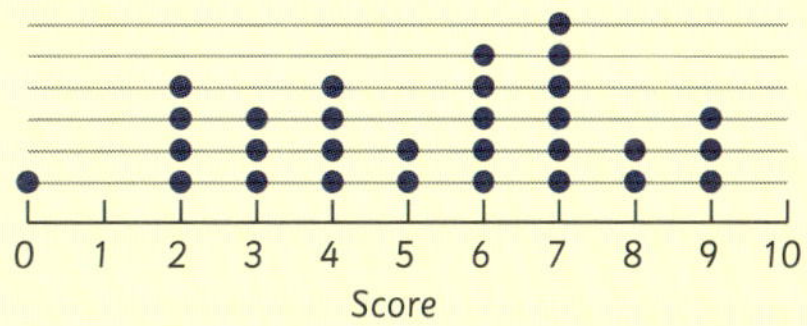

b Answers will vary. A score of 1 occurred 0 times. A score of 7 occurred 6 times. 9 was the highest score. 0 was the lowest score.

3

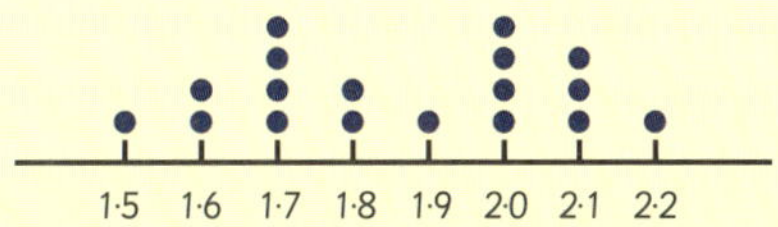

Answers will vary. The smallest finger width was 1·5 cm and the largest finger width was 2·2 cm.

5:08

1 a red $\frac{1}{6}$, yellow $\frac{5}{6}$
b red $\frac{3}{6}$ (or $\frac{1}{2}$), yellow $\frac{3}{6}$ (or $\frac{1}{2}$)
c red $\frac{2}{6}$ (or $\frac{1}{3}$), yellow $\frac{4}{6}$ (or $\frac{2}{3}$)
d red $\frac{5}{6}$, yellow $\frac{1}{6}$
The chance of spinning red plus the chance of spinning yellow is equal to 1.

2 a $\frac{1}{6}$ b $\frac{1}{6}$ c $\frac{1}{6}$ d $\frac{1}{6}$ e $\frac{3}{6}$ (or $\frac{1}{2}$)

3 a $\frac{1}{5}$ b $\frac{2}{5}$ c $\frac{1}{5}$ d $\frac{1}{5}$ e $\frac{3}{5}$

4 a Two balls will be coloured red.
b Three will be red. c Five will be red.
d Three will be red. e Four will be red.
f Two will be red. g Two will be red.
h Five will be red. i Eighteen will be red.

5:09

1 a green b red c blue
d red, yellow, blue, green
e $\frac{3}{20}$ f $\frac{9}{20}$ g $\frac{2}{20}$ h $\frac{6}{20}$

2 a $\frac{12}{15}$ b $\frac{3}{15}$ c $\frac{2}{15}$ d $\frac{13}{15}$
e $\frac{15}{15}$ f 0 g $\frac{1}{15}$

3 Answers will vary.

5:10

1 Graph titles will vary. Sample title: "Naomi and Luke's Birds"
- a i 30 mm ii 20 mm iii 50 mm
- b i 30 ii 20 iii 50
- c i $\frac{30}{100}$ ii $\frac{20}{100}$ iii $\frac{50}{100}$
- d i $\frac{3}{10}$ ii $\frac{2}{10}$ iii $\frac{5}{10}$

2 Graph titles will vary. Sample title: "Easter eggs found"
- a Emma b Kate
- c Answers may vary. Kate found 3, Sarah 4, Emma 7 and Ben 6

3 Our Team

5:11

1 a lions b tigers c 30% d 4

2 a volleyball b tennis c tennis

3 a $\frac{1}{4}$ b 20

4 No, because several sports are grouped together as 'other'.

5 a i Naomi ii Luke b i 20 times ii 10 times
- c No. The results will change because the cubes are randomly drawn.

5:12

1 a 5 b 11 c 3 d 10 e 6 f 7

2 a i blue ii playing sport
- b i pink ii watching sport

3 a

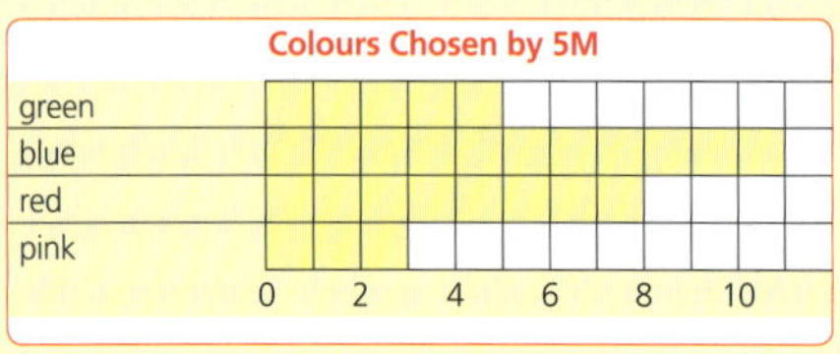

b

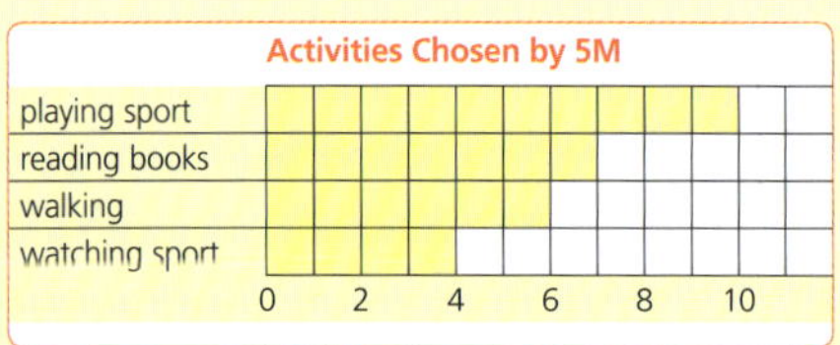

5:13

Answers will vary.

5:14

1 a i 7 cm ii 0 cm iii 2 cm iv 10 cm
- b i day 5 and day 14 ii day 9 iii day 10
  - iv about halfway through day 7
- c day 4 d day 6
- e Students should draw a smooth graph passing through all points.
- f 14 g day 14
- h about halfway through day 6

2 a 7:00 am
- b i 100 km ii 150 km iii 300 km
  - iv 400 km v 200 km vi 200 km
- c 11:00 am d 1 hour e 50 km f 100 km

3

Quote	Fee to arrive	Cost per 15 min	Cost for parts to be replaced	a 3-hour job	b 15-min job	c $1\frac{1}{2}$-hour job
1	$25	$40	$80	$585	$145	$345
2	$100	$15	$115	$395	$230	$305
3	$50	$30	$70	$480	$150	$300

The cheapest quote would be: a 2 b 1 c 3

5:15

1 a i 16°C ii 22°C b i 11°C ii 14°C
- c Thursday

2 Answers will vary.

5:16

1 A coin has 2 sides. Each side is equally likely to be tossed, so the chance is 1 in 2 (or $\frac{1}{2}$).

2 a $\frac{1}{8}$ b $\frac{3}{8}$ c $\frac{5}{8}$ d $\frac{7}{8}$

3 a $\frac{1}{10}$ b $\frac{4}{10}$ or $\frac{2}{5}$ c $\frac{3}{10}$ d $\frac{2}{10}$ or $\frac{1}{5}$

4 a $\frac{4}{10}$ or $\frac{2}{5}$ b $\frac{2}{10}$ or $\frac{1}{5}$ c $\frac{7}{10}$
- d $\frac{1}{10}$ e $\frac{5}{10}$ or $\frac{1}{2}$ f $\frac{9}{10}$
- g $\frac{5}{10}$ or $\frac{1}{2}$ h $\frac{7}{10}$

5:17

1 a $\frac{1}{5}$ b $\frac{1}{5}$ c $\frac{3}{5}$ d $\frac{5}{5}$ or 1 e $\frac{4}{5}$

2 a $\frac{2}{8}$ or $\frac{1}{4}$ b $\frac{6}{8}$ or $\frac{3}{4}$ c yes

3 a 5 b 12 c 22 d 1 e blue
- f $\frac{1}{40}$ g $\frac{39}{40}$ h yes i $\frac{5}{40}$ j $\frac{35}{40}$
- k yes l $\frac{12}{40}$ m $\frac{22}{40}$ n yes

5:18

1 Column graph 1 joins B. Sector graph 2 joins A. Picture graph 3 joins C. Divided bar graph 4 joins E. Line graph 5 joins D.

2 Answers will vary. Tables clearly display all available data using a standard and simple format. Tables are easy to create and the information is easy to access. Tables take up little space.

3 A column graph will be drawn.

5:19

1 a Answers will vary and could include:

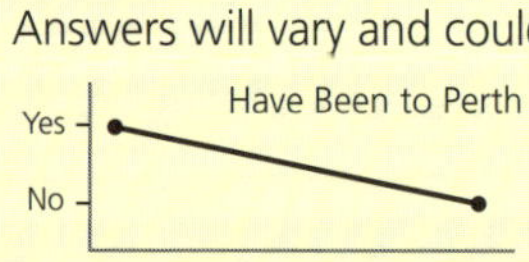

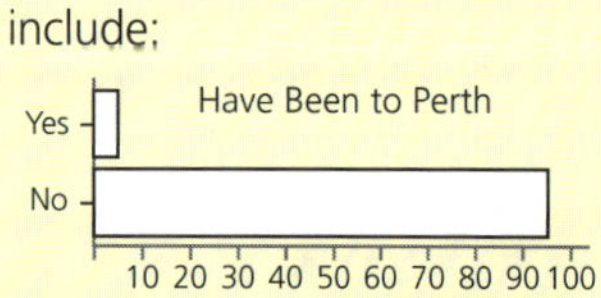

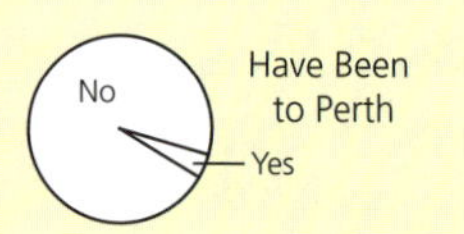

b Answers will vary and could include:

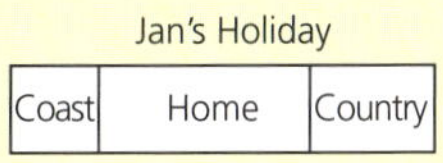

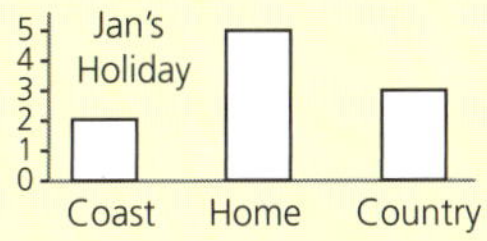

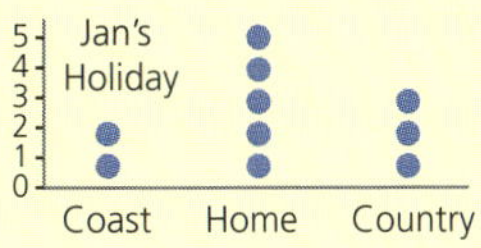

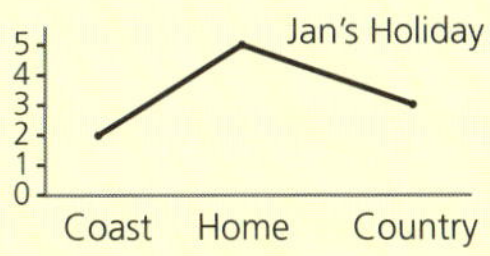

c Answers will vary.

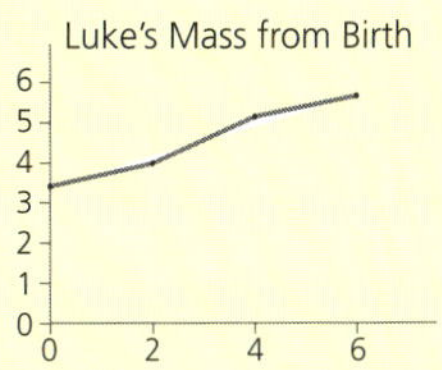

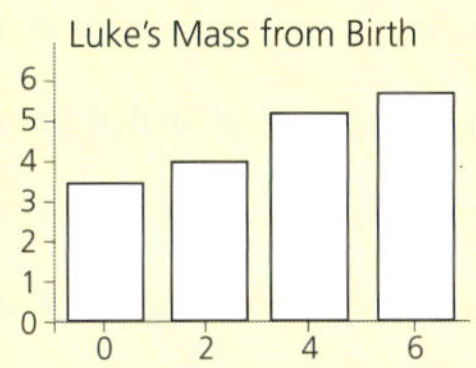

2 Answers will vary.
a A column graph would be best as it allows easy comparisons to be made.
b A sector or divided bar graph would be best as there are only two categories but other graphs could also be used.

5:20

1 Heads total = 9. Tails total = 11.
a 20 **b** **i** $\frac{9}{20}$ **ii** $\frac{11}{20}$
2 **a–c** Answers will vary.
d Different. 'A head and a tail' is most likely to occur.
3 **a** $\frac{5}{9}$ **b** $\frac{3}{9}$ or $\frac{1}{3}$ **c** $\frac{1}{9}$
d Answers will vary.

5:21

1 **a** $\frac{3}{6}$ or $\frac{1}{2}$ **b** $\frac{2}{6}$ or $\frac{1}{3}$ **c** $\frac{1}{6}$
d $\frac{0}{6}$ or 0 **e** 3
2 Answers will vary. Discuss your results.
3 8 ($7\frac{1}{2}$ rounded up). Discuss your results.
4 **a** **i** licorice **ii** chocolate
b **i** $\frac{10}{20}$ or $\frac{1}{2}$ **ii** $\frac{6}{20}$ or $\frac{3}{10}$ **iii** $\frac{3}{20}$ **iv** $\frac{1}{20}$
5 Answers will vary. Discuss your results.
6 Koalas

5:22

1 **a** 25 miles **b** 37·5 miles **c** 20 km **d** 50 km
e 20 miles **f** 10 miles
2 **a** sleep **b** working **c** yes
d play and other **e** work
3 **a** 13 **b** 9 **c** 46
d 151 cm–160 cm
e No. The category is 100 cm to 110 cm. **f** 91
4 **a** 14 **b** 19 **c** 35 **d** 54
e No. The graph only shows scores in groups.

5:23

1 **a**

Plan	Total Cost of Calls	Total Cost of Data	Total Cost of Messages	Total Cost per Month
1	$40.20	$0	$2.50	$42.70
2	$17.28	$0	$1.50	$18.78
3	$31.44	$0	$1.00	$32.44

The cheapest plan for Jake is Plan 2.

b

Plan	Total Cost of Calls	Total Cost of Data	Total Cost of Messages	Total Cost per Month
1	$0.00	$5	$8.75	$13.75
2	$0.00	$10	$5.25	$15.25
3	$0.00	$12	$3.50	$15.50

The cheapest plan for Simon is Plan 1.

c

Plan	Total Cost of Calls	Total Cost of Data	Total Cost of Messages	Total Cost per Month
1	$33.50	$15	$3.00	$51.50
2	$14.40	$30	$1.80	$46.20
3	$26.20	$36	$1.20	$63.40

The cheapest plan for Katie is Plan 2.

d

Plan	Total Cost of Calls	Total Cost of Data	Total Cost of Messages	Total Cost per Month
1	$10.05	$5	$50.00	$65.05
2	$4.32	$10	$30.00	$44.32
3	$7.86	$12	$20.00	$39.86

The cheapest plan for Jasmine is Plan 3.

Diagnostic Test 1

Number

see page 1, 4

1 Write the value of the 6 in:

a 267 304 ☐

b 6 349 721 ☐

Write the numeral for:

c 300 000 + 5 000 + 70 + 2 ☐

d 500 000 + 70 000 + 300 ☐

Round to the nearest million:

e 5 684 271 ☐

f 7 163 395 ☐

Operations

see page 27

2 Complete:

a 4 × 5 = ☐ b 8 × 10 = ☐

c 7 × 4 = ☐ d 8 × 2 = ☐

e 6 × 3 = ☐ f 9 × 5 = ☐

see page 27

3 Write the first five multiples of 4.

☐, ☐, ☐, ☐, ☐

see page 28, 29

4 Complete the table:

×	5	6	8	4	9	7
6						
7						
8						
9						

see page 29, 30

5 a 12 ÷ 2 = ☐ b 15 ÷ 3 = ☐

c 42 ÷ 6 = ☐ d 28 ÷ 7 = ☐

e 35 ÷ 5 = ☐ f 24 ÷ 3 = ☐

see page 31

6 Round to the nearest hundred.

a 642 ☐ b 786 ☐

c 3267 ☐ d 4831 ☐

see page 32

7 a

	Hund	Tens	Ones
		7	2
+		4	5

b

	Hund	Tens	Ones
	4	3	6
+		8	0

c

	Hund	Tens	Ones
	3	6	9
+		1	9

see page 33

8

	Hund	Tens	Ones
	4	5	9
+	2	6	2

see page 33

9 a

	Hund	Tens	Ones
	1	8	3
+	1	0	9

b

	Hund	Tens	Ones
	6	2	5
+		9	7

 • *AUSTRALIAN SIGNPOST MATHS 5* • ISBN 978 1 4886 2190 1

Diagnostic Test 1 (continued)

Measurement

10 see page 95
- a 2 days = ☐ hours
- b 4 hours = ☐ minutes
- c 50 years = ☐ decades
- d 700 days = ☐ weeks
- e 180 mins = ☐ hours

11 How much time has passed from: see page 95
- a 4:10 pm to 4:36 pm? ☐
- b 9 am to 10:49 am? ☐
- c 6:30 pm to 8:36 pm? ☐

12 How many metres are there in: see page 96
- a 4 km? ☐ b 7 km? ☐

How many kilometres are there in:
- c 5000 m? ☐ d 8000 m? ☐

13 Choose the most suitable unit of length (km, m, cm, mm) to measure: see page 97
- a the length of a road ☐
- b the width of your hand ☐
- c the depth of a lake ☐
- d the length of an ant ☐

Geometry

14 What 3D object can be made from the net? see page 120

a
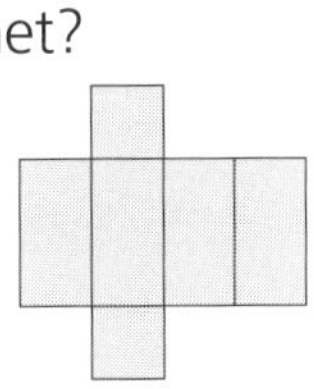

b
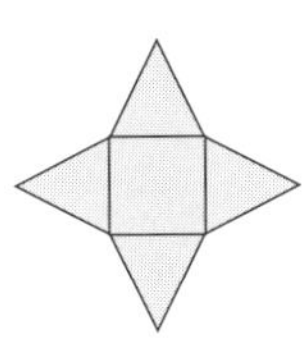

☐ ☐

15 see page 121

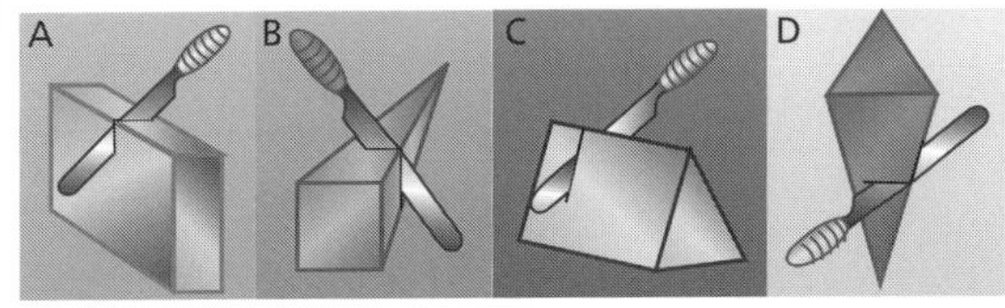

Name the cross-section of:
- a A ☐ b B ☐
- c C ☐ d D ☐
- e Which objects are prisms? ☐
- f Which objects are pyramids? ☐

Statistics

16 Alyssa had 8 fish, 4 mice and 12 birds. Complete this column graph to show the number of each. see page 143

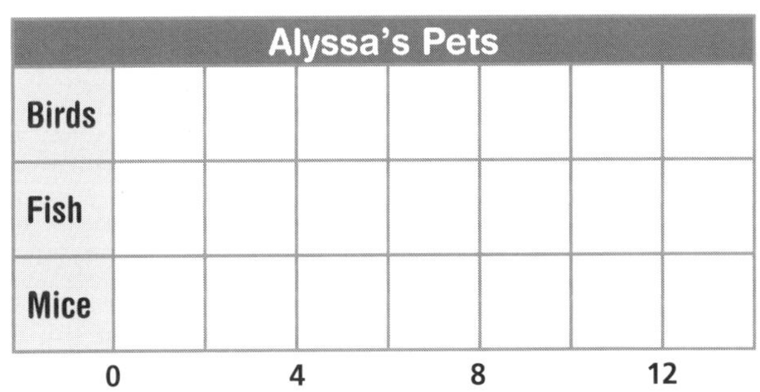

17 see page 144

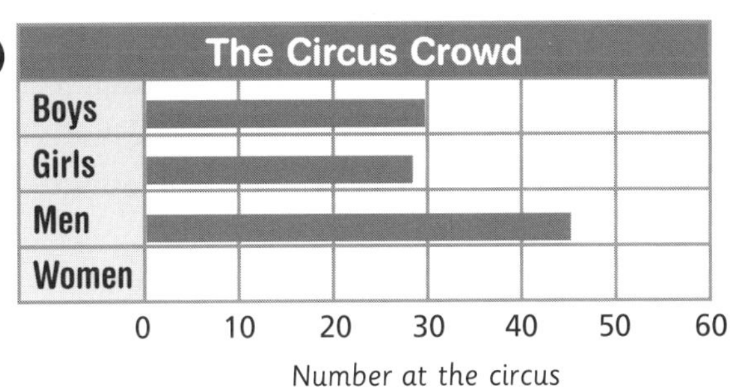

- a How many girls were there at the circus? ☐
- b How many men were at the circus? ☐
- c There were 42 women at the circus. Draw in the column for women.
- d What was the total number of people in the crowd? ☐

 • *AUSTRALIAN SIGNPOST MATHS 5* • ISBN 978 1 4886 2190 1

Diagnostic Test 2

Number

see page

1. Write the numeral for: 3
   a $(4 \times 10^3) + (9 \times 10^2) + (7 \times 10^1) + 8$ ☐
   b $(6 \times 10^3) + (5 \times 10^2) + (2 \times 10^1) + 1$ ☐

2. Write the fraction and the decimal shown in each hundred square. 4

   a 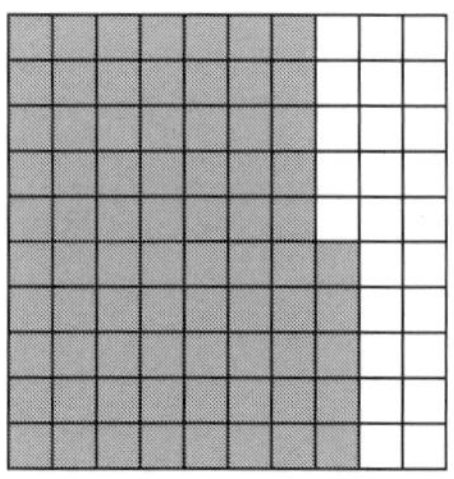
   b 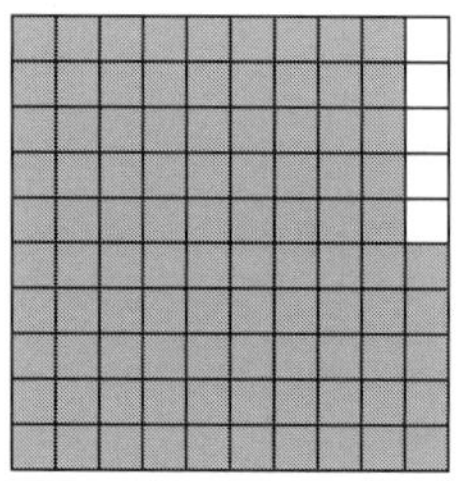

   a ☐ or ☐ b ☐ or ☐

3. 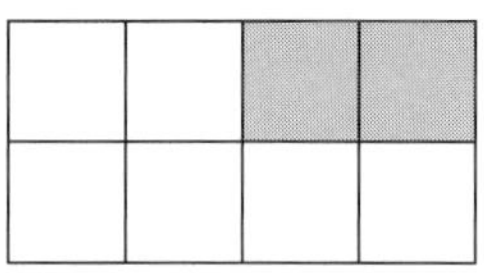 5

   ☐ is coloured.
   ☐ is not coloured.
   $\frac{2}{8}$ and ☐ makes 1 whole.

4. Put the group of fractions in order, from smallest to largest. 6
   $\frac{1}{2}, \frac{1}{5}, \frac{1}{10}$ ☐ ☐ ☐

5. a Match each fraction with a part of the circle. 6

   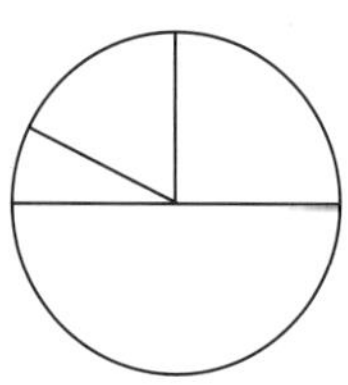

   $\frac{1}{2}$ $\frac{1}{6}$ $\frac{1}{4}$

   Complete using < or >.
   b $\frac{1}{6}$ ☐ $\frac{1}{4}$ c $\frac{1}{2}$ ☐ $\frac{1}{6}$

Operations

see page

6. Write all the factors of 24. 39
   ☐, ☐, ☐, ☐, ☐, ☐, ☐, ☐

7. a Write the first five multiples of 2. 38
   ☐, ☐, ☐, ☐, ☐
   b Write the first five multiples of 6.
   ☐, ☐, ☐, ☐, ☐

8. 34

a

	Hund	Tens	Ones
	5	4	9
−	3	2	6

b

	Hund	Tens	Ones
	7	9	8
−	3	5	4

9. 35

a

	H	T	U
	4	7	5
+	2	6	9

b

	H	T	U
	3	5	8
+	4	7	5

10. 36

a

	H	T	U
	5	4	3
−		9	6

b

	H	T	U
	2	8	0
−		9	6

11. 37

a

	H	T	U
	7	2	8
−	3	6	5

b

	H	T	U
	2	1	5
−	1	8	5

 ISBN 978 1 4886 2190 1

Measurement

see page

12 Find the perimeter. 98 99

a

b

13 Complete the table for rectangles. 99

	Length	Width	Perimeter
a	10 cm	8 cm	
b	12 m	9 m	

c The perimeter of a square with 12 cm sides = ☐

Geometry

14 **a** Write a letter under each coloured shape. 122

A reflection **B** translation **C** rotation

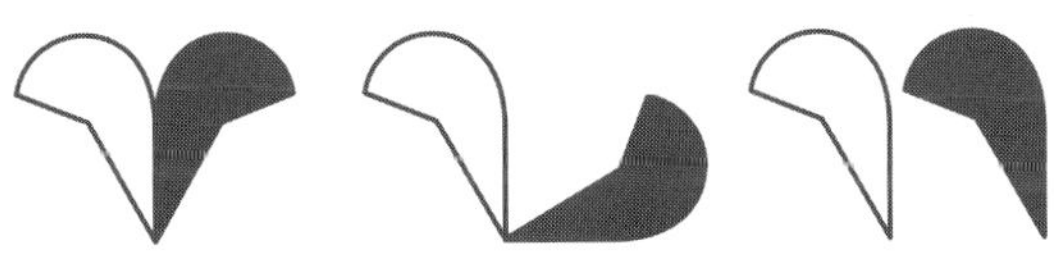

b Continue this pattern.

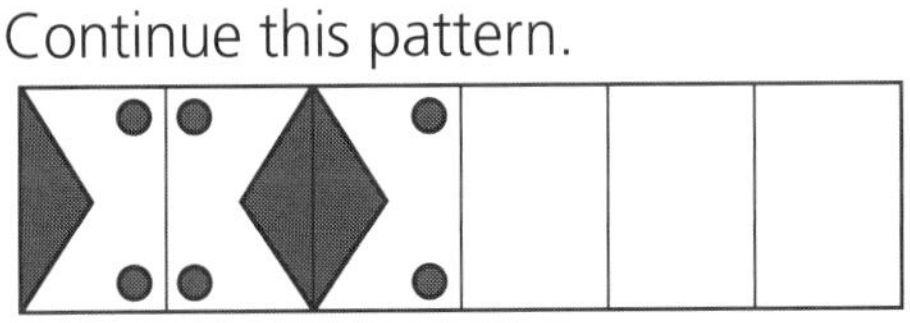

15 Match each object to a net. 122

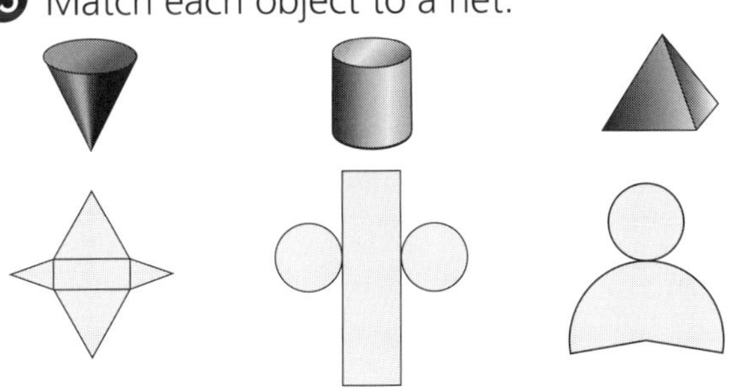

Statistics

see page

16 121

Fish Caught	
Day 1	
Day 2	
Day 3	

 = 4 fish

a How many fish were caught on day 1?

b On day 3, twelve fish were caught. Complete the graph.

17 145

Heads	𝍸 𝍸 𝍸 𝍸 𝍸 𝍸 𝍸 ||
Tails	𝍸 𝍸 𝍸 𝍸 𝍸 𝍸 𝍸

I tossed a coin many times and recorded the results in the table above.

a How many times did I toss the coin?

b What fraction of the time did I toss a head?

18 146

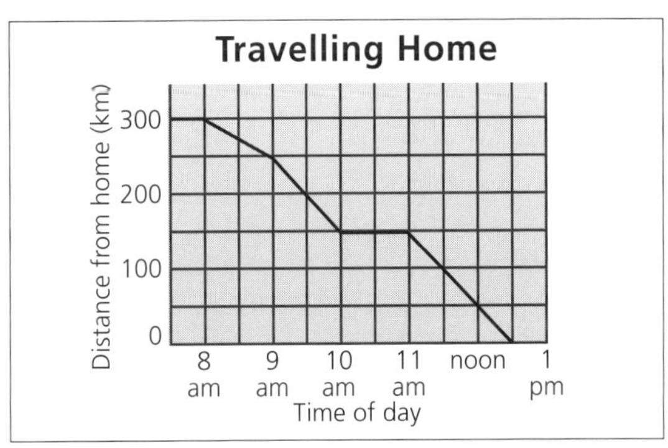

How far from home were we at:

a 8 am?

b 10 am?

c 11 am?

d noon?

e What distance was travelled between 8 am and 12:30 pm?

 • *AUSTRALIAN SIGNPOST MATHS 5* • ISBN 978 1 4886 2190 1

Diagnostic Test 3

Number

see page

1 Write the decimal for: 7

a $\frac{9}{10}$ ☐   b $3\frac{3}{10}$ ☐

Use decimals to write:

c 7 tenths ☐   d 4 tenths ☐

e Match each fraction to the correct decimal.

$\frac{6}{10}$	0·3
$\frac{3}{10}$	0·9
$\frac{9}{10}$	0·6

2 Complete the number line. 7

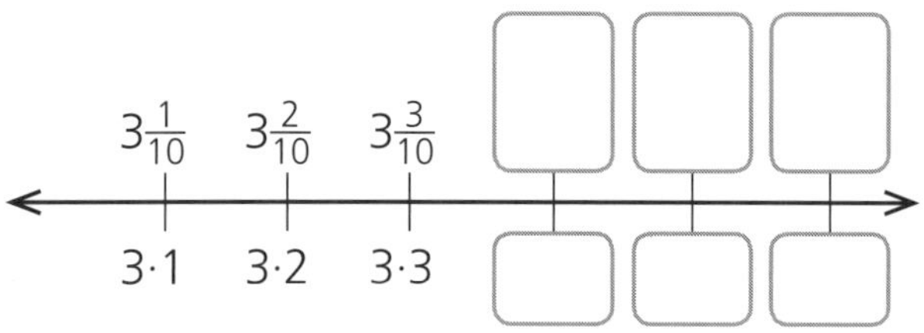

3 Write the fraction and decimal shown. 8

a 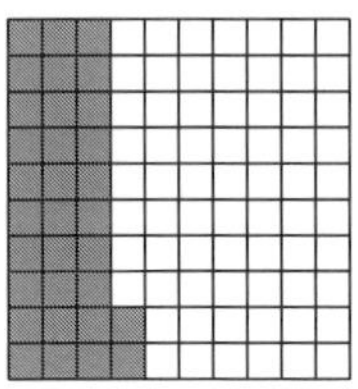   b 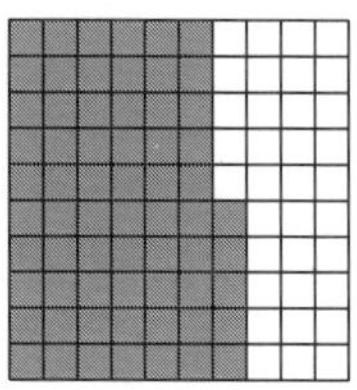

0· ☐ or ☐   0· ☐ or $\frac{}{}$

Operations

4 40

a $0.25 + $0.34 = ☐

b $5.16 + $2.04 = ☐

5 41

a $5.39 − $2.13 = ☐

b $3.52 − $1.61 = ☐

see page

6 a 17 hats are put in groups of 3. How many groups? 44

17 ÷ ☐ = ☐ (r)

★★★★★★★★★
★★★★★★★★★

b 18 ÷ 4 = ☐ groups ☐ remaining

c 18 ÷ 7 = ☐ groups ☐ remaining

d 18 ÷ 10 = ☐ groups ☐ remaining

7 45

a 8) 32   b 9) 45

c 7) 56   d 6) 24

e 4) 21   f 7) 38

g 5) 49   h 6) 32

8 How many teams of 8 can be made from 52 students? 46

☐ teams, ☐ remaining

9 Complete the pattern and write the rule. 47

a 7, 19, 31, ☐, ☐, ☐, ☐

Rule: ☐

b The number of star points:

★, ★★, ★★★, …

5, ☐, ☐, ☐, ☐, ☐

Rule: ☐

Measurement

see page

10

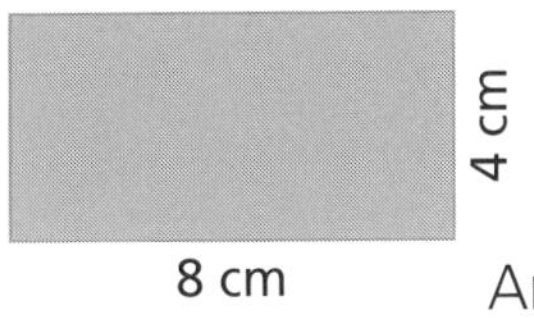

Area = ☐

100

11 Complete for rectangles.

101

	Length	Width	Area
a	6 cm	9 cm	cm^2
b	8 m	7 m	m^2
c	15 cm	5 cm	cm^2

d

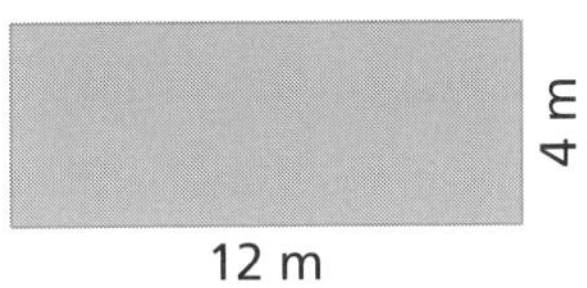

Area = ☐

Geometry

12

125

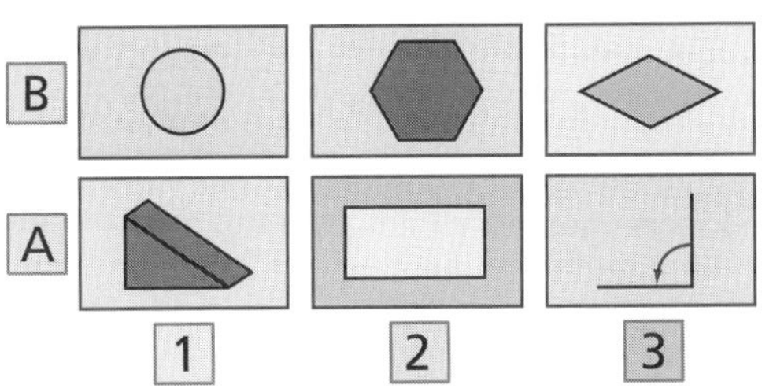

Give the number and letter for:

a the circle ☐

b the rectangle ☐

c the right angle ☐

13 I used geostrips to make these angles.

125

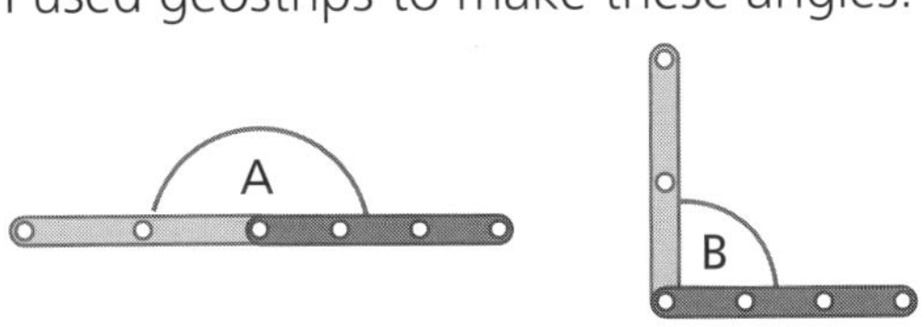

a Which angle is larger? ☐

b What is the name of angle **B**? ☐

see page

14

121

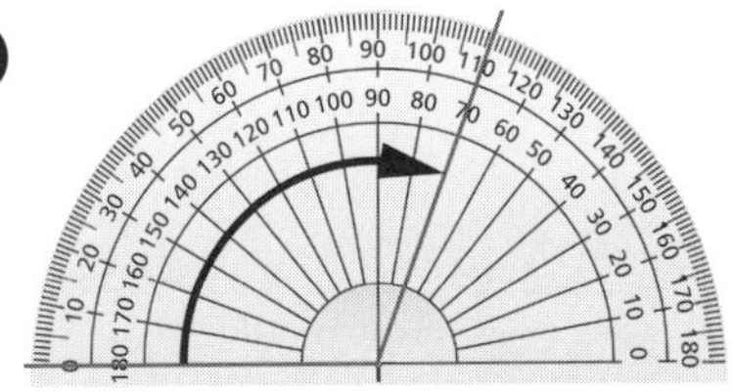

a What is the size of the angle? ☐

How many degrees are in a:

b right angle? ☐

c straight angle? ☐

d revolution? ☐

15

127

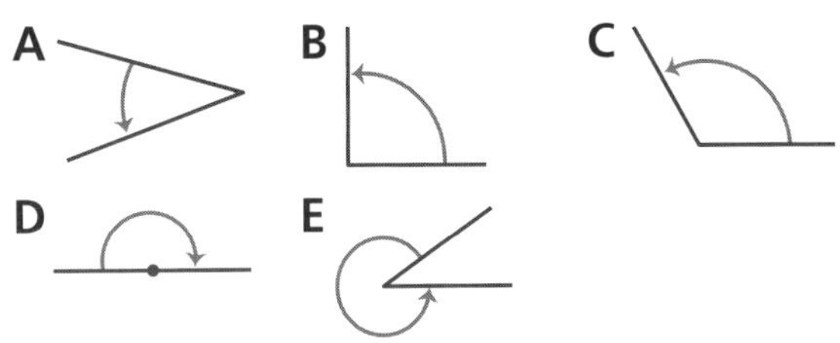

Which angle is:

a a right angle? ☐

b less than a right angle? ☐

c a straight angle? ☐

d more than a straight angle? ☐

e more than a right angle but less than a straight angle? ☐

Statistics

16

147

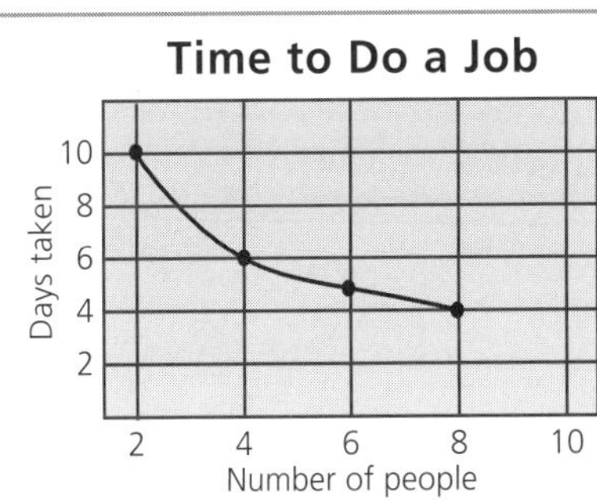

a How long would 3 people take to do the job? ☐

b How long might it take 10 people to finish the job? ☐

Diagnostic Test 4

Number

see page

1 Write the numeral for: 10 11

a

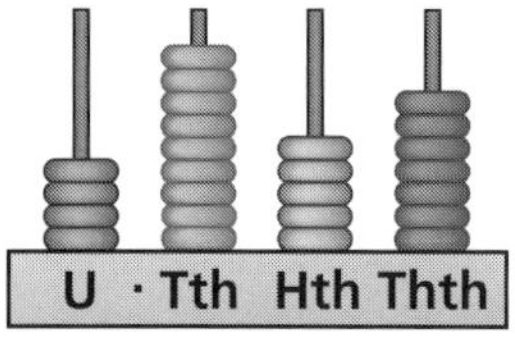

b

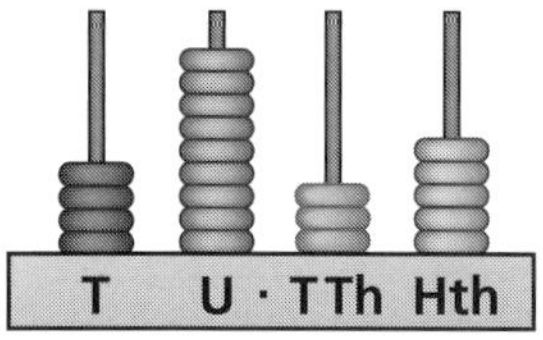

2 a Complete the label for the model. 9

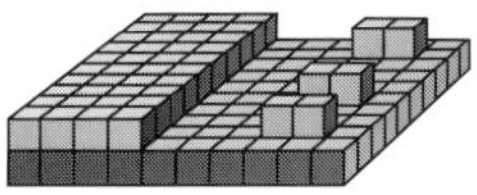

Ones	Tenths	Hundredths
0 •		

b Complete the numeral expanders.

0·58

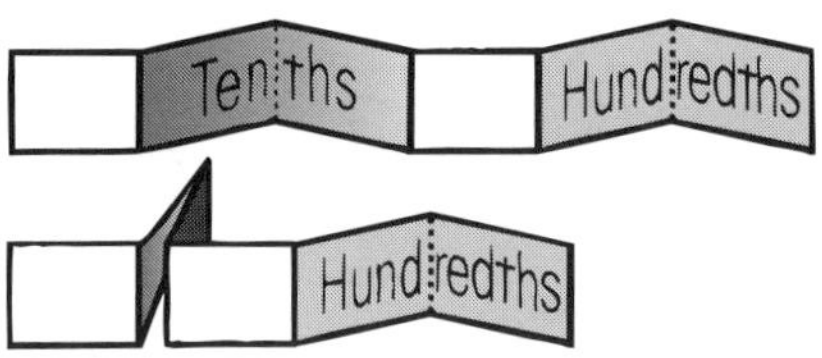

3 Write each number on the place-value chart. 10

a nine point two three

b seven point one oh six

Units •	10ths	100ths	1000ths
•			
•			

4 Complete the following: 11

a 0·89 = ☐ tenths and ☐ hundredths

b 0·07 = ☐ tenths and ☐ hundredths

c Write the number represented by:

6 tens, 2 units, 8 tenths and 1 hundredth.

Operations

see page

5 50 51

a
```
  4 9 5 3
+   6 9 7
```

b
```
    5 1 6
+ 3 7 4 8
```

c
```
    3 0 4
  2 1 3 2
+ 5 4 0 3
```

d
```
  4 2 6 4
  1 7 5 3
  2 3 0 6
+ 1 1 9 5
```

e
```
  $12.67
  $25.38
+ $19.95
```

f
```
  $24.76
  $ 9.84
+ $37.57
```

6 48 49

a
```
  4 2 6
− 1 7 8
```

b
```
  9 0 0
− 4 9 3
```

7 52 53 54

a
```
  5 9 6 7
− 2 4 3 6
```

b
```
  9 4 6 2
− 5 6 9 4
```

c
```
  6 0 0 0
− 2 4 6 3
```

d
```
  9 0 0 0
− 7 1 6 8
```

e Ron had 7695 bricks but used 5977. How many did he have left?

8 Use the jump strategy to find: 55

a 367 + 137 =

b 243 + 325 =

Use a suitable strategy to find:

c 800 − 93 =

d 4000 − 539 =

 ISBN 978 1 4886 2190 1

Measurement

see page

102

9 a About how many soccer fields make one hectare? ☐

How many hectares are there in:

b 20 000 m^2 ☐ c 50 000 m^2 ☐

How many square metres are there in:

d 7 ha ☐

e 4 ha ☐

f A rectangle has length 50 m and width 100 m. Is this more or less than 1 ha?

☐

103

10 Find the area of each shape.

a

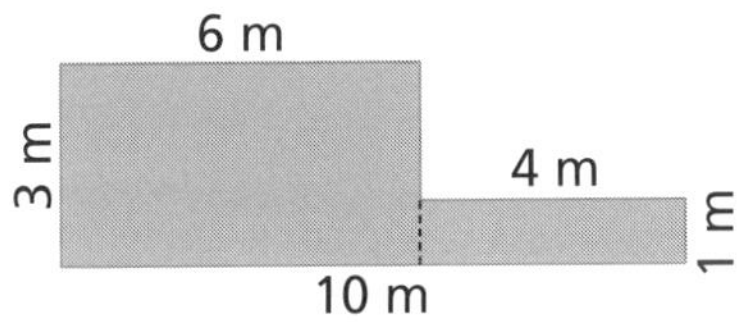

Area = ☐

b

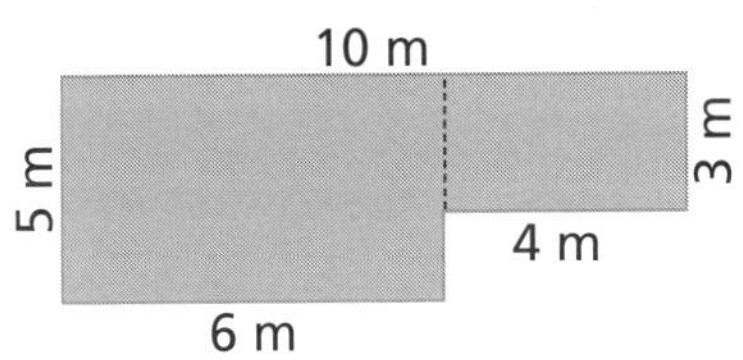

Area = ☐

103

11 Write the most appropriate unit of measure (cm^2, m^2, ha) for:

a a classroom floor 150 ☐

b a farm 250 ☐

c a garden 600 ☐

d a postage stamp 6 ☐

104

12 Find the volume of each prism.

a

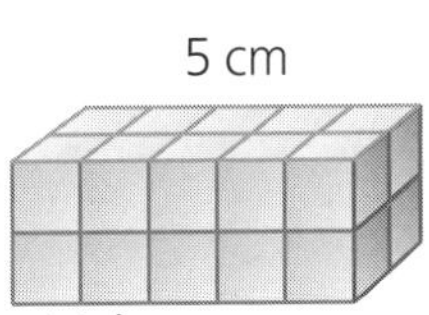

Volume ☐ cm^3

b

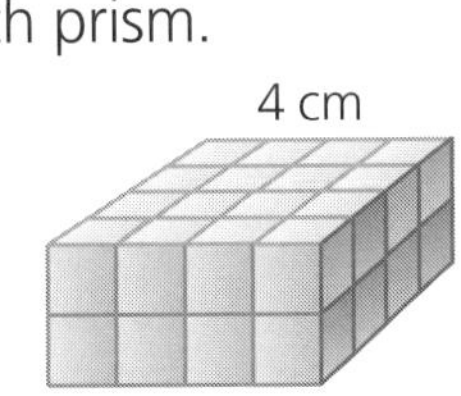

Volume ☐ cm^3

Geometry

see page

129

13 Name each type of angle.

a

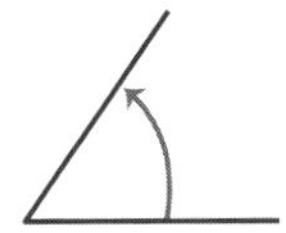

☐

b

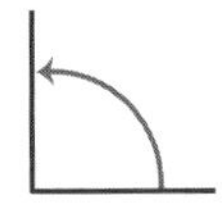

☐

c

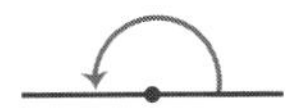

☐

d

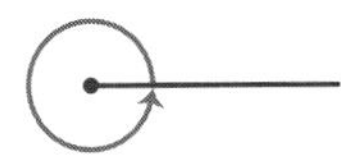

☐

Statistics and Probability

150

14 What is the probability, as a fraction, that the spinner will stop on:

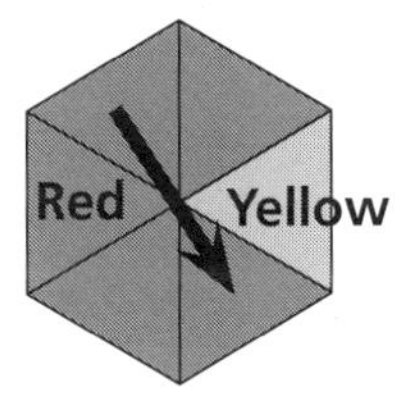

a red? ☐

b yellow? ☐

148

149

15 Jim drew this dot plot of the number of goals kicked in 11 soccer games.

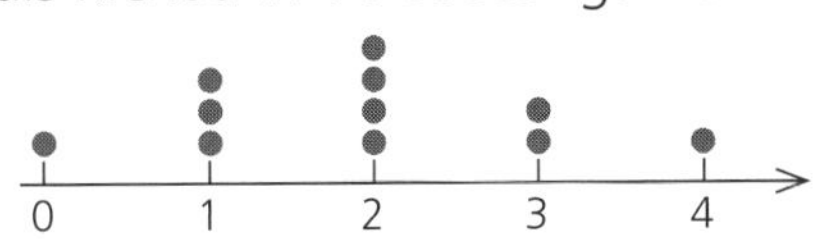

a What number was kicked most often? ☐

b How many times were three goals kicked? ☐

c How many times were zero goals kicked? ☐

152

16 Colour of Rachel's Budgies

Rachel has 50 budgies. How many are:

a blue? ☐

b yellow? ☐

c green? ☐

Diagnostic Test 5

Number

see page

1 Add these fractions. 14

a 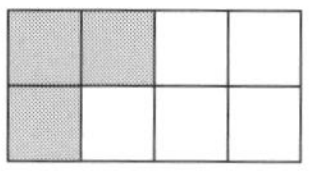 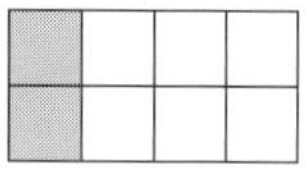

$\frac{3}{8} + \frac{2}{8} = \frac{\square}{\square}$

b 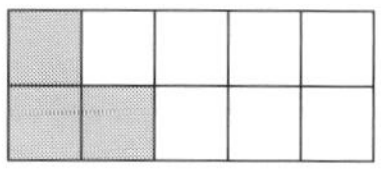 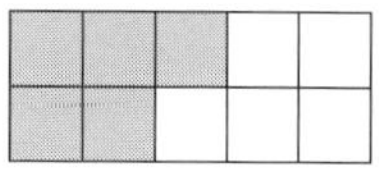

$\frac{3}{10} + \frac{5}{10} = \frac{\square}{\square}$

2 Complete: 15

a 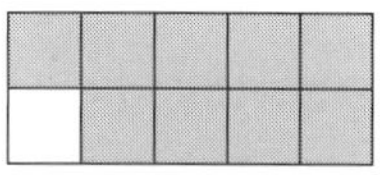 $\frac{9}{10} - \frac{4}{10} = \frac{\square}{\square}$

b 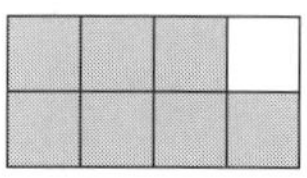 $\frac{7}{8} - \frac{2}{8} = \frac{\square}{\square}$

3 Write the numeral for: 12 13

a 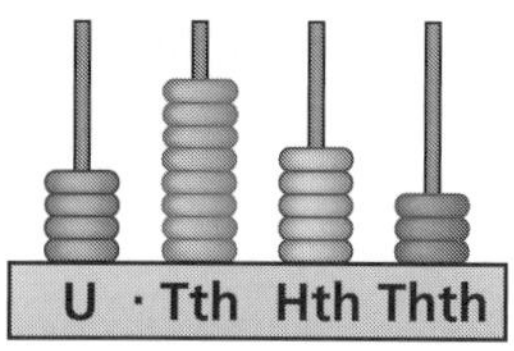

b 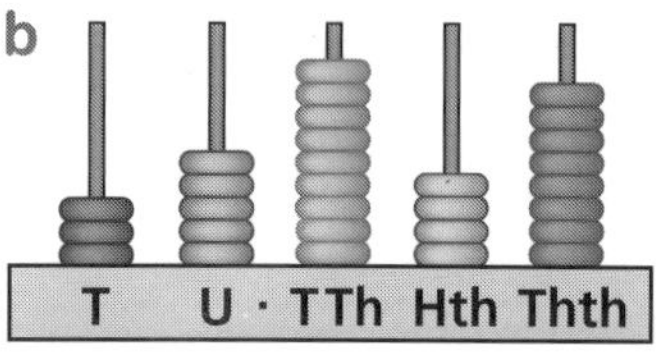

True or false?

c 3·619 < 3·916

d 9·543 > 9·245

4 Write the decimal for: 13

a $\frac{59}{100}$ b $\frac{735}{1000}$

Write the fraction for:

c 0·5 d 0·03

e 0·217 f 0·34

Operations

see page

5 a Write down all the factors of 12. 56 57

☐, ☐, ☐, ☐, ☐, ☐

b Write the first six multiples of 7.

☐, ☐, ☐, ☐, ☐, ☐

6 62 63

a 15 × 10

b 14 × 100

c 5 × 60

d 4 × 700

e

	H	T	U
		7	0
×			8

f

	H	T	U
	2	0	0
×			3

7 58 59 60

a $3\overline{)96}$ b $4\overline{)48}$ c $2\overline{)96}$

d $3\overline{)84}$ e $5\overline{)97}$ f $4\overline{)98}$

8 61

a $3\overline{)396}$ b $4\overline{)804}$

c $5\overline{)135}$ d $3\overline{)234}$

9 61 64

a $5\overline{)918}$ b $6\overline{)859}$

c $6\overline{)800}$ d $4\overline{)700}$

e $10\overline{)980}$ f $10\overline{)647}$

10 65

a $2\overline{)816}$ b $3\overline{)609}$

Measurement

11 Use 24-hour time to write: see page 106

a 8:00 am ☐

b 11:00 pm ☐

12 Write each time in 24-hour time. 106

a 

b 

13 These watches show 24-hour time. 106
Use am or pm to rewrite each time.

a 

b 

14 Write in two other ways. 107

a 5 pm = ☐ or ☐

b 0700 = ☐ or ☐

c 21:00 = ☐ or ☐

Geometry

15 What letter is found: 130

B	C	D
I	A	E
H	G	F

a west of **A**? ☐

b east of **G**? ☐

c north of **E**? ☐

d south-west of **A**? ☐

e north-east of **A**? ☐

f south-east of **A**? ☐

16 see page 131 136

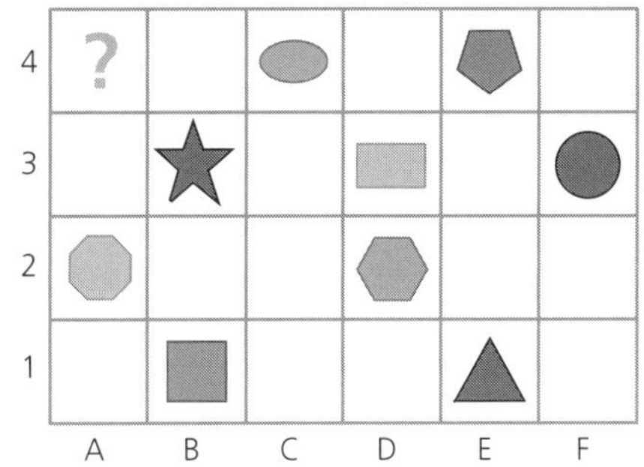

What is found in:

a E1? ☐

b B3? ☐

In which square is the:

c circle? ☐ d question mark? ☐

17 131 136

A 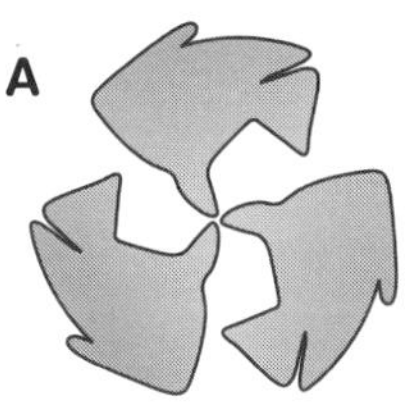

B 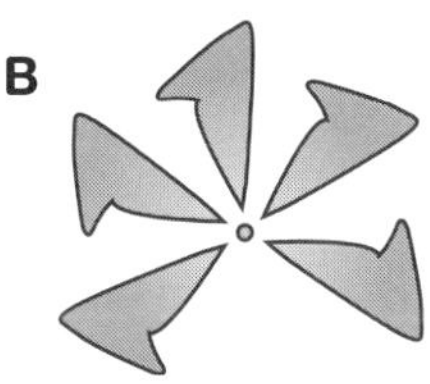

Which picture has rotational symmetry? ☐

18 Use your protractor to measure these angles to the nearest 10°. 133

a 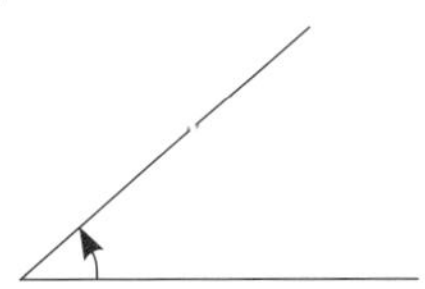 ☐

b 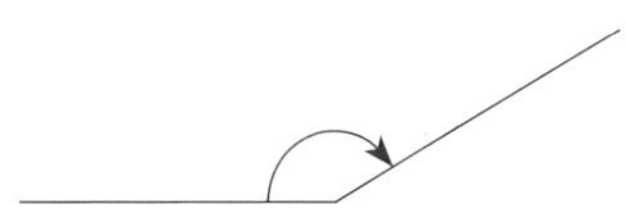 ☐

Probability

19 I took a coloured counter from a bag 100 times. This graph shows the results. 153

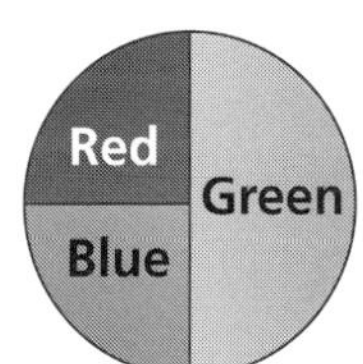

How often did I take:

a a green counter? ☐

b a red counter? ☐

 • *AUSTRALIAN SIGNPOST MATHS 5* • ISBN 978 1 4886 2190 1

Diagnostic Test 6

Number

see page

1 a $\frac{3}{8} + \frac{2}{8} = \frac{\quad}{\quad}$ b $\frac{5}{10} + \frac{4}{10} =$ ☐ 16 17

c $\frac{7}{8} - \frac{3}{8} = \frac{\quad}{\quad}$ d $4 - \frac{3}{4} =$ ☐

2 Write the improper fraction for: 17

a $3\frac{1}{2}$ $\frac{\quad}{\quad}$ b $2\frac{2}{3}$ $\frac{\quad}{\quad}$ c $1\frac{5}{6}$ $\frac{\quad}{\quad}$

Write the mixed numeral for:

d $\frac{9}{4}$ ☐ e $\frac{7}{3}$ ☐ f $\frac{8}{7}$ ☐

3 Circle the faster time shown on each pair of stopwatches. 19

a 06:13 74 b 23:16 81

06:13 59 23:16 85

Operations

4 Write all the factors of: 67

a 24 ☐

b 30 ☐

c Write down the first five multiples of 3.

☐, ☐, ☐, ☐, ☐

5 Factorise the first number then find the answer. 70

a $20 \times 14 = 2 \times$ ☐ $\times$ ☐

= ☐

b $30 \times 11 = 3 \times$ ☐ $\times$ ☐

= ☐

6 Complete the labels. 71 74

$16 \times 8 = (10 \times 8) + (6 \times 8)$

= ☐ + ☐

= ☐

see page

7 75

a

H	T	U
	8	4
×		3

b

H	T	U
	6	5
×		4

c

H	T	U
	3	7
×		6

d

H	T	U
	5	2
×		8

8 Circle the numbers that are divisible by 9. 66

93 45 64 72 81

63 56 36 27 76

9 a Paul has 36 stamps and Ann has 48. What is the average number of stamps held? ☐ 68 69

b The lengths of bugs measured were 16 mm, 19 mm and 25 mm. What was the average length? ☐

10 a Complete the table. 72 73

Input Number	32	42	52	62	72
Output Number	41	51	61		

b The rule is: To the input number we ☐.

c If the input number is 100, then the output number is ☐.

Measurement

see page 110

11 How many millimetres are there in:

a 9·4 cm? ☐ b 10·6 cm? ☐

Write the measurements shown.

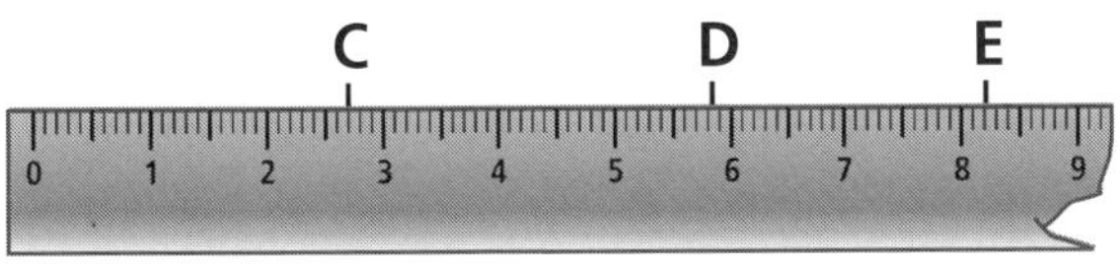

c C: ☐ mm d E: ☐ mm

e D: ☐ mm

111

12 a Perimeter = ☐

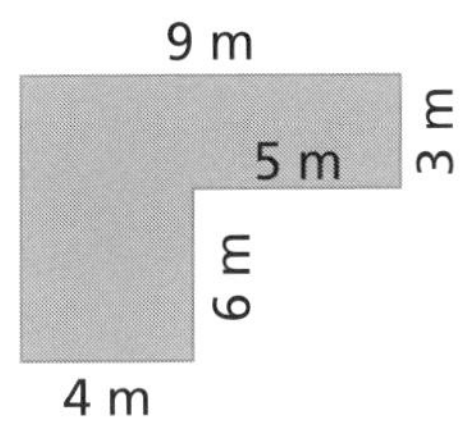

b Measure the perimeter in centimetres.

Perimeter = ☐

108

13 Would you use cm^3 or m^3 to measure the volume of a:

a house? ☐ b lunch box? ☐

c book? ☐ d room? ☐

e Would the volume of a bathroom be more than $1\ m^3$? ☐

109

14

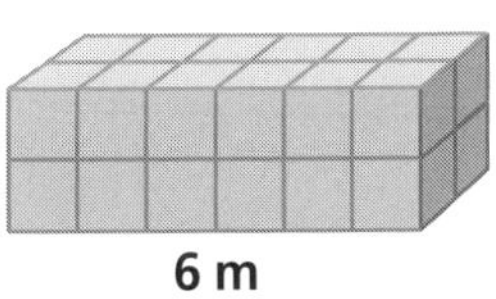

Volume = ☐ m^3

109

15 Complete the table for rectangular prisms.

	Length	Width	Height	Volume
a	6 cm	3 cm	2 cm	
b	8 cm	4 cm	3 cm	

Geometry

see page 135

16 Name solid:

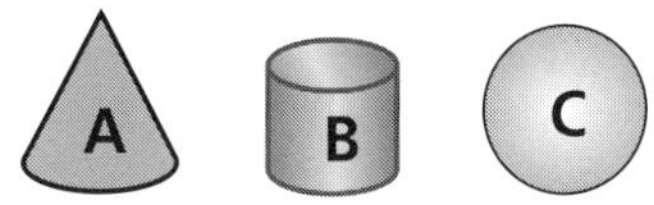

a A ☐

b B ☐

c C ☐

How many flat surfaces are on:

d B? ☐ e C? ☐ f A? ☐

g Which is the net of a cube? ☐

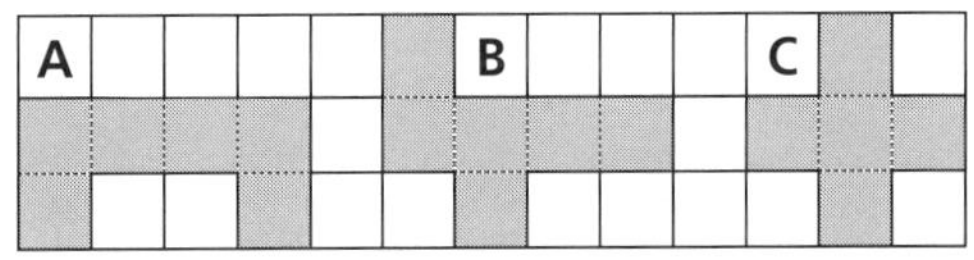

136

17

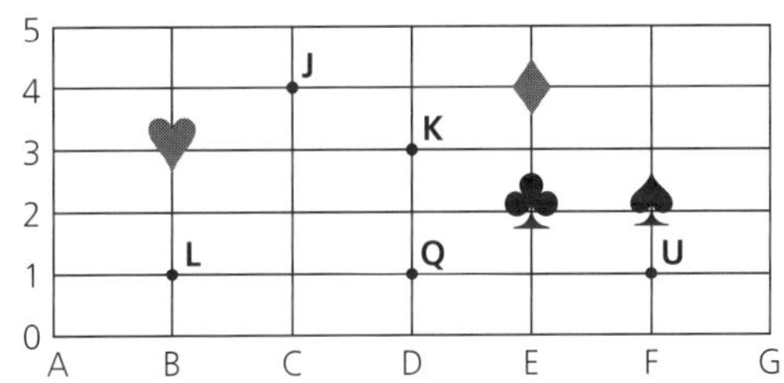

Name the letter found at:

a D3 ☐ b B1 ☐

c C4 ☐ d F1 ☐

Give the coordinates of:

e ♥ ☐ f ♦ ☐

g ♣ ☐ h ♠ ☐

Statistics

156

18

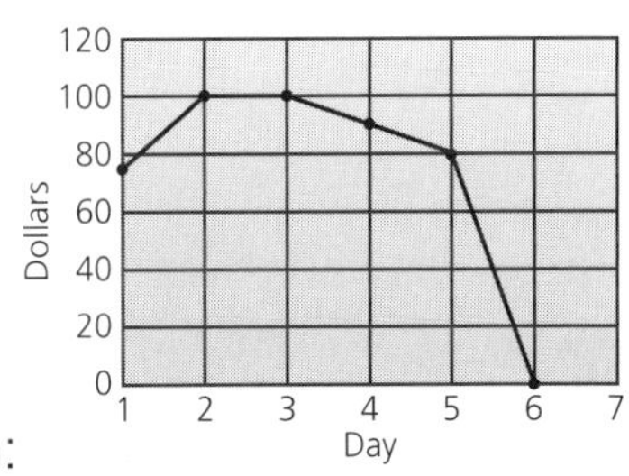

How much did I have in my account on:

a day 3? ☐ b day 5? ☐

c day 1? ☐ d day 6? ☐

Diagnostic Test 7

Number

see page 22

1 Write the equivalent fractions.

a $\frac{2}{8} = \frac{\square}{4}$ b $\frac{2}{4} = \frac{\square}{2}$

c $\frac{6}{8} = \frac{\square}{4}$ d $\frac{4}{8} = \frac{\square}{4}$

23

2 Write an equivalent fraction.

a $\frac{3}{5} = \frac{\square}{10}$ b $\frac{2}{3} = \frac{\square}{6}$

Complete the patterns.

c $\frac{4\ (\times\ 2)}{5\ (\times\ 2)} = \frac{\square}{\square}$ d $\frac{1\ (\times\ 4)}{3\ (\times\ 4)} = \frac{\square}{\square}$

20 21

3 Complete the patterns.

a 3·1, 2·9, 2·7, ☐, ☐

b 1·22, 1·32, 1·42 ☐, ☐

20

4

a 1·1 + 0·3 = ☐ b 4·3 + 0·3 = ☐ c 1·9 + 0·1 = ☐

d 0·9 − 0·3 = ☐ e 1·0 − 0·4 = ☐ f 1·8 − 0·4 = ☐

Operations

76

5

a

H	T	U
	3	5
×		3

b

H	T	U
	7	2
×		6

77

6

a

H	T	U
3	0	8
×		3

b

H	T	U
1	7	9
×		4

see page 78

7 Round to estimate and then calculate.

a 169 + 38

Estimate: ☐

Answer: ☐

b 975 + 263

Estimate: ☐

Answer: ☐

80

8

a 47 × 10 b 206 × 4

c 161 × 6 d $1.90 × 5

e $1.17 × 6 f $1.35 × 7

g 1368 × 6 h $12.19 × 8

82 83

9 In a town with a population of 93 215 there are 46 713 males.
If all the population are either male or female, how many are female?

☐

Measurement

see page

10 Write each time in 24-hour time. 112

a 

b 

11 Write as 24-hour time. 112

a 3:41 pm   b 8:35 am

Write as am or pm time.

c 19:23   d 01:55

e How long is it from 07:29 to 11:16?

12 a On our holiday we drove for $4\frac{1}{2}$ hours. What time did we arrive if we left at 4:30 pm? 113

b We arrived at the shops at 13:30. If we spent $2\frac{1}{2}$ hours at the shops, what time did we leave to go home?

Geometry

13 Use a protractor to draw an angle of 30°. 138

see page

14 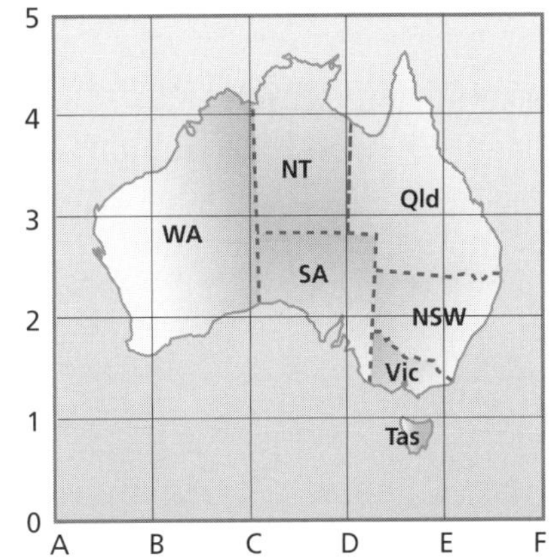 137

Which state has the coordinates:

a B3?   b D2?

What are the coordinates of:

c New South Wales?

d Queensland?

15 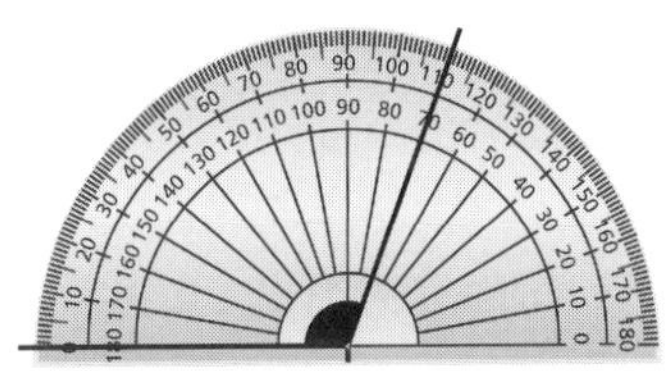 139

What is the size of this angle?

16 What is the size of each reflex angle? 139

a 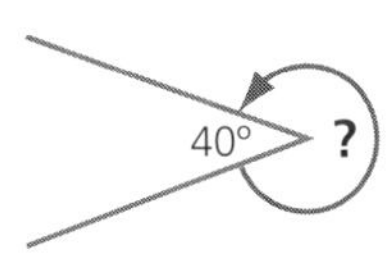

b 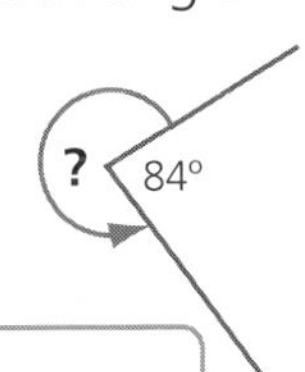

Probability

17 Rate the probability from 0 to 1. 158

a The next ant I see will fit into a car.

b I will swim 100 m in 10 seconds.

c A dice will land on 2, 4 or 6 when it is thrown.

d My next teacher will be ten years old.

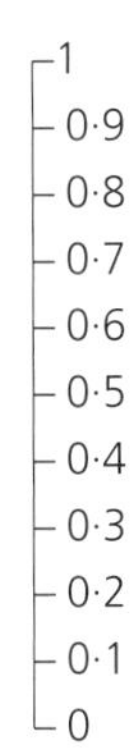

Diagnostic Test 8

Number

see page 24

1 Write an equivalent fraction for:

a $\frac{3\ (\times 3)}{5\ (\times 3)} = \frac{\square}{\square}$ b $\frac{5\ (\times 2)}{6\ (\times 2)} = \frac{\square}{\square}$

c $\frac{1}{2} = \frac{\square}{8}$ d $\frac{3}{4} = \frac{\square}{12}$

e $\frac{4}{5} = \frac{\square}{10}$ f $\frac{1}{3} = \frac{\square}{6}$

Multiply both the numerator and denominator by 2:

g $\frac{1}{2} = \frac{\square}{\square}$ h $\frac{1}{6} = \frac{\square}{\square}$

What number has been used to multiply the numerator and denominator in each pair of equivalent fractions?

i $\frac{1}{3} = \frac{3}{9}$ ☐ j $\frac{3}{4} = \frac{6}{8}$ ☐

Operations

2 Round these numbers to the nearest hundred. 84 85

a 309 ☐ b 749 ☐

c 197 ☐ d 451 ☐

3 Is 48 × 6 closer to 50 × 5, 50 × 10 or 40 × 5? ☐ 84

4 a Eight children each had $11.35. How much money did they have altogether? ☐ 88

b A farmer sold 345 sheep from his flock of 3230. How many did he have left? ☐

c Melinda said that the average of 624, 196 and 32 was 281. Was she correct? ☐

see page 86

5 You need to make up a budget to buy items for a fish tank. Complete the table below using no more than $200.

Item	Number	Cost	Total
Tank			
Stones			
Fish			
Plant			
		Total Spent	

Cost of items:
Tank.......................... $80
Each fish.................... $12
Plant.......................... $15
Each stone................ $10

Measurement

6 How many tonnes are there in: 114

a 3000 kg? ☐ b 9000 kg? ☐

How many kilograms are there in:

c 6 t? ☐ d 5 t? ☐

7 How many grams are there in: 115

a 5 kg? ☐ b 9 kg? ☐

c 3·5 kg? ☐ d 7·5 kg? ☐

Write true or false for each statement:

e 12 kg < 15 000 g? ☐

f 700 kg < 7 t ☐

8 Record the measure shown on each scale. 116

a 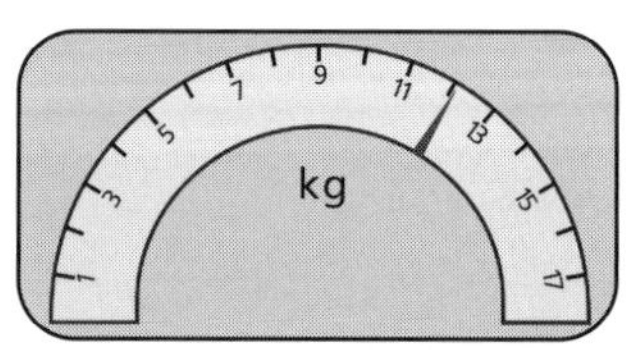

The measure is ☐ kg.

b 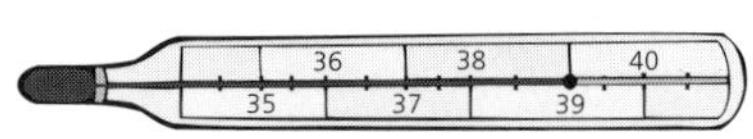

☐ °C.

 • *AUSTRALIAN SIGNPOST MATHS 5* • ISBN 978 1 4886 2190 1

see page 117

❾ How many centimetres are there in:

a 3·14 m? ☐ b 5·69 m? ☐

c 10·75 m? ☐ d 8·03 m? ☐

How many metres are there in:

e 645 cm? ☐ f 923 cm? ☐

g 896 cm? ☐ h 1030 cm? ☐

Geometry

140

❿

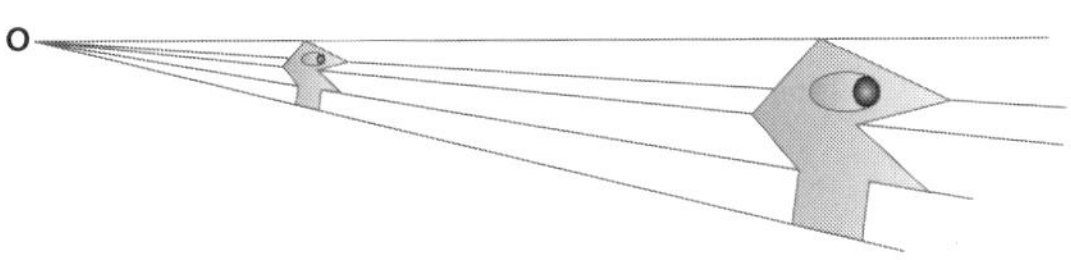

True or false?

a If you double the distance of each point from the centre, the figure becomes larger. ☐

b If you halve the distance of each point from the centre, the figure becomes smaller. ☐

Probability

⓫ If the chance of choosing a red lolly is $\frac{1}{8}$. What is the chance of not choosing a red lolly? ☐ 159

⓬ Three red balls and seven yellow balls are placed in a bag. 162

a What colour ball would more likely be drawn from the bag first? ☐

b If a red ball was drawn first, could the second ball drawn be red? ☐

Statistics

see page 160

⓭ Name each type of graph.

a

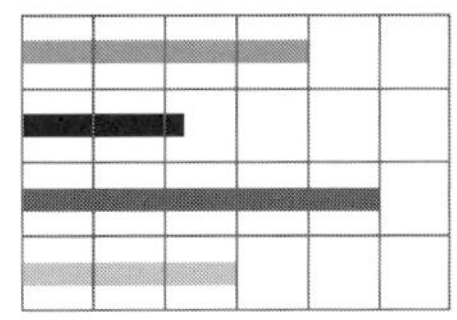

☐

b

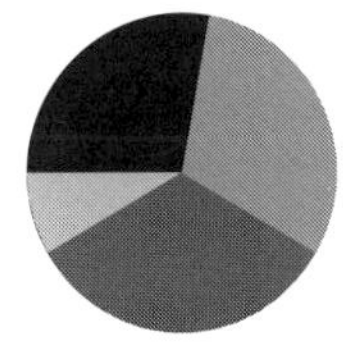

☐

c

☐

164

⓮ 16 students were asked to choose one of three flavours.

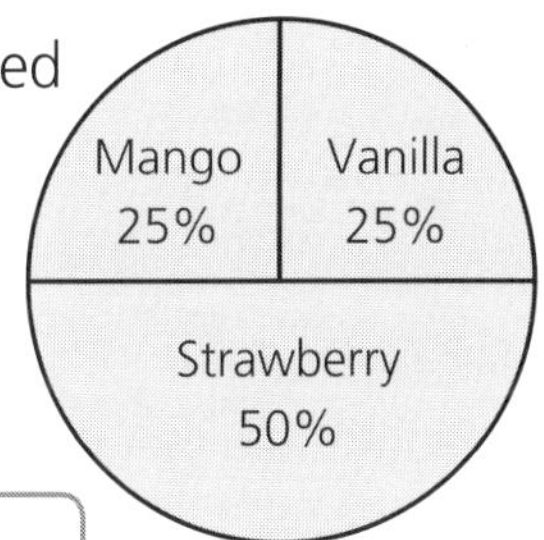

How many chose:

a strawberry? ☐

b mango? ☐

c vanilla? ☐

164

⓯

Heads	𝍸 𝍸 𝍸 𝍸 𝍸 𝍸 𝍸 ||
Tails	𝍸 𝍸 𝍸 𝍸 𝍸 𝍸 𝍸

a I tossed a coin many times and recorded the results in the table above. How many times did I toss the coin? ☐

b How many more heads were there than tails? ☐

c Do these results suggest that tossing a head has the same chance as tossing a tail? ☐

 • AUSTRALIAN SIGNPOST MATHS 5 • ISBN 978 1 4886 2190 1